METAL CUTTING MECHANICS

METAL CUTTING MECHANICS

Viktor P. Astakhov

CRC Press
Taylor & Francis Group
Boca Raton London New York

CRC Press is an imprint of the
Taylor & Francis Group, an **informa** business

CRC Press
Taylor & Francis Group
6000 Broken Sound Parkway NW, Suite 300
Boca Raton, FL 33487-2742

First issued in paperback 2019

ISBN-13: 978-0-8493-1895-5 (hbk)
ISBN-13: 978-0-367-40014-9 (pbk)

Visit the Taylor & Francis Web site at
http://www.taylorandfrancis.com

and the CRC Press Web site at
http://www.crcpress.com

Preface

Except for casting, where the product forms directly from the liquid state, many metal products are subjected to at least one metal-removing process during their manufacture. The current rapid technological progress has forced the theory of metal cutting into the forefront of engineering application and design. The facts of economic life have made the efficient utilization of material, labor, and time and a more efficient approach to design necessities even for less sophisticated industrial applications. Therefore, today there is a greater need for all industrial and manufacturing engineers to be familiar with the fundamentals of metal cutting analysis, not only to be more numerate and hence reduce the extent of empirical approach to problems, but also to close the gap which exists between design and manufacture.

Although the theory of metal cutting has advanced considerably over the last hundred years, rigorous solutions to many practical metal cutting problems are still not available. The common analyses of orthogonal cutting — a simplified version of the practical cutting process, assuming idealized materials and boundary conditions — gives valuable insight into the effects of cutting speed, feed, tool geometry, material properties, and various other parameters on force, energy, and power requirements. However, there are still many more questions than answers in regard to a great number of particular cases.

The main differences between the author's work and the other works on metal cutting include:

- **A system engineering approach is taken to consider the cutting process.** For the first time, the cutting process is considered as taking place within the cutting system, which is defined as consisting of the following components: cutting tool, chip, and workpiece. Although the known models of machining essentially include the same elements, they have never been considered as systems having the system time frame; instead, they have always been considered as assemblages without any time reference. Using the system approach, the basic mechanisms governing the chip formation process are considered.

- **The uniqueness of the cutting process, among other closely related manufacturing processes, is identified by chip formation, which is considered to be caused by the bending stress in the deformation zone.** The bending moment, which develops in the deformation zone due to the interaction between the chip and tool rake face, causes the bending stress in this zone. This bending stress combined with the shear stress due to compression has been proven to be the cause for chip formation and the consequent distinction of metal cutting from other manufacturing processes. The presence, significance, and variation of the bending stress in the deformation zone are proven both theoretically and experimentally, resulting in the chief conclusion that the cutting process has a cyclical nature. The dynamics of the cycles are considered in terms of variation of the cutting force, cutting temperature, chip structure, and other major cutting parameters.
- **A distinctive move towards predictability of the metal cutting process is suggested.** At the present stage of development, the predictability of a metal cutting theory depends entirely on the accuracy with which it accounts for the properties of workpiece material, as the design and geometry of the cutting tool along with the properties of tool material are well known, and the cutting regime can be set at any desirable level and/or can be varied according to any defined sequence. The resistance of workpiece materials to cutting is critical in cutting mechanics; therefore, special attention is paid to the properties of workpiece materials, with the main emphasis being placed on the study of the thermomechanical behavior of these materials in cutting. It has been proven that the shear flow stress, used as the primary mechanical characteristic of workpiece material in cutting, is not a relevant parameter to be used to predict the cutting process, and this conclusion explains significant discrepancies in reported results. The reason for this is that the metal cutting process is always accomplished by fracture — that is, by separation or fragmentation of a solid body into two or more parts under the action of stress. Because fracture in metal cutting processes is the result of plastic deformation, ductile fracture is analyzed in details.
- **Special attention is devoted to experimental methodology.** Because the theory of metal cutting is still in the early stages of development, experimental study prevails as the essential instrument of knowledge to reveal the influence of the cutting parameters on the processes' outcomes as well as to prove certain aspects of developing theories. Therefore, particular significance is given to the theory of similarity and the theory (and practice) of simulation in their new, broader sense, in which they are the basis for experimental procedures and provide guidance as to the formulation of tests, as well as making it possible to condense the resulting experimental information gained. A number of new, original methods for experimental studies in metal

cutting are discussed. Special attention has been paid to the methodology of the cutting force measurements, simulation of orthogonal cutting conditions, microhardness scanning procedure, and fractography.

After a brief introductory chapter, the emphasis in the second chapter is placed on the system approach in metal cutting. The mathematical theory of system is employed to demonstrate that the cutting tool, workpiece, and tool constitute the cutting system. Using the primary system properties, it has been proven that the bending moment rising in the deformation zone due to the interaction between the chip and tool rake face causes chip formation. A definition of metal cutting is suggested to distinguish this process from other closely related manufacturing processes. The third chapter gives a description of the model of chip formation having parallel boundaries. The approach in this chapter is to apply, to the greatest extent possible, continuum mechanics to the analysis of the chip formation process, with this model demonstrating that proper application of the theory leads to the proper velocity diagram and to the conclusion that the deformation process in metal cutting cannot be thought of as simple shearing. Although highly idealized, the model presents a valuable starting point for further development. The fourth chapter considers the relevant properties of workpiece materials. This chapter is mainly devoted to proving that the metal cutting process is purposeful fracture of the workpiece material so that fracture toughness should be considered in the analysis of the energy required. The strain at fracture is suggested to be the main mechanical characteristic of workpiece material in cutting. Chapter 5 presents a finite element analysis of the cutting system. Little emphasis is given to the computation side of the analysis, as it is covered sufficiently in many books and papers. Instead, the main attention is given to physical factors and aspects of analysis. Chapter 6 deals with experimental methodologies, experimental setups, data collection and acquisition procedures, and analyses of the obtained results.

The book should be of particular relevant to researches in the field of metal cutting and machine tool and tool design and to all practicing engineers and metallurgists who are directly concerned with improving design and efficiency in the manufacturing industry. Also, it should be of benefit to graduate students and final year undergraduates specializing in manufacturing processes.

I am deeply indebted to the following people for their help with the work described in the book: Professor S. Shvets, for valuable help understanding the chip formation process and its analysis; Professor Xinran Xiao, for her supportive discussions and criticism of the material aspects of my work and painstaking corrections of my earlier papers; Professor Jawahir, for the encouragement to write this book, constructive criticism, and his important advice; Professor Milton Shaw, for his willingness to help, his open-minded discussions, and his valuable advice; and Professor M.O.M. Osman, for his support. I was greatly assisted by my graduate students, whose names appear in the references throughout the text, and valuable assistance has especially been provided by my graduate student, doctoral candidate Mohammed Hayajneh, who has conducted a number of very sophisticated cutting experiments included in the book. I wish to thank Thomas Hay for his great help in careful checking of the manuscript. A special note of thanks is due to the Manufacturing Engineering Division of the American Society for Mechanical Engineers (ASME) for giving me the opportunities to organize the symposium "Predictive Modeling in Metal Cutting as Means of Bridging Gap Between Theory and Practice" in 1997 and the symposium "Machining Science and Technology", in 1999.

Viktor P. Astakhov

The author

Viktor P. Astakhov, Ph.D., Dr.Sci., received his B.Eng. in manufacturing from Odessa College of Industrial Automation (U.S.S.R.) in 1972, his M.Eng. in 1978 from the Odessa Polytechnic University (U.S.S.R.), his B.Sc. and M.Sc. in mathematics from Mechnicov's State University (Odessa, U.S.S.R.), and his Ph.D. from Tula Polytechnic University (U.S.S.R.) in 1983. He was awarded a Dr.Sc. designation in 1991 for his outstanding performance and profound impact on science and technology. His first teaching appointment was in the department of metal cutting and cutting tools at Odessa Polytechnic University in 1984, where he became a full professor and head of the deep-hole machining center for the aerospace industry. He has served on national scientific and planning committees and has also been an active consultant for the machine tool building, aerospace, nuclear, and gas turbine industries. Currently, he is an adjunct professor with Concordia University and a technical representative and senior consultant with American Heller Corporation.

Dr. Astakhov is the recipient of a "Bronze Medal for Achievements in Development of Industry of the U.S.S.R." in 1984, a "Bronze Medal for Achievements in Development of Industry of the Ukraine" in 1986, a certificate designating him "Professor of Ukraine" in 1994, and an ASME certificate of appreciation and ASME's "BOSS" award in 1997.

Dr. Astakhov's principal research interest is in manufacturing, including metal cutting theory, mechanical metallurgy, materials failure properties, and cutting tool and machine tool design. He is also involved in teaching and research of various manufacturing processes, machine design, and creativity enhancement algorithms. Being active in both fundamental and industrial-related research, he has published two books and more than 100 papers in refereed journals and proceedings. Dr. Astakhov is the author of more than 40 patents. He has served as the organizer and contribution editor of the "Predictable Modeling in Metal Cutting as Means of Bridging Gap Between Theory and Practice" and "Machining Science and Technology" symposiums held at the 1997 and 1999 International Mechanical Engineering Congress and Exposition. He is serving as a board member and frequent reviewer for several international scientific journals.

Contents

Chapter one. Introduction 1
1.1 General 1
1.2 Reasons for studying metal cutting 2
1.3 History of metal cutting 3
1.4. Basic problems requiring solution 6
References 8

Chapter two. System approach 11
2.1 Observation part of existing studies on chip formation 12
2.2 System and assemblage 15
2.2.1 Mathematical definition of system 15
2.2.2 Mathematical definition of an assemblage 20
2.2.3 The relation between a system and an assemblage 21
2.3 Cutting system 26
2.3.1 First step: time dependence of parameters of cutting system and their dynamic interactions 26
2.3.1.1 First stage: observational part 26
2.3.1.2 Second stage: experimental verification 33
2.3.1.3 Third stage: quantitative analysis of the bending stress 39
2.3.1.4 Definition of the metal cutting process 44
2.3.1.5 Difficulties due to high ductility of workpiece material 45
2.3.1.6 Model of chip formation when seizure occurs at tool-chip interface 47
2.3.1.7 Generalized model of chip formation and structure classification 51
2.3.2 Second step: controllability of the cutting system 55
2.3.3 Principle of control of the cutting system 61
References 69

Chapter three. Parallel-sided deformation zone theory 73
3.1 The role of engineering plasticity in metal cutting studies 73
3.2 Velocity diagram: what seems to be the problem? 79
3.3 Role of the velocity diagram in the theory of metal cutting 81
3.4 The discontinuity of the tangential velocity 82
3.5 Real velocity diagram in metal cutting 88
3.5.1 Velocity diagram at the second stage 88
3.5.2 Velocity diagram at the third stage 94
3.6 Real virtual work equation 94
3.7 Analysis of plastic deformation in metal cutting 95
3.7.1 Chip compression ratio and strain — what is a real measure of plastic deformation in metal cutting? 95
3.7.2 Infinitesimal strain 98
3.7.3 The rotational element 102
3.7.4 Infinitesimal strain of a line element 105
3.7.5 Strain compatibility equation 106
3.7.6 Finite displacement of a point in a continuous body 108
3.7.7 Finite strain 109
3.7.8 Principal strains and strain invariants 111
3.7.9 Increments in the strain components 113
3.7.10 Strain in metal cutting 114
3.8 Strain rate in metal cutting 117
3.8.1 Reported strain rates 117
3.8.2 Critical analysis of the reported strain rates 119
3.8.3 Strain rate tensor 121
References 123

Chapter four. Work material considerations 127
4.1 What has to be predicted according to existing theories? 128
4.2 Review of attempts made to predict shear flow stress 129
4.2.1 The first approach 130
4.2.2 The second approach 131
4.2.3 The third approach 132
4.3 What has to be predicted in reality? 132
4.4 Experimental verification 134
4.4.1 Incremental compression 134
4.4.2 Metal cutting 136
4.4.3 Influence of high temperatures 137
4.5 The cause of significant discrepancies in reported results 141
4.6 Nature of plastic deformation of polycrystalline materials 143
4.6.1 Fracture and its correlation with deformation 143
4.6.2 Plasticity and stress relaxation 145
4.6.3 Defect population in metals 147
4.6.4 Influence of the state of stress 150

4.6.5 Generalizations ... 154
4.7 Fracture of ductile polycrystallines ... 156
4.8 Mechanism of fracture in metal cutting ... 156
4.8.1 Possible regions of fracture ... 157
4.8.2 Experimental evidence ... 158
4.8.3 Type of fracture ... 163
4.9 Fracture strain determination ... 167
4.9.1 Existent criteria of ductile fracture ... 169
4.9.2 Basic starting concepts for the analysis of ductile fracture ... 171
4.9.3 Relationship between strain at fracture of a material and parameters of current plastic deformation ... 174
References ... 178

Chapter five. Finite element simulation ... 183
5.1 General ... 183
5.1.1 FEM in metal cutting ... 183
5.1.2 Aims and layout ... 184
5.2 Computational details of simulation ... 184
5.2.1 Program structure ... 185
5.2.2 Workpiece and tool materials modeling ... 187
5.3 Applications of FEM ... 189
5.4 Results and discussion ... 191
5.4.1 Modeling of the deformation zone ... 192
5.4.2 Modeling of the tool/chip interface ... 197
5.4.2.1 Known results ... 197
5.4.2.2 Simulation of stress distribution at tool/chip interface ... 198
5.4.2.2.1 Shear stress distribution ... 198
5.4.2.2.2 Normal stress distribution ... 199
5.4.3 Contact processes at the tool/chip interface ... 202
5.5 Comparison with machining experiments ... 204
References ... 210

Chapter six. Methodology of experimental studies in metal cutting ... 213
6.1 Similarity methods in metal cutting ... 213
6.1.1 General ... 213
6.1.2 Basics of similarity ... 215
6.1.3 Applications in metal cutting studies ... 218
6.1.3.1 Optimum cutting temperature — Makarow's law ... 218
6.1.3.2 Similarity numbers ... 227
6.1.3.3 Use of similarity numbers in metal cutting ... 229
6.1.3.3.1 Energy balance ... 229

6.1.3.3.2 Optimum cutting speed and machinability ... 234
6.1.3.3.3 Quality of the machined surface and tool wear ... 238
6.2 Temperature measurements in metal cutting ... 245
6.2.1 Conventional thermocouples ... 245
6.2.2 Tool-work (natural) thermocouple ... 249
6.2.3 Semi-artificial and running thermocouples ... 253
6.3 Cutting force measurements ... 255
6.3.1 General ... 255
6.3.2 Procedure used in the current study ... 257
References ... 267

Appendix. Cutting tool geometry ... 271
A.1 Tool-in-hand system ... 271
A.1.1 Planes ... 271
A.1.2 Angles ... 272
A.2 Tool-in-machine system (setting system) ... 276
A.3 Tool-in-use system ... 279
A.4 Determination of the uncut chip cross-section for non-free and non-orthogonal cutting conditions ... 283
References ... 287

Index ... 289

Chapter one

Introduction

1.1 General

Over the last hundred years, an extensive study has been carried out on the machining of metals. Most of this has focused on the down-to-earth reduction of machining costs and a pragmatic approach to the manufacture of parts of acceptable dimensional accuracy and surface quality. Unfortunately, a much smaller volume of research was devoted to discovering the fundamental mechanisms underlying metal machining processes in general, as opposed to seeking case solutions for particular machining problems. The real boom in fundamental metal cutting research, in the 1960s, brought the field a recognition of the need for an applicable metal cutting theory as well as a reputation for being extremely complex. Since then, the practice has advanced by its own costly way of trial and error, while fundamental research has experienced a decay after producing huge amounts of data that only occasionally match the practical results.

The modern history of metal cutting began in 1945 when Merchant published his vision of metal cutting phenomena.[1] As recorded in an excellent survey presented by the CIRP (International Institution for Production Engineering Research) working group on chip control,[2] numerous attempts to improve the theory proposed by Merchant failed to improve its predictive ability. Moreover, the original objectives of metal cutting research became somewhat obscure.[3] Instead of the original objective of establishing a predictive theory, the center of gravity has been shifted to developing theories of a descriptive nature which only explain post-process phenomena and thus have no prediction ability. As a result, no significant progress has been made, and, after many years of study, theory is still lagging behind practice. Shaw,[4] in a book summarizing his lifetime of experience in the field, came to the discouraging conclusion that it is next to impossible to predict metal cutting performance.

Nevertheless, university courses on metal cutting and, thus, the corresponding textbooks (Reference 5, for example) continue to teach Merchant's theory, because it offers the simplest explanations for metal cutting phenomena, although no physical background is provided.

Nowadays, industry relies completely upon empirical data as presented by tool and machine tool manufacturers, as well as by professional engineering associations through handbooks and workshops. Because these recommendations do not originate from a common theory, they provide only a good "starting point" which forces users, at their own cost, to determine the optimal values of cutting parameters for separate applications. And, for an outside observer with an obscure knowledge of the field, it may appear that industry is adapting well to this situation.

At this point, one may ask a logical question: "At the present stage of development, do we really need a realistic metal cutting theory?" The answer to this question is given in a quote from a recent CIRP working paper:[6] "A recent survey by a leading tool manufacturer indicates that in the U.S.A. the correct cutting tool is selected less than 50% of the time, the tool is used at the rated cutting speed only 58% of the time, and only 38% of the tools are used up to their full tool-life capability. One of the reasons for this poor performance is the lack of the predictive models for machining." The same was found in an earlier survey of cutting regime selection on CNC (computer numerical control) machine tools in the American aircraft industry,[7] which showed that selected speeds are far below the optimal economic speeds. If we recall that the U.S. now spends more than $115 billion annually to perform its metal removal tasks using conventional machining technology,[8] the price of the lack of the predictive models for machining is evident.

1.2 *Reasons for studying metal cutting*

Although the necessity for the study of metal cutting seems to be clear, it is believed that the original objective of metal cutting studies was to establish a predictive theory or analytical system which would enable us, without any cutting experiment, to predict cutting performance such as chip formation, cutting force, cutting temperature, tool wear, and surface finish.[3] In the author's opinion, two different groups of parameters are mixed in this definition. First is the group of external parameters which may be used to optimize the cutting process, such as the cutting speed. Second is the group of internal parameters which are necessary for study but cannot be used directly for process optimization. For example, the cutting temperature and tool wear are typical internal parameters which have no meaning for a process planner.

The ultimate objective of the science of metal cutting is to solve practical problems associated with efficient material removal in the metal cutting process. To achieve this objective, the principles governing the cutting process should be revealed. A knowledge of these principles makes it possible to predict the practical results of the cutting process and thus to select the optimum cutting conditions for each particular case.

The optimum cutting conditions are achieved by the proper selection of the cutting regime, cutting tool, and machine. These constitute the machining system[9] performance which defines the efficiency of parts' production.

Three major reasons define the necessity to study the metal cutting process vigorously:

- A predictive metal cutting theory would be beneficial to process planners by providing sufficient knowledge of process efficiency.
- A predictive metal cutting theory would be beneficial to the tool designers, producers, and users, as it constitutes a proper basis for making correct decisions. At present, it is usual to use experience, empirical relations, or even common sense for this purpose. Common books on cutting tool design do not sufficiently address the cutting process, even though the tools are designed to perform this process. Today, the geometry of cutting tools is a matter of technological convenience rather than of process considerations. That is to say that the tool geometry is chosen empirically to suit a particular tool material, chip-shape concern, tool holder, etc.
- A predictive metal cutting theory would be beneficial to the designers and users of machine tools by providing them with real process parameters such as the cutting force (direction and magnitude), heat generation rate, energy consumption, sources and energy of the vibrations, etc.

1.3 History of metal cutting

Machining research has been carried out for well over 100 years, resulting in a vast amount of literature on the topic. Though a number of reviews are known,[10–11] the most comprehensive today is probably the review by the CIRP.[2] It is worthwhile to point out here that the book by Zorev[11] is the most extensive experimental work in the field. No single study offers the results of so many reliably conducted experiments using a number of different workpiece materials, tools, and cutting conditions. It is understood that it is next to impossible to accomplish all these by a single researcher. Basically, the book summarizes experimental results conducted over many years by leading scientists in the former Soviet Union research institute which coordinated manufacturing activities in the industry. Professor Zorev used to be a director of the institute and a chairman of its scientific committee in which the main results presented in the book have been discussed in detail. As a result, the book is a scholarly treatment of the mechanics of metal cutting and represents a valuable source for researchers. The main points in the book have been presented clearly and logically and are well supported by multiple experimental results, such that the book deservingly could be referred to as the Bible of metal cutting.

Unfortunately, this is not the case, even though each serious study in the field is sited as a reference. It is this author's opinion, however, that the book has been poorly translated from Russian, which makes it very difficult to follow, particularly for the North American reader who must spend a great deal of time trying to understand the meaning of "tangential stress", "acute-angle cutting", "angle of cut", "shear rate vector", "blue shortness", "plastic shear force", "loss of strain hardening", etc. Moreover, the designations for Russian workpiece and tool materials are used throughout the book with no mention of the fact that practically all of these materials have AISI (American Iron and Steel Institute) analogs with which a North American reader is much more comfortable. Regardless of these shortcomings, the book is still a valuable source of information on metal cutting and deserves to be re-evaluated by many researchers. To facilitate this, we would like to list major points that have never been acknowledged by the researchers, as, in this author's opinion, the book was much ahead of its time:

- The book attempts to prove that the single shear plane model, proposed as early as the 19th century, is not a good approximation for the real cutting process. Two reasons for this conclusion have been analyzed in detail. First, the stress gradient in this plane must be infinite. Second, the acceleration of a particle of material being machined past the shear plane must be infinite. Consequently, there is a transition zone of plastic deformation between the chip and material being machined, the limits of which the shear stresses change continuously, as do the degree of plastic deformation and the speeds of movement of the particles of material being machined. This transition zone can conveniently be called the "chip formation zone" or the "plastic zone". The book points out that, as early as 1896, Bricks[12] justly criticized the single shear plane method and suggested a more realistic shape of the deformation zone which may be represented as a family of planes arranged fanwise and passing through the edge of the cutting tool. The book, when analyzing the drawbacks of Brick's scheme, introduces the unique configuration of the deformation zone based on a slip-line field consideration. This consideration is in agreement with the theory of engineering plasticity and offers the analytical determination of the initial and outer boundary of the deformation zone. For the first time, the principal axes of deformation and true deformation of the chip have been considered, and the explanation is offered that the orientation of the resultant chip texture does not coincide with the direction of plastic shear. Almost 20 years later, Shaw[4] acknowledged the fact that the material flow in metal cutting will not occur at the maximum shear stress; however, no explanation of this fact, which is at odds with common sense and the material test practice, has been provided.

- The experimental part of the book includes the study of more than 40 workpiece materials carefully chosen to represent all groups of steels, including plane low, medium, high carbon steels, low and high alloys, stainless steels, and heat-resistant alloys. It is worthwhile to point out here that the chemical composition and mechanical and physical properties for all workpiece materials were actually tested prior to cutting experiments, and the results are presented in the book. The separate studies at low and the high cutting speed are presented. For the first time, the angle of action is considered to be of prime importance. Unfortunately, this has been completely ignored by later studies.
- It is the fact that the temperature at the tool/chip contact interface cannot be considered as the single universal factor which determines the influence of the cutting conditions on the chip formation process (even for a given rake angle and a given work material). The assumption that the thermal strain hardening of steel as the temperature in the chip formation zone raises leads to a reduction of the deformation and the cutting ratio has no basis and is not consistent with experiment.
- The book demonstrates that the influences of different cutting conditions on the chip formation process are interconnected and can be discussed in isolation only to the first approximation without making any general conclusions. The existing theoretical formulas for the chip formation process do not take into account the direct influence of the cutting speed and incorrectly reflect the laws governing the chip formation at high cutting speeds. Most of these formulas are incorrect, even at low cutting speed, as they do not take into account the influence of the resistance of the workpiece material to plastic deformation.
- For the first time, the nature of the forces on the clearance face is considered, and the influence of the machining conditions on these forces has been analyzed. The study reveals the significant influence of the properties of workpiece materials and process parameters on these forces. In light of these findings, the experimental measurements of the cutting forces and their dependance on the process parameters conducted later by other researches have not considered or even mentioned the discussed results. To date, the forces on the clearance face have not interested researchers of metal cutting (see References 13 to 15, for example), although they claim they are trying to develop a predictive model for machining. When it comes to the analysis of tool wear, it is recognized that the flank wear is of prime concern, so that all tool life testing has primarily considered this wear. Therefore, it is only logical that one should recognize that flank forces play a significant role in metal cutting, as all other tool-workpiece (chip) contact

areas wear less than flanks (under commonly used cutting conditions). The underestimation of these values and the role of the flank forces is one of the barriers impeding the development of a reliable, predictive model for machining.

- The book offers a very good model for the mechanics of three-dimensional cutting process. Unfortunately, the very poor translation resulted in incorrect terminology being used throughout the book. As a result, this particular model is referred to in the book as a model for cutting at an acute (?) angle, instead of a model for oblique or three-dimensional cutting. It is not a surprise, then, that the succeeding researchers did not acknowledge the model.[4,16–18]

1.4. *Basic problems requiring solution*

As pointed out by Oxley,[13] if attention is limited to machining with a continuous chip, then the process is essentially one of plastic deformation, and the appropriate theory for its solution is plasticity theory. The plane-strain, slip-line field approach is used in this book to reveal the principles of the shearing process in metal cutting, as the plastic deformation of ductile materials takes place by shearing even under combined stress fields.

The applicability of this approach is limited by the rate of plastic strain in the deformation zone. It is known that, in general, when a metallic material experiences high strain rates, twinning seems to play a significant role in its plastic deformation.[19] Although the mechanism of twinning occurrence is not well understood, a great deal of experimental work on the nature of plastic deformation under extremely high rates (more than 10^3 s^{-1}) of strain reveals that twins form more copiously than during slow deformations.[20] At the strain rates believed to occur in cutting (10^6 and even higher),[13,15] all the plastic deformation observed in cutting should by governed by twinning; however, this is not the case in metal cutting because (1) the maximum contribution which the twinning could make to the plastic strain is about 17%,[19] which is only a tiny fraction of the strains observed in cutting; and (2) the microstructure of twins (see Figure 4.26 in Dieter[19]) is very different from that of slipping and has never been observed in cutting, or at least it has never been reported in work in which the traditional qualitative metallography at magnifications of 100× to 1000× and scanning electron microscopy were used.[1,4,13,14,21,22] Therefore, this problem has yet to be solved.

The problem of uniqueness of machining is the chief problem to be solved. It has been suggested by Hill[23] that the machining process is not uniquely determined by the input conditions so that no steady-state mode exists during the process. Trying to establish the principle of minimization of work of plastic deformation, Hill[24] also concluded that this principle is not applicable in metal cutting, as the geometry of deformation is not known *a priori*. As suggested by Oxley,[13] it is doubtful if a complete solution to the problem, as it is formulated in the known model for chip formation, can be found even with the powerful numerical methods now available. One of the main objectives of this book is to resolve this problem by showing ways to obtain a unique, exact solution.

Another major obstacle in predicting metal cutting performance is the modeling of the behavior of workpiece material in cutting. Numerous contradictory results are present in the published works. These works have aimed to find a material mechanical characteristic(s) responsible for its resistance to cutting. In particular, the so-called shear flow stress has been considered exclusively, and several different approaches to determine this stress under a given cutting conditions are suggested. To date, the usual approach to the problem has been to propose a chip formation model based on experimental observations and to develop an approximate machining theory from this.[13] However, the resistance of workpiece material to cutting as well as the mechanical properties of this material should not depend on the selected model. Unfortunately, this is not the case, so that a number of different models are in existence.[1,4,11,13,14] The objective of this book is to clarify this problem.

Finite element analysis (FEA) is maturing into a promising analysis tool to enhance understanding of machining and for prediction of the machining process output; however, it should not be forgotten that the accuracy of FEA is defined by how adequate the selected physical model is for the analysis. Because the results of the analysis depend primarily on the selected model, it is not very appropriate to compare the results of FEA obtained using different physical models. Moreover, a criterion of separation of the chip and the workpiece should be adopted in FEA, rendering the results of different studies even more incompatible. Therefore, FEA is a useful method if and only if an adequate physical model for chip formation is selected and a separation criterion, which directly stems from this model, is used.

To put the analysis of the metal cutting operation on a qualitative basis, a properly designed and conducted experiment along with the use of proper methodology for analysis of the experimental results play vitally important roles in metal cutting studies. Unfortunately, the space limitation found in many scientific journals forces the authors to sacrifice a great deal of description of their experimental procedures, leaving only the final results and their brief comparison with those obtained theoretically. Furthermore, common books on metal cutting do not consider working methodologies for the determination of many important cutting parameters such as the metallographic chip structures (sample selection, preparation, treatments, etc.), chip compression ratio, cutting forces, etc. These books do not suggest any methods for designing metal cutting experiments, data collection and acquisition techniques, design and calibration of special transducers and sensors, and workpiece and machine preparation for experiments. The combination of all these factors makes it next to impossible to compare the experimental results obtained by different researchers and unavoidably slows down the progress in metal cutting development. In the author's opinion, the experimental study of metal cutting is as important as the theoretical study. This book presents some of the basic methodologies of experimental study in metal cutting.

References

1. Merchant, M.E., Mechanics of the metal cutting process, *J. Appl. Phys.*, 16, 267, 1945.
2. Jawahir, I.S. and van Luttervelt, C.A., Recent developments in chip control research and applications, *CIRP Ann.*, 42, 659, 1993.
3. Usui, E., Progress of "predictive" theories in metal cutting, *JSME Int. J., Series III*, 31(2), 363, 1988.
4. Shaw, M.C., *Metal Cutting Principles*, Clarendon Press, Oxford, 1984, pp. 199, 200.
5. Boothroyd, G. and Knight, W.A., *Fundamentals of Machining and Machine Tools*, 2nd ed., Marcel Dekker, New York, 1989.
6. Armarego, E.J.A., Predictive modelling of machine operations — a means of bridging the gap between theory and practice, CSME Forum SCGM "Manufacturing Science and Engineering," Hamilton, Canada, May 7–9, 1996, p. 18.
7. Armarego, E.J.A., Jawahir, L.S., and Ostafiev, V.K., Working paper, STC "C", Modelling of Machining Operations working group, 1996.
8. King, R.I., Ed., *Handbook of High-Speed Machining Technology*, Chapman & Hall, New York, 1985.
9. Jawahir, I.S., Balaji, A.K., Stevenson, R., and van Luttervelt, C.A., Towards predictive modeling and optimization of machining operations, in *Manufacturing Science and Engineering*, Vol. 6(2), Proc. 1997 American Society of Mechanical Engineers International Mechanical Engineering Congress and Exposition, November 16–21, 1997, Dallas, TX, 1997, p. 3.
10. Bailey, J.A. and Boothroyd, G., Critical review of some previous work on the mechanics of the metal-cutting process, *ASME J. Eng. Industry*, 89, 54, 1968.
11. Zorev, N.N., *Metal Cutting Mechanics*, Pergamon Press, Oxford, 1966.
12. Briks, A.A., *Metal Cutting*, Dermacow Press, St. Petersburg, 1896 (in Russian).
13. Oxley, P.L.B., *Mechanics of Machining: An Analytical Approach to Assessing Machinability*, John Wiley & Sons, New York, 1989.
14. Trent, E.M., *Metal Cutting*, Butterworth-Heinemann, Oxford, 1991.
15. Stephenson, D.A. and Agapionu, J.S., *Metal Cutting Theory and Practice*, Marcel Dekker, New York, 1997.
16. Arsecularatne, J.A., Mathew, P., and Oxley, P.L.B., Prediction of chip flow direction and cutting forces in oblique machining with nose radius tools, *Proc. Inst. Mech. Engrs.*, 209, 305, 1995.
17. Stephenson, D.A. and Wu, S.M., Compute models for the mechanics of three-dimensional cutting process. Part I. Theory and numerical methods, *ASME J. Eng. Industry*, 110, 32, 1988.
18. Shi, H.-M. and Wang, J.-L., A model for non-free-cutting, *Int. J. Mach. Tools Manufact.*, 35(11), 1507, 1995.
19. Dieter, G., *Mechanical Metallurgy*, 3rd ed., McGraw-Hill, New York, 1986.

20. Cahn, R.W., Survey of recent progress in the field of deformation twinning, in *Proc. Conf. Deformation Twinning*, Vol. 25, Gainesville, FL, March 21–22, 1963, Gordon and Breach Science Publishers, New York, 1964, p. 1.
21. Vyas, A. and Shaw, M.C., Chip formation when hard turning steel, in *Manufacturing Science and Engineering*, Vol. 6(2), Proc. 1997 American Society of Mechanical Engineers International Mechanical Engineering Congress and Exposition, November 16–21, 1997, Dallas, TX, 1997, p. 13.
22. Kishawy, H.A. and Elbestawi, M.A., Effect of process parameters on chip morphology when machining hardened steel, in *Manufacturing Science and Engineering*, Vol. 6(2), Proc. 1997 American Society of Mechanical Engineers International Mechanical Engineering Congress and Exposition, November 16–21, 1997, Dallas, TX, 1997, p. 21.
23. Hill, R., The mechanics of machining: a new approach, *J. Mech. Phys. Solids*, 3, 47, 1954.
24. Hill, R., On the state of stress in a plastic-rigid body at yield point, *Phil. Mag.*, 42, 868, 1951.

chapter two

System approach

Modern technological concepts make it possible to define the present stage of development as the system era. Management makes use of terms such as "system concept", "system philosophy", and "system approach". Engineers and physical scientists speak of "system analysis", "system engineering", and "system theory". Even in medicine or biology the specialists speak of "nervous system", "homeostatic system", "gene system", etc. However, the picture is not so bright today as it seemed to be in the 1960s, when the system approach began to boom. Only in certain fields (for example, computer science) has the system concept been developing rapidly with great practical significance. As a result, it is only in this field that system specialists (system analyzers, system programmers, and system managers) are found.

System engineering is beginning to emerge in concept as a generalization of traditional engineering, a generalization in three important aspects:[1]

1. The scope of engineering projects is considerably enlarged and expanded to include many more system interfaces, such as the man/machine interfaces, man/man interfaces, even system/society interfaces.
2. The scientific bases for engineering decisions are necessarily broadened.
3. Materials from engineering products that are fabricated can no longer be limited and circumscribed.

With the emergence of the concept of system engineering, the traditional role of mathematics in engineering has been broadened or even completely changed. Traditionally, mathematics has been the vehicle by which physical concepts are applied to engineering problems, and, traditionally, engineers have been scornful of mathematical rigor, secure in the confidence (often misguided) that physical intuition obviates the need for mathematical rigor. At the system level, however, an engineer is not so much concerned with physics as he is with organization, information, and communication, as well as with the mathematical nature of relationships, whether they are physical or not. At this level, his principal enemy is always the complexity of a system under consideration.

System problems are often aptly described as a "can of worms", for it is difficult to discriminate between the different elements of the problem, such as the system boundaries, components and their various levels, organization, and interrelationships between the levels. The whole problem seems to be constantly in motion; the components are hopelessly intertwined, so much so that there may be only one indivisible component. It is difficult to grasp any one of the slippery components, and the problem is partly immersed in obscuring debris.

Only if the mathematical rigor is adhered to can system problems be dealt with effectively. Therefore, if an engineer is to apply the system concept, he must at least develop an appreciation for mathematical rigor, if not also mathematical competence, as rigorous mathematics is more than a tool by which precisely to apply the concepts from other scientific disciplines to engineering problems. It plays the role of descriptive, inductive, comparative, and, above all, experimental science, providing a scientific basis upon which engineering decisions can be made. The process is somewhat as follows: the system engineer, faced with a problem derived from some system phenomenon, attempts to describe the phenomenon mathematically; he attempts to construct a mathematical model which represents a mathematical structure of some kind. He compares this structure with those existing in mathematics; he experiments with his model both mathematically and deductively, all the while checking the results of such comparison and experiments with the requirements of the problem and experimental or heuristic evidence concerning the phenomenon itself. He modifies the model and experiments some more. Finally, he arrives at a satisfactory model that he can proceed to analyze with various mathematical and computational techniques in order to arrive at an engineering decision.

Though the above-discussed procedure looks relatively simple and logical, the chief problem here is to distinguish the system to be analyzed, its boundaries and components. Intuition and experience at this stage are essential. This chapter illustrates that the cutting process takes place within the cutting system and applies the major system property to analyze this process.

2.1 *Observation part of existing studies on chip formation*

An overall review of previous work on the chip formation theory reveals that the authors have observed essentially the same picture of chip formation regardless of their differences in the analytical modeling. The observations seem to have led to an idealized picture which is known as the model for orthogonal cutting (Figure 2.1). The diagram in Figure 2.1 shows a tool

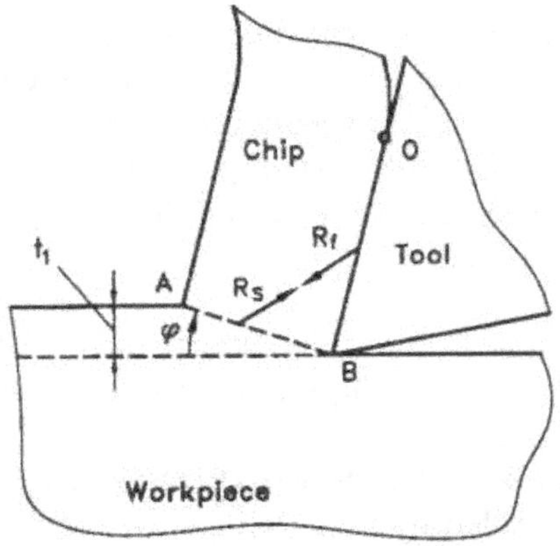

Figure 2.1 The known model for orthogonal cutting. (From Astakhov, V.P. et al., The bending moment as the cause for chip formation, in *Manufacturing Science and Engineering*, Vol. 6(2), presented at the 1997 American Society of Mechanical Engineers International Mechanical Engineering Congress and Exposition, November 16–21, Dallas, TX, pp. 53–60. With permission.)

removing the stock of thickness t_1 by shearing it (as assumed) ahead of the tool in a zone which is rather thin compared to its length, thus it can be represented reasonably well by the shear plane *AB*. The position of the shear plane is customarily defined using the shear angle φ, as shown in Figure 2.1. After being sheared, the layer to be cut becomes a chip which slides first along the tool rake face, following the rake's shape (a straight portion of the chip in Figure 2.1), and then, beyond a certain point *O* on the tool face, curls away. In this figure, derived from visual observation, is included a force system which consists of resultant force R_f on the tool face and resultant force R_s on the shear plane. The most essential assumption for this model is that these forces are collinear; thus, in order to satisfy the requirement of static equilibrium, they must be equal in magnitude and opposite in direction, as shown in Figure 2.1.

The described picture has been used so far almost without exception. Some works have attempted a few amendments to consider the plastic deformation in the shear zone, to account for the presence of a built-up edge, to establish the boundaries of the shear zone, etc. However, the two most essential features of the model (Figure 2.1) have never been questioned and remained unchanged: (1) the chip forms by the process of simple (pure) shearing, and (2) resultant force R_f on the tool face and resultant force R_s on the shear plane are equal in magnitude and opposite in direction.

The discussed model constitutes the very core of the metal cutting process which has been represented, in simple terms, as a cutting tool removing a certain part of the workpiece by means of simple shearing. However, there was no attempt to consider the metal cutting process in the context of other closely related manufacturing processes known as the shearing press operations,[3] which have been modeled essentially using the same assumption/mechanism. The conclusion that the comparison would be reasonable can be substantiated by the list of features common to the shearing press operations and the metal cutting process:

- They both include a tool with a sharp cutting edge(s).
- They both include the workpiece with an allowance to be removed by the cutting edge(s).
- The interaction between the workpiece and the tool takes place at a certain speed limited by tool wear.
- The part to be removed is deformed and then separated from the rest of the workpiece by means of simple shearing.

Therefore, for the present model of orthogonal cutting (Figure 2.1), metal cutting does not have any specific technical feature to distinguish it from other closely related manufacturing processes. From the above consideration it may be concluded that metal cutting has been attracting much more attention than other shearing press operations only because it is a more important component in the overall manufacturing activity.

For further understanding, the following questions, which are often avoided in literature on metal cutting, should be answered:

- What is the difference between metal cutting and cutting?
- If a polymer or any other non-metal (wood, stone) material is cut by turning, milling, or even drilling, how should this process be classified?
- What kind of separation process is taking place when one cuts a piece of bread with a knife or paper with a pair of scissors?

It is only logical that, at this point, one should ask the question: "What is metal cutting all about?" The proper answer to this question would have enormous implications on both the theory and the metal cutting practice. The answer could bring us closer to a starting point for developing a realistic metal cutting theory of a predictive (instead of descriptive) nature.

To be able to answer the question, we should first understand why metal cutting cannot be distinguished among other shearing press processes based on the existing model for chip formation. In the author's opinion, two significant aspects have been overlooked in the analysis of the existent model. First, the model does not account for a number of workpiece materials which cannot be deformed by shearing — cast irons, for example. Second, even when applied to the study of the cutting process of ductile materials, the theory cannot possibly explain chip formation as accomplished by simple shearing. The latter becomes obvious when a number of related manufacturing processes such as blanking, punching, etc. are considered, as these are also accomplished by simple shearing; however, no chips are produced. Moreover, the indentation of a semi-infinite mass of a rigid, perfectly plastic material by a rigid, straight-sided, or acute-angled indenter or by a flat punch causes extensive shearing; however, the chip does not form even if an extremely high force is applied.[4]

Keeping in mind these two aspects and the fact that the theoretical results derived using the existing model exhibit, in general, poor agreement with experiments, the author would like to present a system approach in metal cutting studies. The proposed approach should lead to a better understanding of the metal cutting process and its unique features and help explain its well-known paradoxes. As a result, it is hoped that it will help bridge the gap between theory and practice.

2.2 *System and assemblage*

In techniques, aggregates are classified as either system or assemblage.[1] Because the properties, characteristics, and methods for analysis are considerably different for these two, it is important to define to which group an aggregate under consideration belongs. Unfortunately, this stage has never been considered in the known analysis of the metal cutting process. Therefore, it is necessary to study this stage in detail.

If the professionals of different specialities are asked what the word "system" means, many diverse answers utilizing the various professional languages would be obtained. If these answers were translated into simple English and compared to each other, the notion common to them all might be stated as such: "A system is something which accomplishes an operational process; that is, something is operated in some way to produce something." That which is operated upon is usually called *input*, that which is produced is called *output*, and the operating entity is called a *system*.

In more strict language, a system is a device, procedure, or scheme which behaves according to some descriptions, its function being to operate on information and/or energy and/or matter in a time reference to yield information and/or energy and/or matter.

2.2.1 *Mathematical definition of system*

A significantly simplified mathematical definition of a system which supports the above-discussed intuitive definition follows. A system is a set:

$$Z = \{S, P, F, M, T, \sigma\} \tag{2.1}$$

Here S is a set (not empty); P is a set (not empty); F is an admissible set of input functions with values in P; M is a set of functions each defined on S

with values in S; T is a subset of R containing 0; σ is function defined on $F \times T$ with values in M such that σ is "onto", and:

1. The identity mapping $\omega \in M$ and for every $f \in F$, $\sigma(f, 0) = \omega$.
2. If $f \in F$, s, t, and $s + t \in T$, then $\sigma(f \to s, t)\sigma(f, s) = \sigma(f, s + t)$.
3. If f and $g \in F$, $s \in T$, and $f(t) = g(t)$ for all $t \in R(s)$, then $\sigma(f, s) = \sigma(g, s)$.

If Z is a system, then S is called the set of states of the system Z; P is called the set of input states of the system Z; F is called the set of output functions for the system Z; M is called the set of transition function of the system Z; T is called the time scale of the system Z; and σ is called the state transition function of the system Z. If $f \in F$, $x \in S$, and $t \in T$, then the state of the system at time t given the input function f and initial state x is $(\sigma(f, t))(x)$. The time trajectory of Z determined by $f \in F$ and $x \in S$, denoted *timetraj(f, x)*, is a function, defined on T with values in S, as follows. For every $t \in T$: $(timetraj(f, x))(x) = \sigma(f, t)(x)$. The input trajectory of Z determined by $t \in T$ and $x \in S$, denoted *inputtraj(t, x)*, is a function, defined on F with values in S as follows. For every $f \in F$: $(inputtraj(t, x))(f) = \sigma(f, t)(x)$. If Q is an arbitrary not empty set, then any $\zeta_y \in \psi(S,Q)$ is an output function for Z with values in Q. The output trajectory of Z determined by $f \in F$, $x \in S$, $Q \neq \phi$, and $\zeta \in \psi(S, Q)$ is the function $\zeta_y\,(timetraj(f, x))$: $(\zeta(timetraj(f, x))(t) = \zeta_y\,(\sigma(f, t)(x))$ for every $t \in T$.

The set S, formally representing the set of possible states of the system, describes the internal workings of the machine or system. The set P represents the set of possible input states or input conditions for the system. The set F represents the class of possible or admissible input schedules, or input histories, or input functions for the system. Sufficient input functions are ensured by insisting that in F there are at least one input function and all the functions that can be generated from it by segmentation and translation. The set M represents the set of all modes of behavior available to the system in the following sense. The actual behavior of the system can be thought of as being represented by the states in S or, rather, the trajectory in S generated be the system in operation. Given that the system is in state x at a particular time, the set of possible states of the system at some later time is $\{\alpha(x):\alpha \in M\}$. Thus, the set of mapping in M determines the behavior of the system. Each mapping in M has coordinates, provided at least one input function and time t through the mapping σ: ... "σ is *onto*". Thus, M is the range of σ in $\psi(S, S)$; that is, $M = \Gamma(\sigma, F \times T, \psi(S, S))$. $M = \Gamma(\sigma)$ can be written for short because σ is always a function on $F \times T$ with values in $\psi(S, S)$.

In specifying particular systems, it will be convenient, quite often, to define M as $\Gamma(\sigma)$ and then give σ a very careful and detailed definition. In fact, if M and σ are defined independently, for a particular system, it must be shown that σ is "onto" M. Note that the coordinatization of M through the function σ has the form of a function of the variable (f, t). This is more general than a function of the variable $(t, f(t))$. The form of the independence of the state of the system on the input function is not specified. It might be as simple as a functional dependency on $(t, f(t))$, or it may be, at least in principle, so complicated that it must depend on integration over the past. Actually, the form of dependence may be any of an infinite number of possibilities. The essence of the postulate in this respect is that the state of the system depends only on the states at time 0, the input, and how long the system has been operating on a given input function, however complicated the dependency on the input functions might be.

The subset T of R represents the period of time over which the system exists, or over which the system is observed, or over which a given phenomenon exhibits system-like behavior. Any of these representations might be valid. Most interesting will be the cases for which $T = R$ and $T = R^{++}$. Intuitively, it is to be expected that T would be connected (no holes in the time scale) and that T would be an interval. These are the cases of most concern, but in general no such restrictions are placed on T in order, for example, to make the initial discussion of time-discrete systems somewhat more convenient.

Through mapping σ, the dynamic behavior of the system is modeled. If the system at time $t = 0$ is in a state $x \in S$ with an input function $f \in F$, then for every $t \in T$, the state of the system at time t is designated $(\sigma(f, t))(x)$. From now on, the parentheses about the symbol $\sigma(f, t)$ will be left out, for it should be understood that this composite symbol always stands for mapping of S into itself.

In Condition 1 it is required that no matter what input function is considered, the mapping corresponding to $t = 0$ for that function is the identity. This is consistent with the postulation of the dynamic behavior of the system. Given an input function f and the system in state x at the time $t = 0$, $\sigma(f, t)(x)$ will be the state of the system at time t. In particular, for $t = 0$, it is required that the state of the system be $\sigma(f, t)(x)$, which, under the original assumption, is again x. So, $\sigma(f, 0)$ has to be identity mapping. This means, further, that M is not empty; M contains the identity mapping.

In Condition 2 the dynamic behavior of the system is required to be consistent with respect to transformations of input functions and with the orientation of the time scale. Time $t = 0$ is thought of as being the present moment, positive times as being in the future, and negative times as referring to the past. Given an input function, f, the state of the system at time t, for any $t \in T$, is designated $\sigma(f, t)(x)$ if the state of the system now (at time 0) is x, regardless of whether t is positive, negative, or 0.

A past state of the system fulfills a certain consistency requirement demanded by intuition. The system could have been in state y at time $-t$ (where t is positive), given the input f and the present state x, only if the system initially in state y, experiencing the successive values of the input function f, beginning with $f(-x)$, arrives in state x at time t. This means that y could have been the state of the system at time $-t$ only if $\sigma(f \to -t, t)(y) = x$. Equivalently, in order that y be the state of the system at time $-t$, it is necessary that y be an element of $(\sigma(f \to -t, t))^{-1}(\{x\})$. To put it still another way, if y was the state of the system at time $-t$ with input function f, then the present state of the system is $\sigma(f \to -t, t)(y)$. Thus, states of the system in the past can be imagined as being subjected to experimental verification.

Any $y \in S$, proposed as a possible system state at time $-t$, given the input f and the present state x, can be qualified by the following experiment: start the system in state y, apply the successive values of the input function f beginning with $f(-t)$, and observe the state of the system at time t. In this experiment, time is measured with 0 indicating the time at which the system is placed in state y. On this time scale, the function f is now the function $f \to -t$, so that the state of the system at time t is designated $\sigma(f \to -t, t)(y)$. This state, of course, must be equal to x if y is a possible past state of the system under the given circumstances.

It is important to note that the introduced definition does not insist that the state of the system in the past be included in the definition of the system; it might be that T contains no negative values of time. In this case, the definition of the system does not include the past history of the system. The point of view taken in this discussion is that past states of the system can be included in the definition only if they are unique so that the mapping $\sigma(f, -t)$ can be single valued.

Summarizing the foregoing consideration of the possible state of the system in the past, the following may be concluded. In general, there are many states of a system from which the system, experiencing input f from time $-t$, would have arrived at the state x by this time (now). Therefore, it makes little sense to speak of the state of the system at time $-t$. In this case, $\sigma(f, -t)(x)$ could not be defined uniquely; the only recourse is to exclude the value $-t$ from the set T, because, if $-t$ is in T, then $\sigma(f, -t)(x)$ has to be defined as an element of M for every f. One way to exempt $y = \sigma(f, -t)(x)$ from the kind of verification indicated above is to exclude t, itself, from T, and then $\sigma(f \to -t, t)$ would not be defined. This still, however, does not absolve the requirement that $\sigma(f, -t)$ be a single-valued transformation.

States of the system in the future also can be imagined as being subjected to verification by experiment. For $s > 0$, and $t > 0$, Condition 2 of the definition can be interpreted as postulating a relation between the outcomes of two experiments performed on a system. In the first experiment, the system is started in state x, input function f is applied, and the state of the system is observed at time s and again at time $s + t$. In the notation of the definition, these states would be designated $\sigma(f, s)(x)$ and $\sigma(f, s + t)(x)$, respectively.

In the second experiment the system is started in the state $\sigma(f, s)(x)$, the successive values of the input function f beginning with $f(s)$ are applied, and the state of the system is observed at time t. On the time scale of the second experiment, the input function would be designated $f \to s$; hence, in the notation of the definition, this last observed state would be denoted $\sigma(f \to s, t)(\sigma(f, t)(x))$. Condition 2 can now be interpreted as insisting that for every system of interest in this discussion the second state observed in the first experiment and the state observed in the second experiment — no matter what f, no matter what x, no matter what s and t, so long as s, t, and $s + t$ are all elements of T — must be equal; that is,

$$\sigma(f \rightarrow s,t)\ \sigma(f,s) = \sigma(f,s+t) \tag{2.2}$$

Because this relation must hold for every $s \in S$, the following equation holds for composition of the corresponding mappings in M:

$$\sigma(f \rightarrow s,t)(\sigma(f,s)(x)) = \sigma(f,s+t)(x) \tag{2.3}$$

It is within the framework of these kinds of considerations made above that the remainders of the cases for Condition 2 are justified. There are six cases that might be separately justified:

1. $s \geq 0, t \geq 0$
2. $s \geq 0, t < 0, s + t \geq 0$
3. $s \geq 0, t < 0, s + t < 0$
4. $s < 0, t \geq 0, s + t \geq 0$
5. $s < 0, t \geq 0, s + t < 0$
6. $s < 0, t < 0$

It is seen that Case 1 has been discussed so far.

In general, it is possible to discuss (and imagine) both past and future states of the system at the outcomes or results of computations, and Condition 2 can be interpreted as requiring that, if the state of the system at time s is computed, assuming the input function f and the initial state x, obtaining $\sigma(f, s)(x)$ and the successive input values of the function f beginning with $f(s)$, obtaining $\sigma(f \rightarrow s, t)(x)\ (\sigma(f, s)(x))$, the result should be the same as if the state of the system at time $s + t$ had been computed assuming the input f and the initial state x, for every $f \in F$, $x \in S$, and for every $s, t \in T$, such that $s + t \in T$ whether s or t is positive, negative, or 0.

To discuss one more case, consider that s, t, and $s + t > 0$ all belong to t but that $s < 0$, $t > 0$, and $s + t > 0$; that is, $0 < -s < t$. This case can be interpreted as follows. Compute the state of the system at time s assuming the input function f and the initial state x to obtain $\sigma(f, s)(x)$, then perform two experiments. First, start the system in state $\sigma(f, s)(x)$, applying the successive values of the input function f beginning with the value $f(s)$, and observe the state of the system at time t to obtain $\sigma(f \rightarrow s, t)(x)(\sigma(f, s)(x))$. Then, start the system in state x, apply input f, and observe the state of the system at time $s + t$ to obtain $\sigma(f, s + t)(x)$. Condition 2 requires that these two experiments have the same outcome for every $f \in F$, $x \in S$: $\sigma(f \rightarrow s, t)(x)\sigma(f, s)(x) = \sigma(f, s + t)(x)$, and rightly so. The reasons are as follows. If $\sigma(f, s)(x)$ is taken to be a legitimate past state of the system in the first experiment at time $-s$ (s is positive, recall), the state of the system will be x, and from there until time t the system will experience the same input values with the same state as in the second experiment. In other words, the first experiment from time $-s$ to time t is the same as the second experiment from time 0 to $s + t$, the only difference being the origin from which time is being measured. Therefore, these two experiments should have the same outcome.

The effect of Condition 3 is to eliminate from the discussion mathematical constructions that do not correspond to real system phenomena. It is desirable, for example, to be able, at least in principle, to compute the state of the system at time t knowing only the state of the system at some prior time s and the input function between s and t. Thus, it is desirable to eliminate from the discussion systems such as mathematical constructions whose state at time t depends on the future of the input beyond t. This is not to exclude anticipatory systems from discussion, for to anticipate and attempt to predict the future is not the same as acting or behaving on the basis of sure knowledge of the future and it is not the same as requiring knowledge of future input to compute the present state. Similarly, it is desirable to eliminate from the discussion systems such as mathematical constructions for which the computation of the state at time t depends on the values of the input function prior to s as well as the state at time s and input between s and t. This is not to exclude from the discussion those systems that are affected by or remember the input from other times, for in these cases the effect or memory of prior input must be part of the description of the state at time s. The requirement that accomplishes such control on discussion is the one that states that if two input functions f and g agree between 0 and t, then it makes no difference which input is specified; with the system initially in an arbitrary state x and either f or g applied to the system, the state of the system at time t will be the same in either case.

It has been pointed out that if past state is to be defined it must be unique. Condition 3 goes just a bit further. If the past state has been defined, it must be computable strictly on the basis of the preset state and the values of the input the system must have experienced between the time of interest in the past and now. This requirement provides some protection, by means of consistency, for those cases in which $-s$ might belong to T but s does not, thus exempting the state at time $-s$ from verification of the prerequisite characteristic for states in the past.

2.2.2 *Mathematical definition of an assemblage*

An assemblage is a set:

$$Z = \{S,\ P,\ F,\ M,\ T,\ \sigma\} \tag{2.4}$$

Here, S is a set (not empty); P is a set (not empty); F is an admissible set of input functions with values in P; M is a set of mappings of S into itself; T is a subset of R containing 0; σ is mapping from $F \times T$ onto M. The sets S, P, F, M, T, σ, and their elements are called the constituents of Z.

2.2.3 The relation between a system and an assemblage

Because any set of elements arranged to satisfy Equation (2.4) constitutes an assemblage, it is very important to make a clear distinction between these two aggregates. Such a distinction may be very helpful when one tries to prove that a given assemblage is a system.

Comparison of Equations (2.1) and (2.4) shows that an assemblage $Z = \{S, P, F, M, T, \sigma\}$ is a system if and only if Conditions 1, 2, and 3 of the system definition are satisfied by Z. Though great lengths have been traversed in discussion of these conditions in the above section, this section will provide the weight of evidence in a formal way by introducing a new property with some intuitive justification of its own and by showing that this is formally equivalent to Conditions 2 and 3. The property in question is if $f, g \in F$; s, t and $s + t \in T$, then

$$\begin{aligned} \sigma\left(\left(f \to s \mid g \to s\right) \to -s, s+t\right) &= \sigma\left(f \to s, t\right)\sigma\left(h, s\right) \text{ if } t < 0 \\ &= \sigma\left(g \to s, t\right)\sigma\left(h, s\right) \text{ if } t \geq 0 \end{aligned} \tag{2.5}$$

where $h = g$ if $s < 0$ and $h = f$ if $s \geq 0$.

The rather complicated-looking function $(f \to s \mid g \to s) \to -s$, involved in this property, is merely the input function which is equal to f from $-\infty$ to time s and is equal to g from there on:

$$\begin{aligned} \left(\left(f \to s \mid g \to s\right) \to -s\right)(t) &= \left(f \to s \mid g \to s\right)(t-s) \\ &= \left(f \to s\right)(t-s) \text{ if } t-s < 0 \\ &= \left(g \to s\right)(t-s) \text{ if } t-s \geq 0 \\ &= f(t) \text{ if } t < s \\ &= g(t) \text{ if } t \geq s \end{aligned} \tag{2.6}$$

This function corresponds to the act of an experimenter having input functions f and g available to him and programming or generating a new input function by cutting the lefthand tail off g at s and the righthand tail off f at s and joining the remnants together at s. The mathematical operation of segmentation is possible only at the origin; for this reason, in describing the experimenter's action mathematically, it is necessary to translate the origin on both curves to the point s, obtaining $f \to s$, $g \to s$, then to perform the segmentation obtaining $f \to s \,|g \to s$, and to translate the origin back to the original position obtaining $(f \to s \,|g \to s) \to -s$. This new property is justified in much the same way as Condition 2 was justified, in terms of experiments and computations.

The main crux of the argument stems from the question: In a given time interval, which input function was the system experiencing? Answering this question in two ways and setting the resulting states equal yields the property.

In the theory of systems, an alternative statement of system consistence requirements helps to prove that an assemblage is a system, showing that the assemblage satisfies Conditions 2 and 3. This statement is as follows.[1] Let Z be an assemblage such that $\omega \in M$ and for every $f \in F$, $\sigma(f, 0) = \omega$. Then, Z is a system if and only if Z has the following property. If $f, g \in F$, s, t, and $s + t \in T$, then:

$$
\begin{aligned}
&\sigma\left(\left(f \to s\,|g \to s\right) \to -s,\, s+t\right) = \\
&= \sigma\left(f \to s,\, t\right)\sigma\left(h, s\right) \textit{ if } t < 0 \\
&= \sigma\left(g \to s, t\right)\sigma\left(h, s\right) \textit{ if } t \geq 0 \\
&\textit{where} \\
&h = g \textit{ if } s < 0 \\
&h = f \textit{ if } s \geq 0
\end{aligned}
\tag{2.7}
$$

The first image of a system is the usual black box diagram. Think of a box with k input terminals, each lead having some sort of a meter on it, the box itself having m dials on the side giving information about the internal state of the system (which can be a machine or a process), and n output terminals, each with some sort of meter, as shown in Figure 2.2.

In the example, P would be identified as some subset of R^k, the real Euclidean vector space vector with k dimensions, and the input functions would all be of the form of a function f defined on R with values in R^k:

$$
f(t) = \left(f_1(t), f_2(t), \ldots, f_k(t)\right) \tag{2.8}
$$

that is, a vector-valued function. Then S would similarly be some subset of R^m, and the mapping $\sigma = (f, t)$, determining the state of the system at time t given the input f, would be thought of as acting on the state $x(0) = (x_1(0), x_2(0), \ldots, x_m(0))$, so that $\sigma = (f, t)(x(0)) = (x_1(t), x_2(t), \ldots, x_m(t)) = x(t)$. The output set Q would be some suitable subset of R^n, and the output function ζ_f would be defined as follows:

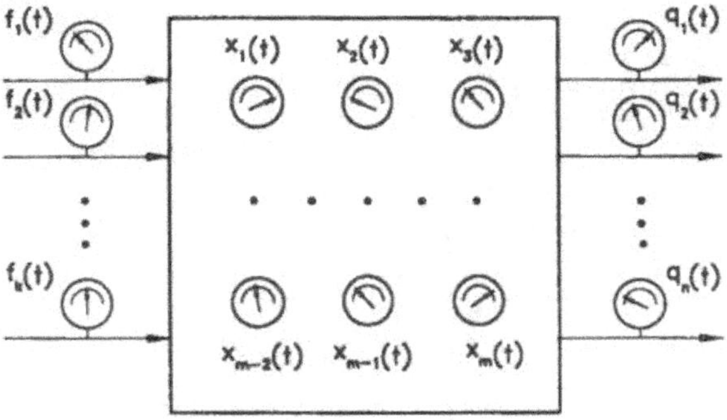

Figure 2.2 The first image of a system.

$$\zeta_f(x(t)) = \zeta_f(x_1(t), x_2(t), \ldots, x_m(t)) = (q_1(t), q_2(t), \ldots, q_n(t)) \qquad \textbf{(2.9)}$$

Alternatively, the set of states might be defined as a suitable subset of R^{m+n} and the state of the system at time t as the vector:

$$y(t) = (x_1(t), x_2(t), \ldots, x_m(t), q_1(t), q_2(t), \ldots, q_m(t)) \qquad \textbf{(2.10)}$$

so that

$$\sigma(f, t)(y(0)) = y(t) \qquad \textbf{(2.11)}$$

Then the set of output states would be some suitable subset of R^n as before, but the function ζ_f would be defined as:

$$\zeta_f(y(t)) = (q_1(t), q_2(t), \ldots, q_n(t)) \qquad \textbf{(2.12)}$$

and it would all amount to the same thing.

To facilitate the above mathematical formalization, the following graphical representation might be useful. If input f is represented as a real-valued ($P = R$) function of time, then a representational sketch for S and the trajectory in S given by f and initial state x can be drawn as illustrated in Figure 2.3. This kind of diagram is particularly useful when trying to keep the effects of translations and segmentation straight. Translations, of course, imply an amount of change in the origin of the time scale.

Figure 2.4 illustrates all the postulated parts of the system theory. In this figure, all discussed interdependencies are indicated by arrows. The point $(f, 0)$ in $F \times T$ determines mapping ω in M corresponding to initial state x in S and output value $\zeta_f(x)$ in Q; similarly, the point (f, t) in $F \times T$ determines

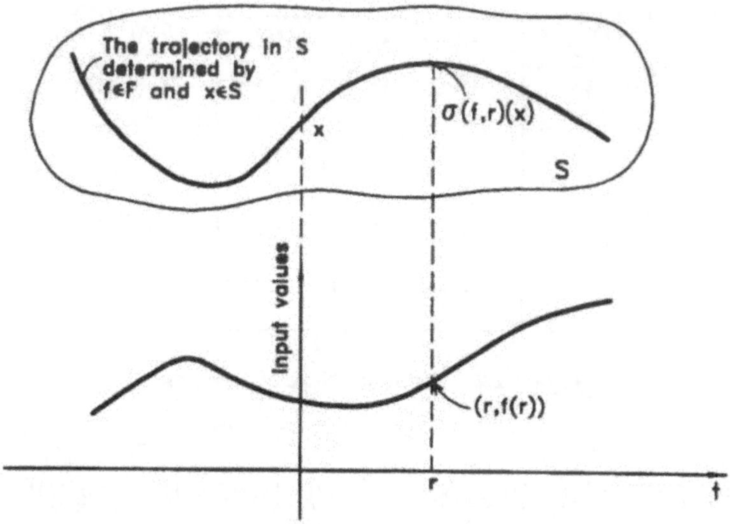

Figure 2.3 Real-valued functions as input to general system.

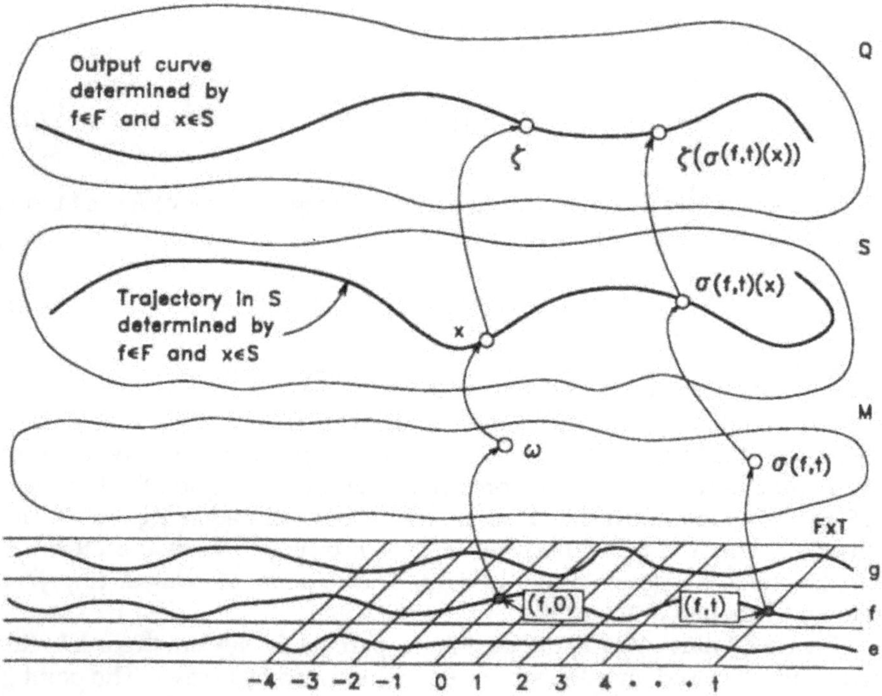

Figure 2.4 The principal parts of system theory.

mapping $\sigma(f, t)$ in M, which in turn determines the point or state $\sigma(f, t)(x)$ on the corresponding trajectory of Z in S, and, finally, determines the output ζ_f $(\sigma(f, t)(x))$ in Q.

The definition given here of a system implies the problems and processes of system design. In any engineering design situation, the first step is always to define and delineate the problem. In the context of system design, with the definition given here this means to specify the set P of input states and the class F of output functions, for these mathematical entities represent the environment (or, better, the input interface) with which the system must cope. Set P may also include the possibility for control of the system outside the system.

With set P and class F defined, it is necessary also to state, as part of the problem definition, the objectives or purpose of the system. In general, this will involve the definition of a set Q (of desirable output states) and function ζ_f defined on $F \times T$ with values in Q. Then, $\zeta_f(f, r)$ symbolizes, in general, what is expected of the system at time r given the input function f. The design process now proceeds ultimately to define the set S of internal states, the class M of mappings of S into itself (the class of models of behavior, heuristically speaking), the time set T, and the mapping of σ from $F \times T$ onto M to define the dynamic behavior of the system in time with respect to each of the possible or admissible input functions. This is carried out in such a way that there exists a function η_f on S with values in Q such that for every $f \in S$ there exists $x \in S$ such that $\eta_f(\sigma(f, t)(x)) = \zeta_f(f, t)$ for every $t \in T$.

There are many steps to the admittedly oversimplified system design process described above. The definitions of set P of input states and discernment of class F of input functions may not be an easy task. Even so, the definition of S, M, and T and the mapping $F \times T$ onto M is not without difficulties.

As follows from strict mathematical definitions of systems vs. assemblages and from their comparison, there are three principal system properties that distinguish a system from an assemblage (as a simple set of elements):

1. The most important property, but one always overlooked in metal cutting studies, is the system time. System time is a new variable in the analysis of a system so that the system's component must be time dependent.
2. Dynamic interaction among system components.
3. Possibility for control of the system outside the system.

In light of these considerations, it is evident that the model shown in Figure 2.1 is not a system, as it does not possess the major system properties; therefore, it is recognized as an assemblage.[5] As a result, the input and output parameters in the known analysis of this model (such as the cutting forces, stress and temperature distributions at the tool/chip interface, rate of strain) are considered to be time invariant. For example, the cutting tool is always considered separately, and its interactions with the workpiece and the chip are substituted by the static loads, stresses, and heat sources. In the author's opinion, this is the principal problem in the known studies on metal cutting.

2.3 Cutting system

To prove that the assemblage shown in Figure 2.1 is a system, the three principal system properties should be examined.[6–9] To do this, a two-step approach has been used. The first step aims to show time dependence of the parameters of the cutting system and their dynamic interactions. The second step is to verify the outside controllability of the system.

2.3.1 First step: time dependence of parameters of cutting system and their dynamic interactions

Let us now consider the changes which the system approach will bring into a conventional picture of the chip formation process. All these changes will be introduced step by step and each one will be illustrated by a separate figure.

As argued before, a simple comparison with other shearing processes shows that chip formation cannot be explained by shearing, regardless of the applied load level. Therefore, the model shown in Figure 2.1 which assumes pure shearing to be the controlling mechanism for the plastic deformation of workpiece material has an internal contradiction.

It seems only logical that another factor should be considered, in addition to shearing, to explain why the chip is formed at all. It has been assumed that the bending moment raised in the deformation zone due to the interaction between the chip and the tool rake face may be considered as the cause for chip formation.[8,9] To illustrate the significance of the bending moment in the chip formation process, a three-stage approach has been used. At the first stage, the metal cutting phenomena are simply observed and discussed qualitatively. The second stage provides a reasonable experimental verification corroborating the conclusions drawn at the first stage. The aim of the third stage is a mathematical treatment to estimate the observed effects with reasonable accuracy.

2.3.1.1 First stage: observational part

Consider the cutting tool starting to advance into the workpiece (Figure 2.5a). As a result, the stresses grow in the workpiece and, as might be expected, the maximum stress occurs in front of the cutting edge. When this maximum stress reaches a certain limit, the following may happen:

- If the workpiece material is brittle, a crack appears in front of the cutting edge (Figure 2.5b). Later on, this crack leads to the final fracture of the layer being removed.

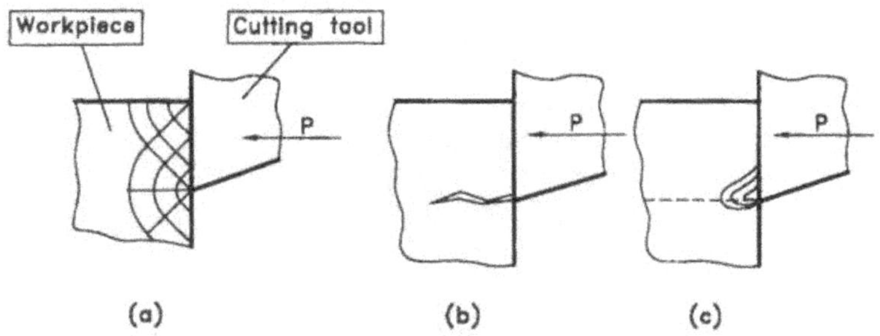

Figure 2.5 Cutting tool starting to advance into the workpiece. (From Astakhov, V.P. et al., The bending moment as the cause for chip formation, in *Manufacturing Science and Engineering*, Vol. 6(2), presented at the 1997 American Society of Mechanical Engineers International Mechanical Engineering Congress and Exposition, November 16–21, Dallas, TX, pp. 53–60. With permission.)

- If the workpiece material is ductile, a visible crack would not be observed because of the "healing" effect of plastic deformation. Instead, a certain elastoplastic zone forms in the workpiece (Figure 2.5c). The dimensions of the plastic and elastic parts of this zone depend on ductility of the workpiece material. It is understood that for a perfect plastic material, the elastic zone would not form at all, while for a perfect brittle material the plastic zone would never form.

This simple consideration shows that the properties of workpiece material play an important role from the beginning of chip formation. Because the general behavior of materials can be classified as ductile or brittle depending upon whether or not material exhibits the ability to undergo plastic deformation, this fact should be incorporated in the chip formation theory.

It is apparent from the above discussion that the system concept along with the concept of the bending moments will be applied to the analysis of chip formation for a variety of cutting conditions. Consider the machining of a brittle material using the proposed system approach. Two basically different cases are possible here. The first takes place when the components of the cutting system are arranged so that resultant force R acting on the chip from the tool rake face intersects the conditional axis of the partially formed chip (Figure 2.6). As such, the bending moment causes chip fracture along Section 1-1, which represents the root of the partially formed chip. Figure 2.7 illustrates a system consideration of the model shown in Figure 2.6. Phase 1 shows the initial state where the tool is in the contact with the workpiece. The application of the penetration force P leads to the formation of a stressed zone ahead of the tool (Phase 2). When stress in this zone reaches a certain level, a crack forms in front of the cutting edge (Phase 3). Further increase in the applied load leads to the development of the crack (Phase 4). A part of the

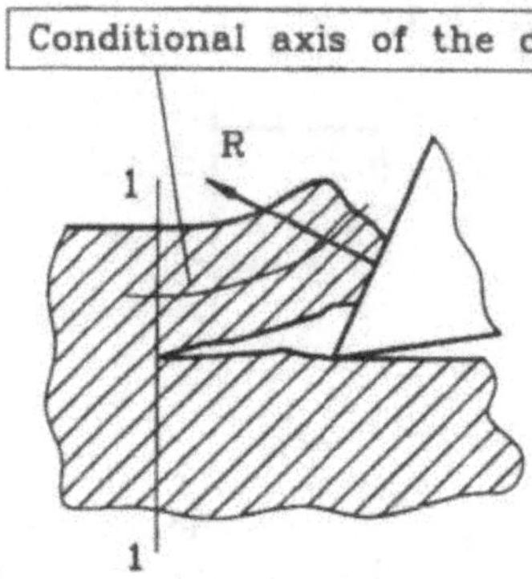

Figure 2.6 Resultant force *R* intersects the conditional axis of the partially formed chip.

separated workpiece material located above the crack now serves as a cantilever. When the applied force reaches a certain limit, the fracture of the workpiece material takes place at the cantilever support (Phase 5). As such, the separate, almost rectangular chip elements are produced (Phases 6, 7). This was observed in the cutting of cast irons using cutting tools with positive rake angles.[10]

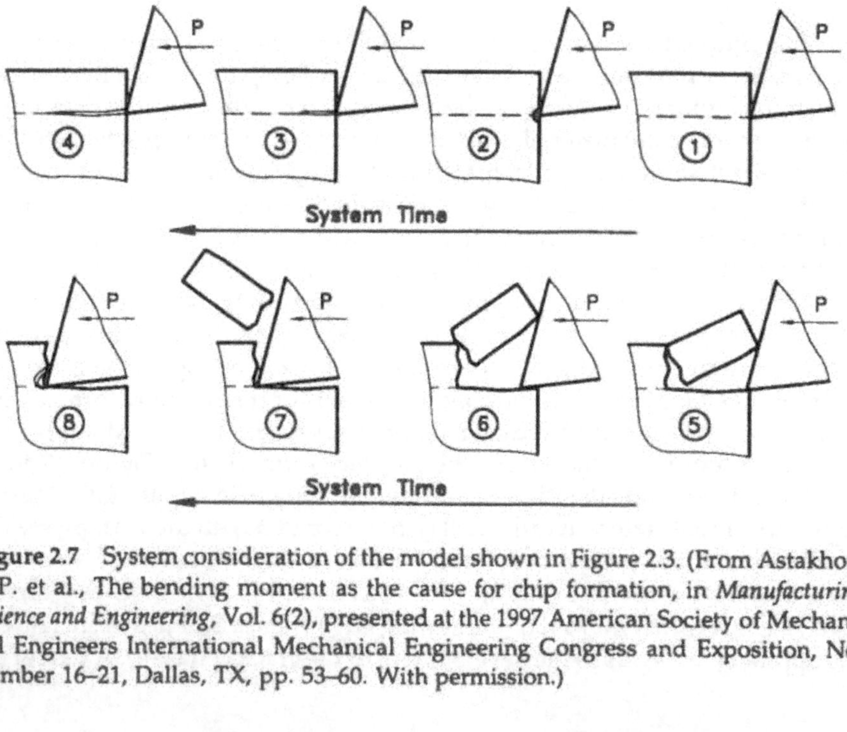

Figure 2.7 System consideration of the model shown in Figure 2.3. (From Astakhov, V.P. et al., The bending moment as the cause for chip formation, in *Manufacturing Science and Engineering*, Vol. 6(2), presented at the 1997 American Society of Mechanical Engineers International Mechanical Engineering Congress and Exposition, November 16–21, Dallas, TX, pp. 53–60. With permission.)

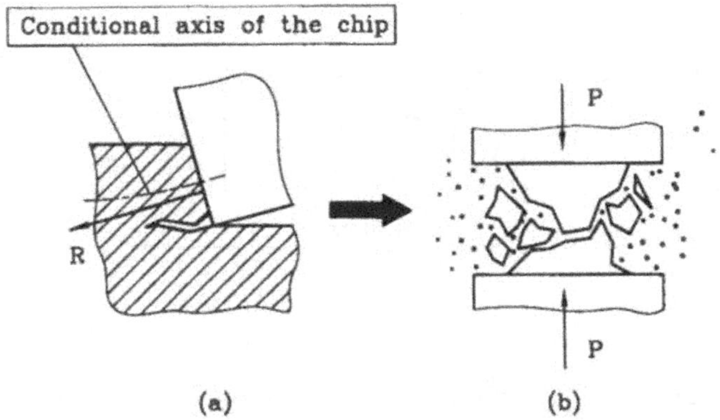

Figure 2.8 Resultant force R does not intersect the conditional axis of the partially formed chip (a) that results in common fracture of a brittle material under compression (b).

The second case takes place when the components of the cutting system are arranged so that the resultant force R does not intersect the conditional axis of the partially formed chip (Figure 2.8a). As such, there is no bending stress in the machining zone. As a result, the final fracture of the layer to be removed takes place due to pure compression. The common fracture of a brittle material under compression takes place (Figure 2.8b). The formed chip consists of irregularly shaped fragments of work material and dust. Such dust is an inherent feature of machine shops dealing with machining of cast irons. Figure 2.9 presents a system consideration of the model shown in Figure 2.8a.

When considering ductile workpiece materials, several different cases are possible here. In this section, the simplest one will be considered for the sake of clarity. The other cases will be presented in later sections.

The basic case of chip formation in cutting ductile materials takes place when no seizure occurs at the tool/chip interface. At the initial stage of chip formation, an elastoplastic zone in front of the cutting edge forms as the result of pure compression. In effect, the plastic deformation of workpiece material takes place by pure shearing at this stage. As the tool advances further, the plastically deformed part of the workpiece material gradually comes into close contact with the tool rake face. When full contact is achieved, this part serves as a cantilever subjected to penetration force P from the tool rake face (Figure 2.10a). Penetration force P can be resolved into two components — namely, compressive force Q, acting along the axial direction, and bending force S, acting along the transverse direction (Figure 2.10b). Therefore, the chip cantilever is subjected to the mutual action of the compressive force Q and the bending moment M (= $S\ l$). As the state of stress becomes complex and includes a combination of bending and compressive stresses, the deformation process would shift from one mechanism to another. As a

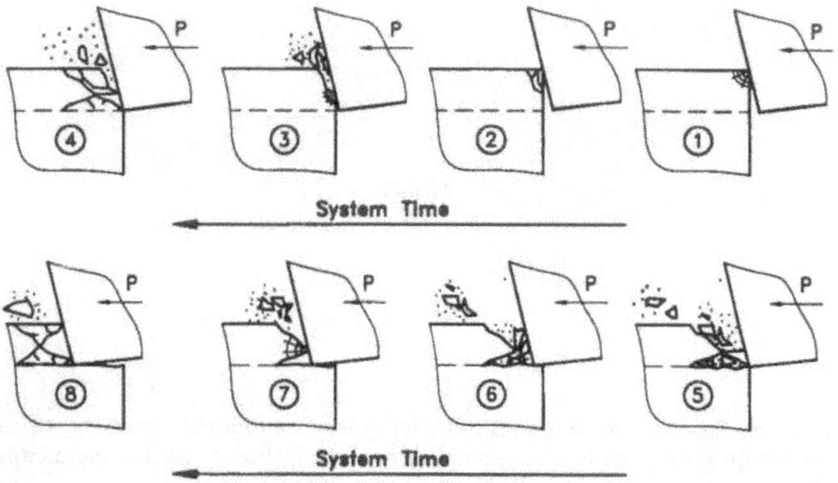

Figure 2.9 System consideration of the model shown in Figure 2.8a. (From Astakhov, V.P. et al., The bending moment as the cause for chip formation, in *Manufacturing Science and Engineering*, Vol. 6(2), presented at the 1997 American Society of Mechanical Engineers International Mechanical Engineering Congress and Exposition, November 16–21, Dallas, TX, pp. 53–60. With permission.)

result of the mutual action of compression and bending, the maximum stress occurs in the vicinity of the chip cantilever support, and its deformation takes place along Section 1–1, which appears to be the plane of the maximum combined stress (Figure 2.10a).

Figure 2.11 presents a system consideration of the model shown in Figure 2.10. In this picture, Phase 1 shows the initial state. When the tool is in contact with the workpiece, the application of the penetration force *P* leads to the formation of the deformation zone ahead of the cutting edge. As might be expected, the workpiece at first deforms elastically (Phase 2). When the

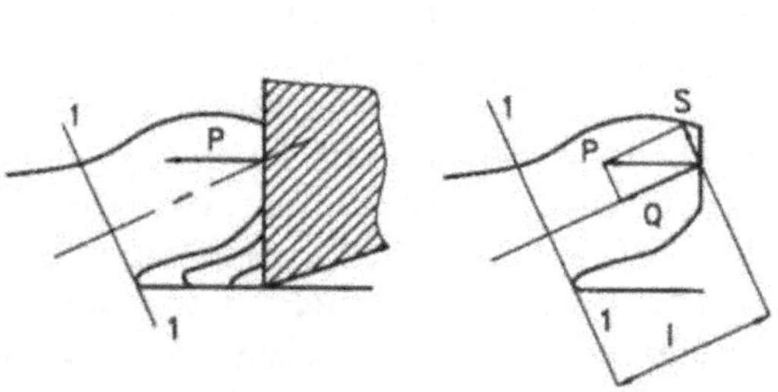

Figure 2.10 The interaction between tool rake face and the chip: **(a)** penetration force *P* acting on the chip, and **(b)** two components of the penetration force *P*, namely, the compressive force *Q* and the bending force *S*.

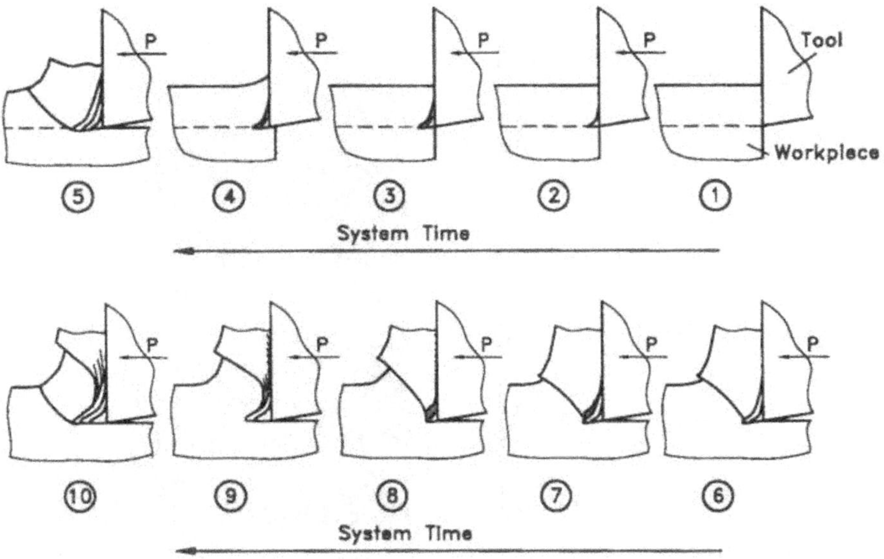

Figure 2.11 System consideration of the model shown in Figure 2.10a. (From Astakhov, V.P. et al., The bending moment as the cause for chip formation, in *Manufacturing Science and Engineering*, Vol. 6(2), presented at the 1997 American Society of Mechanical Engineers International Mechanical Engineering Congress and Exposition, November 16–21, Dallas, TX, pp. 53–60. With permission.)

load from the tool exceeds a value corresponding to the yield strength, the workpiece undergoes plastic deformation. As a result, a certain elastoplastic zone forms in the workpiece ahead of the tool that allows the tool to advance further into the workpiece so that a part of the layer to be removed comes into close contact with the tool rake face (Phase 3). When full contact is achieved, this part serves as a cantilever subjected to the penetration force P from the tool rake face. Therefore, the state of stress in the deformation zone becomes complex due to a combination of bending and compressive stresses. As a result, further deformation is controlled by the combined stress. As such, the dimensions of the plastic part of the deformation zone will increase with increasing the applied load P (Phase 4). When the combined stress in this zone reaches the limit (for a given workpiece material), a sliding surface forms in the direction of the maximum combined stress (Phase 5). This instant may be considered as the very beginning of chip formation. As soon as the sliding surface forms, all the chip cantilever material starts to slide along this surface and thus along the tool face (Phase 6). Upon sliding, resistance to the tool penetration decreases, leading to a decrease in the dimensions of the plastic part of the deformation zone (Phase 7). However, the structure of the workpiece material, which has been deformed plastically and now returns to the elastic state, is different from that of the original material. Its appearance corresponds to the structure of the cold-worked

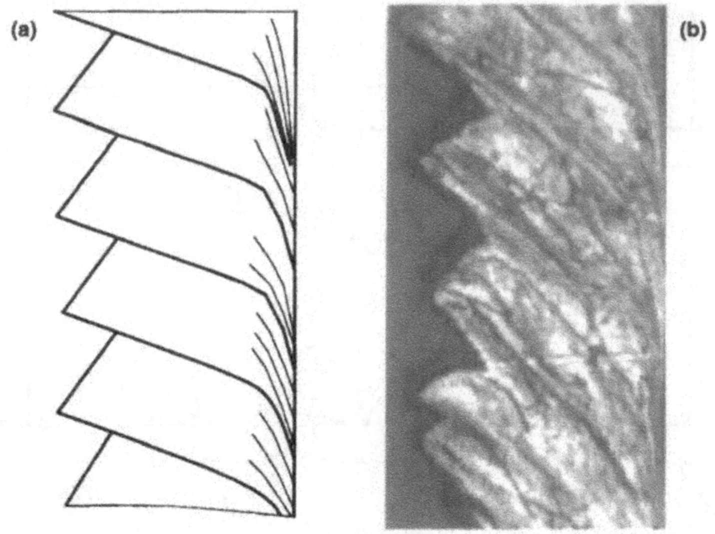

Figure 2.12 Structure of the continuous fragmentary chip: **(a)** prediction, **(b)** experimental. (From Astakhov, V.P. et al., The bending moment as the cause for chip formation, in *Manufacturing Science and Engineering,* Vol. 6(2), presented at the 1997 American Society of Mechanical Engineers International Mechanical Engineering Congress and Exposition, November 16–21, Dallas, TX, pp. 53–60. With permission.)

material. In effect, the hardness of this material is much higher than that of original material. The results of our experimental study using a computer-triggered, quick-stop device have suggested that this material spread over the tool/chip interface by the moving chip constitutes the well-known chip contact layer (Phase 8), which is now believed to be formed due to severe friction conditions in the so-called secondary deformation zone.[11–16] The sliding of the chip fragment continues until the force acting on this fragment from the tool reduces, because a new portion of work material is entering into contact with the rake face. This new portion attracts a part of penetration force *P*. In effect, the stress along the slide plane diminishes, becoming less than the limiting stress that ceases the sliding (Phase 9). A new fragment of the chip starts to form (Phase 10).

The chip formed in this way is referred to as the continuous fragmentary chip. It has a saw-toothed free side (Figure 2.12a). For comparison, Figure 2.12b illustrates the experimentally obtained chip structure which is quite similar to that modeled.

It is worthwhile to notice here two essential facts. The first one is that the resistance to the tool penetration into the workpiece varies within a chip formation cycle. Therefore, it should be expected that the bending moment and thus the bending stress in the deformation zone should vary over this cycle. The second is that the chip contact layer forms in the deformation zone rather than due to severe friction conditions in the so-called secondary deformation zone. Our experimental results have suggested that there is no such phenomenon in metal cutting as the secondary deformation zone, which, in the author's opinion, was invented to explain the formation of the chip contact layer.

2.3.1.2 Second stage: experimental verification

The discussed models of chip formation in cutting of brittle materials can be supported by a number of known experimental observations.[10–12,17] Moreover, the discussed results can be directly examined by a simple cutting experiment. However, the model of chip formation in cutting of ductile materials has to be proven. Here, to prove that the model is equivalent to the real cutting process, the presence, significance, and variation of the bending stress in the deformation zone should be verified.

The presence of the bending stress in the deformation zone was confirmed by comparison of the modeled structure of the deformation zone with that obtained experimentally. The slip-line structure of the deformation zone in the workpiece has been modeled using the following facts. First, it is known that if a cantilever with a curved contour is subjected to the bending moment then the maximum plastic deformation occurs in the vicinity of its external surface adjacent to the support with the known slip-line field.[18] Second, the plastic deformation occurring under compression is the result of a series of shears. As such, all the planes of the maximum shear stress are parallel and inclined at an angle of 45° to the direction of the compressive force so that the slip-line field is also well known.[19] Because it was assumed that the plastic deformation of the workpiece is governed by the combined stress, the general field of plastic deformation in the chip formation zone was modeled by graphical superposition[11] of the slip-line fields due to compression and bending. The result of the modeling is shown in Figure 2.13a. An analysis of the obtained resulting slip-line structure of the deformation zone shows that:

- The obtained shape of the lower boundary of the plastic zone is in excellent agreement with reported experimental data (Figures 5a, 36, 37, and 47 in Zorev;[11] Figures 9.37, 9.6, and 10.15 in Trent[12]).
- The maximum deformation occurs in two distinctive regions. The first one is in the vicinity of the cutting edge, and the second is adjacent to the transition surface of the workpiece where deformation is even higher than in the first region. One can conclude that it complies with the results of modeling shown in Figure 4.12 where fracture in the second region is observed.

A special experiment was carried out to obtain the structure of the deformation zone. A specimen made of AISI 1020 steel was cut on the standard cutting regime with a carbide cutter when the components of the

(a)

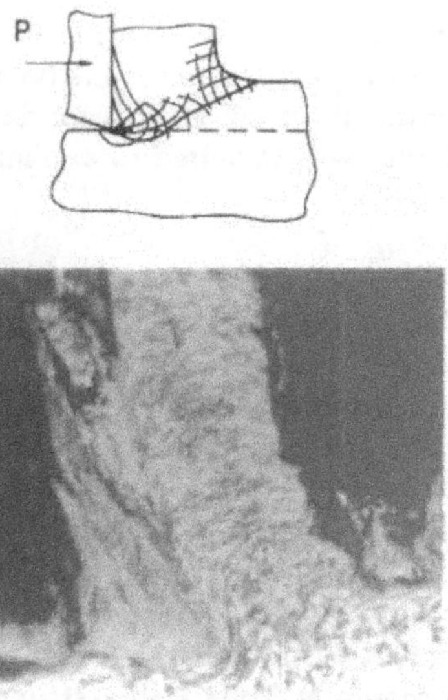

(b)

Figure 2.13 Structure of the deformation zone: **(a)** modeled, **(b)** experimental. (From Astakhov, V.P. et al., *J. Mater. Process. Technol.*, 71(2), 247–257, 1997. With permission.)

cutting system were disengaged almost instantly by a quick-stop device. A micrograph in Figure 2.13b shows a "frozen" picture of the deformation zone. As can be seen, the experimentally obtained structure of the deformation zone is in close agreement with the model that confirms the presence of the bending stress.

To prove that the bending stress in the deformation zone plays a significant role in chip formation and varies in a cycle of chip formation within certain limits, a special experiment was performed. The essence of the experiment was analysis of the chip structure, which should reflect the influence and the variation of bending stress, if this is the case. The experimental methodology, setup, and data collection and acquisition procedures are discussed in Chapter 6. A specimen made of a nickel-based high alloy was cut using the following cutting regime: cutting speed, 50 m/min (experimentally chosen to be the optimum cutting speed, as discussed in Chapter 6); feed, 0.45 mm/rev; depth of cut, 2 mm. Cutting tool parameters: tool material, carbide M30; rake angle, 0°; flank angle, 8°. The time constant (a system lag) for the quick-stop device was determined, and the triggering was shifted by programming the data acquisition system. Then, the part containing a partially formed chip was cut off from the rest of the specimen, polished and etched, and finally examined by optical microscopy. To determine the variation of resistance to the tool penetration within a chip formation cycle, a microhardness test was carried out.

Figure 2.14a summarizes the results of the experiment. In this figure, the system state at the beginning of a chip formation cycle is shown by dashed

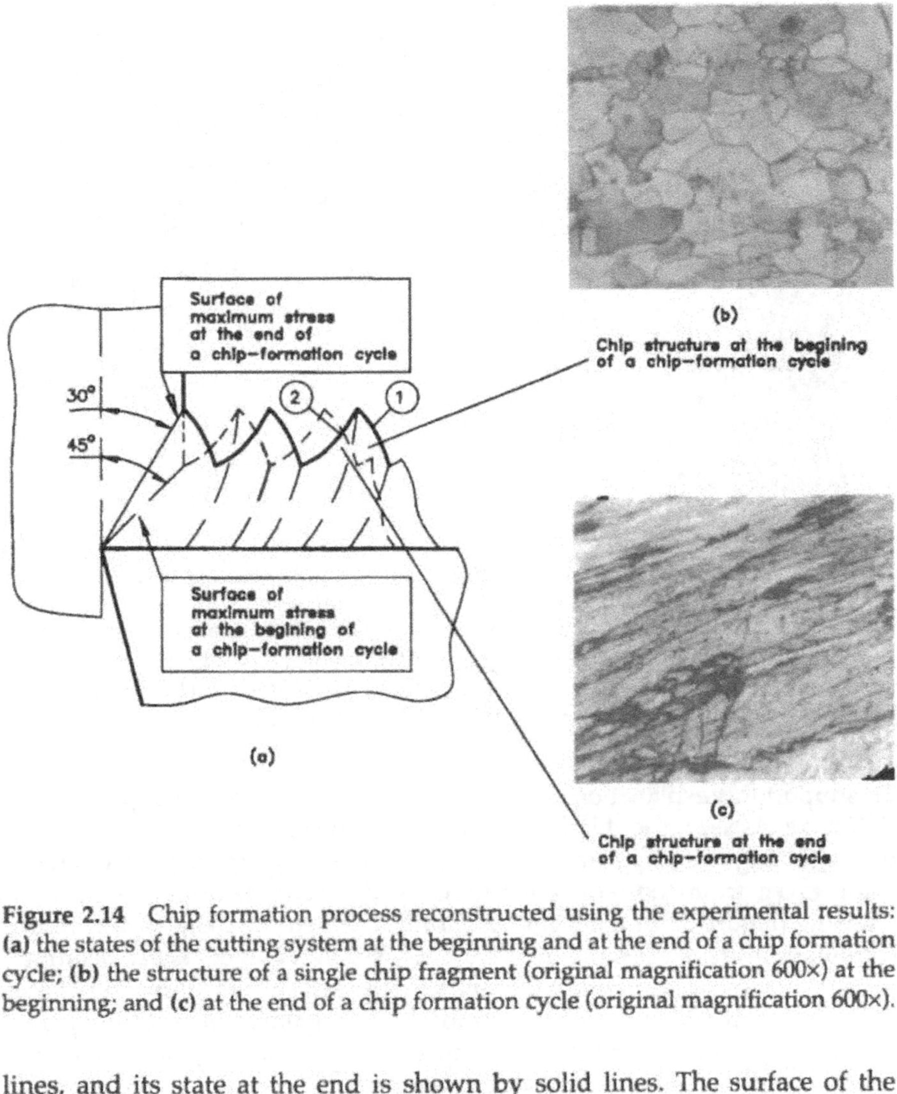

Figure 2.14 Chip formation process reconstructed using the experimental results: **(a)** the states of the cutting system at the beginning and at the end of a chip formation cycle; **(b)** the structure of a single chip fragment (original magnification 600×) at the beginning; and **(c)** at the end of a chip formation cycle (original magnification 600×).

lines, and its state at the end is shown by solid lines. The surface of the maximum combined stress is approximated by a plane extended from the cutting edge to the workpiece free surface. At the beginning of a chip formation cycle, the surface of maximum stress is located at angle $\varphi_1 = 45°$ that corresponds to pure shearing, as the compressive force is the only force applied. At the end of a chip formation cycle, the plane of maximum stress is located at angle $\varphi_2 = 30°$ that corresponds to the maximum contribution of the bending stress to the combined stress. As the inclination of the plane of maximum stress changes from φ_1 to φ_2, the front slope 1 (Figure 2.14a) of the motionless chip forms as the reaction of workpiece material to the applied compressive force. A convex shape of this slope is readily explained by barreling which takes place during plastic deformation under pure compression.

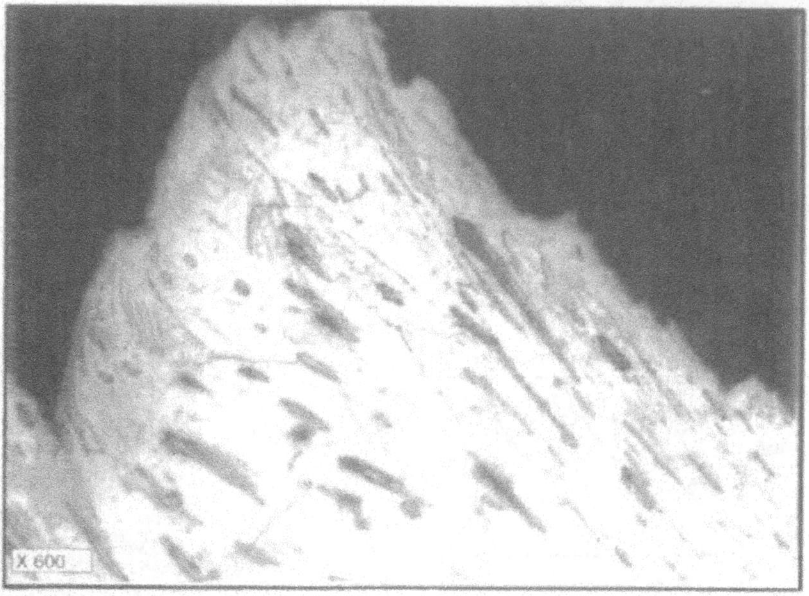

Figure 2.15 Micrograph of a single chip fragment of AISI 303 stainless steel. Cutting conditions: tool material, M30; rake angle, 0°; flank angle, 12°. Cutting regime: cutting speed, 60 m/min; feed, 0.2 rev/min; depth of cut, 5.0 mm. Etched with 40 mL hydrofluoric acid (HF), 20 mL nitric acid (HNO_3), 40 mL glecren.

To support this explanation, Figure 2.14b shows a chip micrograph taken at this stage. As seen, the chip has a structure with slightly deformed grains at the beginning of a chip formation cycle, and this structure is quite similar to those taken from samples that experienced plastic deformation by pure shearing (Figure 32 in the *Metal Handbook*[20]). When the plane of maximum combined stress reaches the inclination of 30°, the ratio of "shear-bending" stress becomes large enough to form a sliding surface in the direction of that plane. All of the chip fragment starts to slide along the formed sliding plane. At this stage, the plastic deformation of the chip takes place under the combined action of bending and compression. As a result, a rare, concave slope of a tooth (slope 2 in Figure 2.14a) forms during the sliding. Figure 2.14c illustrates a chip microstructure at this stage of deformation. As can be seen, at the end of a chip formation cycle the chip has a structure with severely deformed grains which is similar to those taken from the samples that experienced plastic deformation under combined stress.[21]

The mechanism discussed takes place in the machining of different materials. For example, Figure 2.15 shows a micrograph of a single tooth of the fully developed chip of AISI 303 stainless steel. The discussed differences in the slopes and grain deformation can be easily distinguished.

The analysis of the structure of the continuous fragmentary chip shows that this structure is non-uniform at the macro- and microlevels. At the

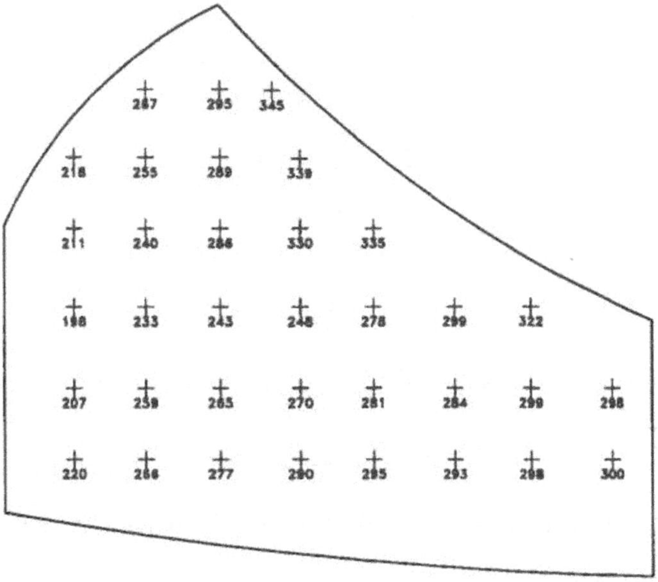

Figure 2.16 Microhardness *(HV)* distribution in the chip fragment shown in Figure 2.15.

macrolevel, the continuous fragmentary chip has non-uniform strength along its length. That is to say that the shear strength of the chip fragments is much higher than that of the fragment connectors. At the microlevel, where the structure of a single chip fragment is considered, the chip has a structure with slightly deformed grains at the beginning of a chip formation cycle; at the end of a chip formation cycle, the chip has a structure with severely deformed grains.

To demonstrate the variations of strain and stress within a chip fragment shown in Figure 2.15, a microhardness study was carried out. The microhardness (*HV*) of the plastically deformed material is uniquely related with the preceding deformation (ε)[22] and with the shear stress τ gained at the last stage of deformation as follows:[23]

$$HV = 0.208\,\varepsilon^{0.219} \qquad 0.05 \le \varepsilon \le 2.00 \tag{2.13}$$

$$\tau = 0.185\, HV \tag{2.14}$$

This makes it possible to obtain information about the levels and distribution of deformation and shear stress on the basis of microhardness measurements (the experimental methodology is discussed in Section 5.5). The experimental results are shown in Figure 2.16. As seen, the variation of microhardness is in the range from 218 *HV* at the beginning of a chip formation cycle to 339 *HV*

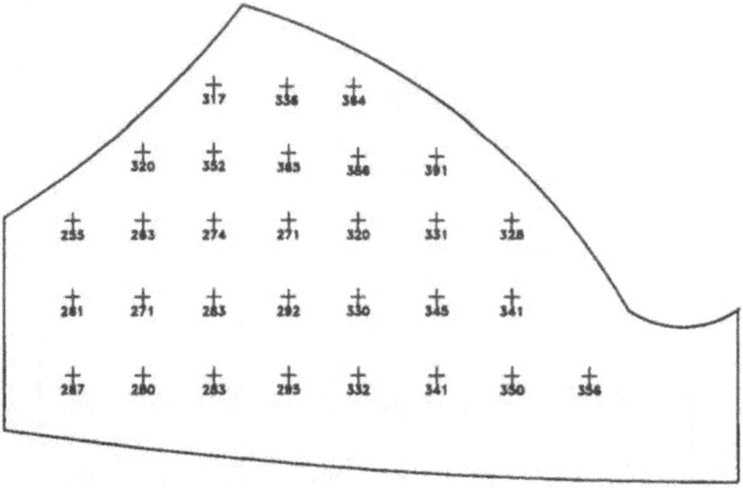

Figure 2.17 Microhardness *(HV)* distribution in the chip fragment shown in Figure 5.13b.

at the end of formation. This quantitatively demonstrates significant variation of the stress and deformation over a cycle of chip formation.

When the initial structure of workpiece material contains very fine grains, like the martensite structure shown in Figure 5.13a, a change in the grain structure after cutting cannot be distinguished readily with an optical microscope (Figure 5.13b). However, the results of a microhardness test shown in Figure 2.17 reveal the variation of microhardness in a chip fragment that proves the discussed model (Figure 2.14).

As the chip formation process appears to be cyclic, its frequency is of interest. The study reveals that the frequency of the chip formation process depends on the cutting speed and workpiece material (Figure 2.18). In the author's opinion, this fact should be incorporated in a dynamic analysis of the cutting process. To prove this point, Figures 2.19 to 2.21 present the frequency autospectra of cutting force obtained in machining of different materials at the same cutting regimes and dimensions of the workpieces. As seen from these figures, the dynamic response of the machining system, including the machine tool, depends not only on the geometry and cutting regime of the system, but also on a particular workpiece material used. The results of the study show that when the experiments are conducted properly (i.e., when the noise due to the misalignment of the workpiece and machine inaccuracy is eliminated from the system response), the amplitudes of peak at the frequency of chip formation are the largest in the corresponding autospectra. The results also show that the frequency of chip formation can easily be determined by measuring the average pitch on the sawtoothed chip (under high magnification).

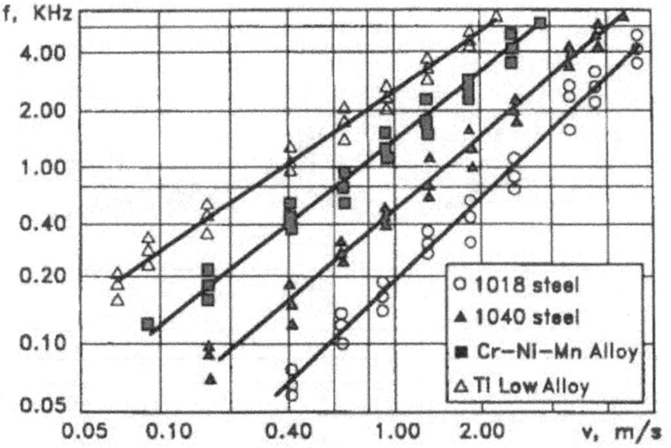

Figure 2.18 Effect of the cutting speed on the frequency of chip formation.

Unfortunately, this fact is not considered in the dynamic analysis of the machining systems where the cutting process as the main source of vibration is totally ignored. The reasons for that are discussed in Chapter 6, where the analysis of cutting force measurement is considered. The experimental results presented clearly show the presence, significance, and variation of the bending stress in the deformation zone.

2.3.1.3 *Third stage: quantitative analysis of the bending stress*

Here, the contribution of bending stress to combined stress will be analyzed quantitatively. To do this, the ratio of bending stress/shear stress due to compression has to be estimated. As soon as the presence and variation of the bending stress in the deformation zone has been experimentally established, the question about significance of the bending moment for the chip formation process should be answered. To be able to address this question, a suitable criterion is needed. For this purpose, the stress induced in the plane of the maximum combined stress by the bending moment is reasonable. The tensile (compressive) bending stress σ_b due to the bending moment can be compared to the shear stress τ caused by compression. If the ratio σ_b/τ is found to be reasonably large, the bending moment can be considered to play a significant role in chip formation and vice versa. A comparison of the bending and shear stress will be made for a representative case using some experimental data obtained in the experiment reconstructed in Figure 2.12.

Consider the state of the cutting system at a certain instant of a chip formation cycle, as shown in Figure 2.22. There, the plane of the maximum combined stress designated *AB* is located at a certain angle φ_i. As such, the average shear stress in the plane *AB* can be computed by using the known equation:[11,15]

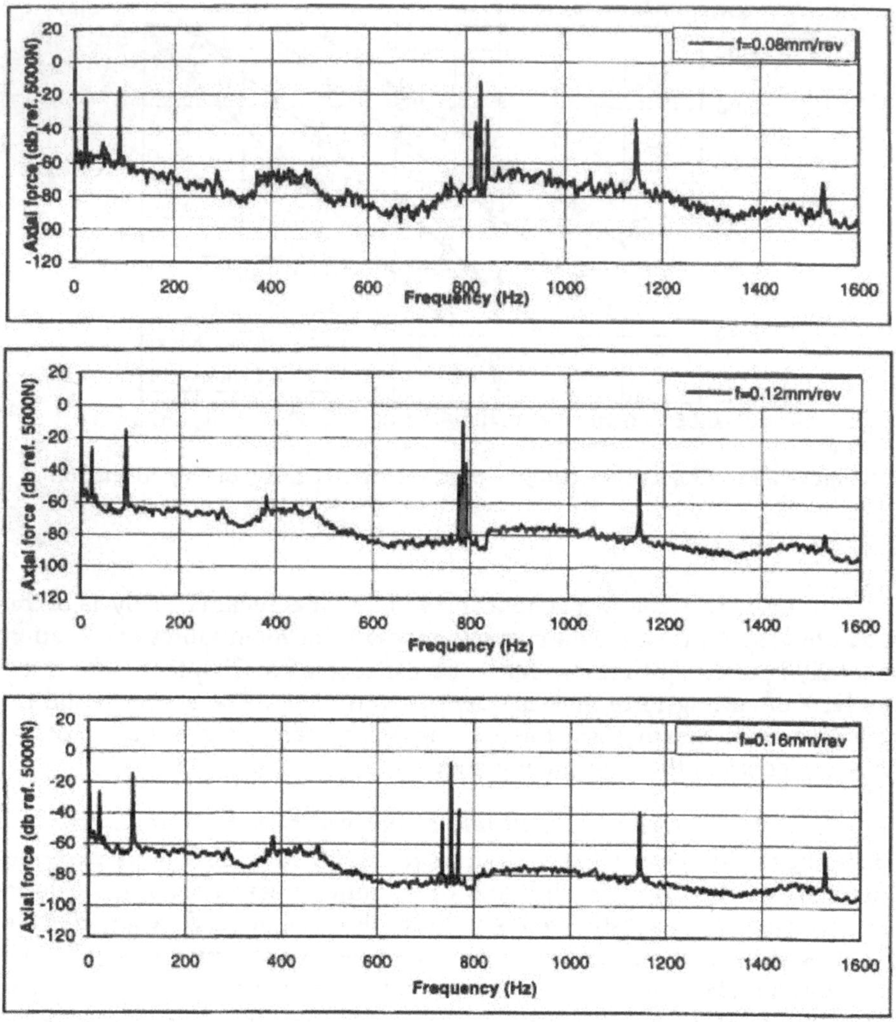

Figure 2.19 Autospectra for the cutting force. Workpiece material, AISI 1045; spindle rotation speed, 1253 rpm.

$$\tau = \frac{R\left(\cos\left(\beta_R - \gamma_n\right)\sin\varphi_i - \sin\left(\beta_R - \gamma_n\right)\cos\varphi_i\right)}{a_1\, b_1}\sin\varphi_i \qquad (2.15)$$

Here, β_R is the angle between the normal to the tool rake face and the direction of the cutting force R (Figure 2.22); γ_n is the rake angle.

The bending stress on the same plane can be calculated using the following:

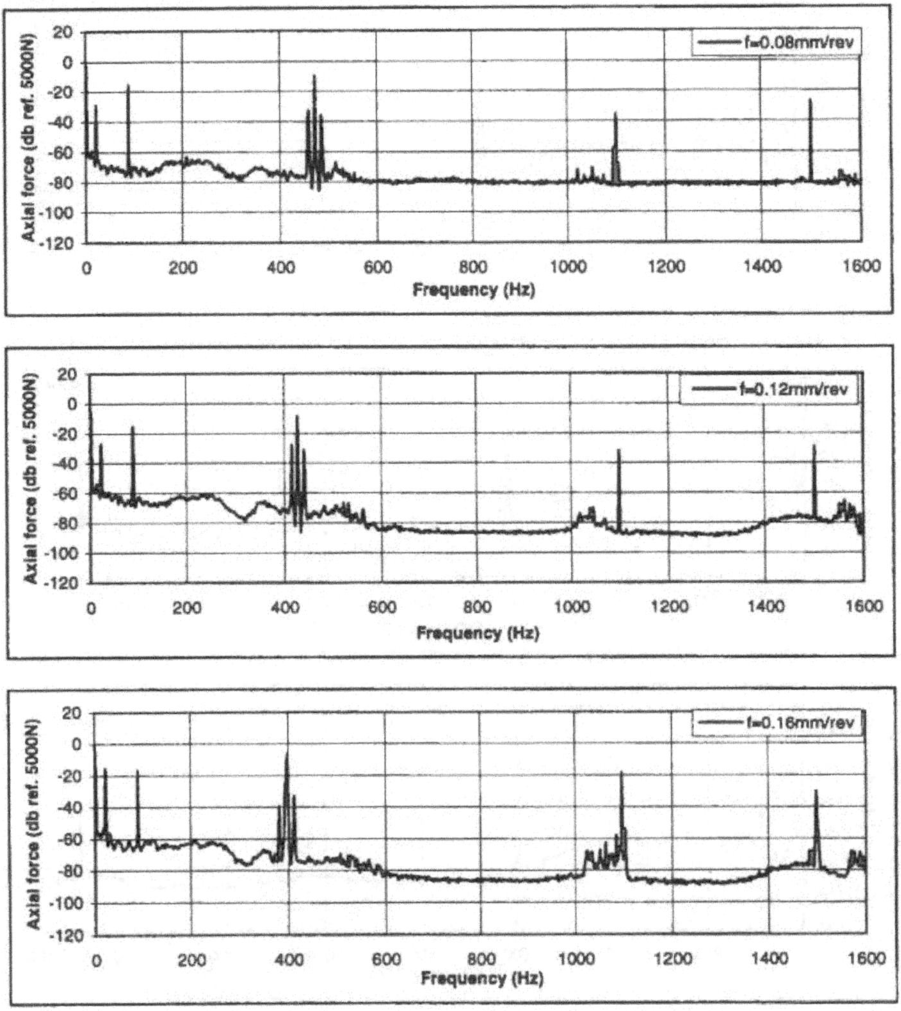

Figure 2.20 Autospectra for the cutting force. Workpiece material, AISI 4340; spindle rotation speed, 1253 rpm.

- The root of the chip cantilever is in the plastic state. Thus, the bending stress distribution along the surface of maximum combined stress is as shown in Figure 2.22.[24] As such, the value of the bending moment, which corresponds to a fully plastic support condition, is called the plastic moment which for a rectangular member made of an elastoplastic material is[24]

$$M_p = \frac{3}{2} M_Y = \frac{3}{2}\left(\frac{I}{c}\sigma_Y\right) \tag{2.16}$$

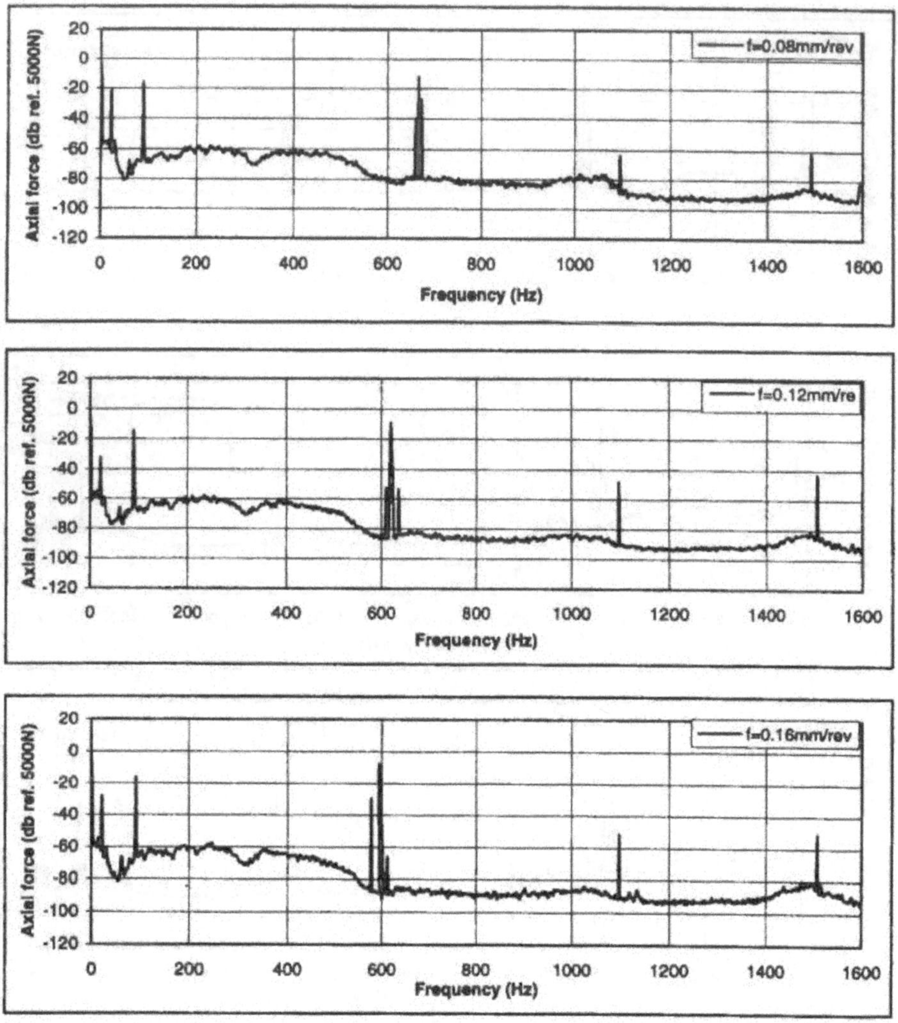

Figure 2.21 Autospectra for the cutting force. Workpiece material, AISI 303; spindle rotation speed, 1253 rpm.

Here, M_Y is the bending moment at the onset of yield; I is the moment of inertia of the cross-section with respect to the neutral axis (NA in Figure 2.22); σ_Y is the shear strength of workpiece material; and $c = a_1/2$.

- The surface of maximum combined stress is approximated by a plane. Therefore the root of the formed chip cantilever has a rectangular cross-section $b_1 \times a_1/\sin\varphi_i$ where b_1 is the width of cut, and a_1 is the uncut chip thickness. Therefore:

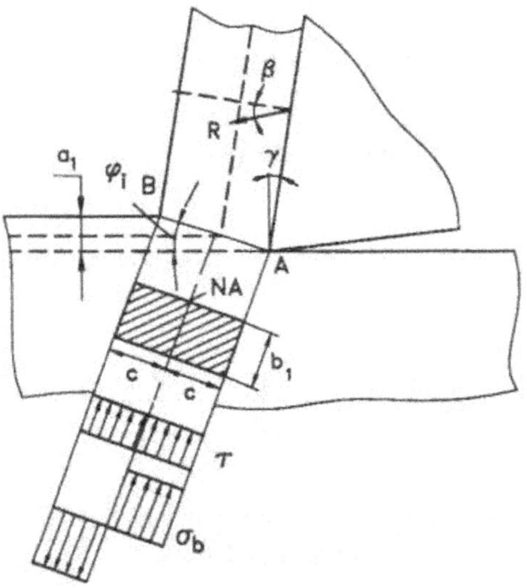

Figure 2.22 A model to estimate the contribution of the bending stress. (From Astakhov, V.P. et al., The bending moment as the cause for chip formation, in *Manufacturing Science and Engineering*, Vol. 6(2), presented at the 1997 American Society of Mechanical Engineers International Mechanical Engineering Congress and Exposition, November 16–21, Dallas, TX, pp. 53–60. With permission.)

$$\frac{I}{c}=\frac{b_1(2c)^3}{12c}=\frac{2}{3}b_1c^2=\frac{1}{6}\frac{b_1a_1^2}{\sin^2\varphi_i} \tag{2.17}$$

Combining Equations (2.16) and (2.17), one may obtain the equation for the bending stress:

$$\sigma_b=\frac{M_p}{b_1a_1^2}\sin^2\varphi_i \tag{2.18}$$

Taking the most conservative approach possible, we will not consider the bending moment in the deformation zone due to the weight of the partially formed chip as suggested by Nakayma[25] or due to the chip interaction with any other obstacles on the tool rake face as suggested by Jawahir.[26,27]

Moreover, we will consider that cutting force R is applied at the middle of tool/chip contact length c which can be calculated knowing the chip compression ratio ζ as follows:[28]

$$c=a_1\,\zeta^{1.5} \tag{2.19}$$

The expression for the bending moment, therefore, is

$$M_p = R\left(\frac{a_1\zeta^{1.5}}{2}\cos(\beta_R - \gamma_n + \varphi_i) - \frac{a_1}{2}\right) = \\ = 0.5\,R\,a_1\left(\zeta^{1.5}\cos(\beta_R - \gamma_n + \varphi_i) - 1\right) \tag{2.20}$$

By substituting Equation (2.20) into Equation (2.18), one may obtain the final expression for the bending stress:

$$\sigma_b = \frac{0.5\,R\left(\zeta^{1.5}\cos(\beta_R - \gamma_n - \varphi_1) - 1\right)}{b_1\,a_1}\sin^2\varphi_i \tag{2.21}$$

Finally, the ratio of bending stress/shear stress due to compression may be obtained using Equations (2.21) and (2.15) as:

$$\frac{\sigma_b}{\tau} = \frac{0.5\left(\zeta^{1.5}\cos(\beta_R - \gamma_n + \varphi_i) - 1\right)\sin\varphi_i}{\cos(\beta_R - \gamma_n)\sin\varphi_i - \sin(\beta_R - \gamma_n)\cos\varphi_i} \tag{2.22}$$

Data for the calculation of this ratio have been drawn from an experiment in which the tangential and radial components of the cutting force and chip compression ratio under instantaneous shear angle were measured (see Chapter 6). Substituting the experimentally obtained value of the chip compression ratio, $\zeta = 3.45$, and directional angle $\beta_R = 11°$ into Equation (2.22), one can obtain $\sigma_b/\tau = 3.38$. Such a high ratio indicates that the stress due to the bending moment at the plane of the maximum combined stress is large enough to affect chip formation significantly.

According to Equation (2.22), the bending stress varies within a cycle of chip formation from practically zero, when the surface of maximum combined stress is inclined at an angle of 45°, to a value which is 3.38 times greater than the average shear stress on this surface, thus showing that the bending stress in the deformation zone significantly affects chip formation.

2.3.1.4 *Definition of the metal cutting process*

On the basis of the foregoing considerations, we may suggest the following definition of the metal cutting process:

> Metal cutting is a forming process of components that are so arranged that by their means the applied external energy causes the purposeful fracture of the layer to be removed. This fracture occurs due to the combined stress including the continuously changing bending stress that is the cause of the cyclic nature of the process.

The bending stress combined in the deformation zone with the shear stress due to compression causes chip formation. The presence of the bending stress in the deformation zone distinguishes the process of metal cutting from other deforming and separating manufacturing processes. Regardless of the workpiece material (wood, stone, glass, plastic, metal); type, shape, geometry, etc. of the cutting tool used; and the kinematics of the process, a forming process possessing this distinguishing feature should be called metal cutting.

From the above definition, it follows that regardless of the workpiece material used, cutting by a knife or a pair of scissors is splitting by shearing and thus cannot be called metal cutting even if a thin metal sheet can serve as the workpiece, because there is no bending stress in the deformation zone and the chip does not form.

System consideration of the metal cutting process reveals that the variation of the bending stress in the deformation zone constitutes a cyclical character of the chip formation process. As a result, the parameters of the cutting system vary over each chip formation cycle. A sliding surface, formed in cutting of ductile materials, does not exist throughout the entire chip formation cycle. Rather, it forms at its end as the result of the stress redistribution in the considered cycle and appears as the surface of maximum combined stress.

2.3.1.5 Difficulties due to high ductility of workpiece material

Having established the presence and significance of the bending stress in the deformation zone, we are ready now to consider the known difficulties in cutting of highly ductile materials. It is well known that an increase in the ductility of workpiece material lowers its machinability. By this is meant that both the cutting temperature and the power per unit volume of metal removed will increase.[29] This phenomenon is explained by an increase in the spread between yield and fracture strength with ductility. The energy required, therefore, to machine, for example, a low carbon steel is higher than that for a medium carbon steel (as discussed in Chapter 4). This conclusion, however, stems from the practice of metal cutting rather than from the theory of chip formation. Because the known model recognizes simple shearing as the only cause for chip formation,[30–32] it cannot explain why machining of a material with a lower shear strength requires more energy than the machining of one with a higher shear strength.

To understand the phenomenon, a special experiment was carried out. Two specimens — the first made of AISI 1040 steel (ultimate tensile strength $\sigma_b = 576$ MPa, yield strength $\sigma_y = 367$ MPa), the second made of much more ductile steel AISI 1018 ($\sigma_b = 538$ MPa, $\sigma_y = 367$ MPa) — were machined using the same cutting regime (cutting speed was 90 m/min; feed, 0.12 mm/rev; depth of cut, 1.5 mm; no cutting fluid; a P10 carbide cutter with the rake angle of $-8°$). In the experiment, the chip compression ratio ζ was measured and the structure of the obtained chip was analyzed. For the first specimen it was

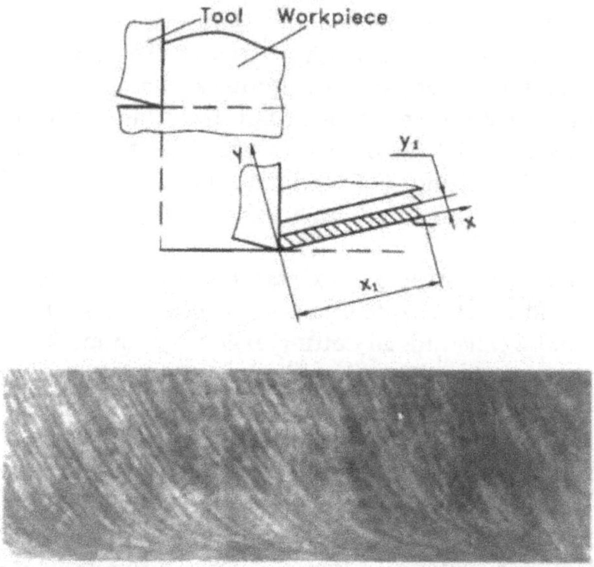

Figure 2.23 A model for chip formation to distinguish the influence of the ductility of workpiece material on the structure of the continuous fragmentary chip (top) and a micrograph of the structure of the chip obtained in cutting of a highly ductile material (original magnification 200×). (From Astakhov, V.P. et al., *J. Mater. Process. Technol.*, 71(2), 247–257, 1997. With permission.)

found that $\zeta_1 = 3.8$, while for the second $\zeta_2 = 6.2$. To analyze the changes in the structure of the continuous fragmentary chip with ductility, consider a model for chip formation shown in Figure 2.23a. Here, the righthand coordinate system is described as follows:

- The *x*–axis coincides with the direction of the sliding plane in the current chip formation cycle.
- The *y*-axis is perpendicular to *x*-axis, as shown in Figure 2.23a.
- The *z*-axis (not shown) is perpendicular to the *x*- and *y*-axes.

Because it was found experimentally that in metal cutting the change in volume in the plastic region is very small,[15] the following expression for strains is valid:

$$\varepsilon_x + \varepsilon_y + \varepsilon_z = 0 \tag{2.23}$$

Here, ε_x, ε_y, and ε_z are strains along the corresponding coordinates.

Consider a volume of the work material located between two successive sliding planes (Figure 2.23a). The plastic deformation of this volume may be considered in terms of corresponding strains as:

$$\varepsilon_x = \ln\frac{x_1}{x_0};\ \varepsilon_y = \ln\frac{y_1}{y_0};\ \varepsilon_z = \ln\frac{z_1}{z_0} \tag{2.24}$$

where x_0, y_0, z_0 and x_1, y_1, z_1 are the dimensions of the volume along the corresponding axes before and after deformation, respectively.

It is known that, when properly measured, the chip width is practically equal to the width of cut,[28] which yields $z_0 = z_1$ and $\varepsilon_z = 0$. Substituting Equation (2.24) into Equation (2.23) and ignoring signs, one can obtain:

$$\ln\frac{x_1}{x_0} + \ln\frac{y_1}{y_0} = 0 \quad or \quad \frac{x_1}{x_0} = \frac{y_1}{y_0} \tag{2.25}$$

As has been shown,[11,29,33] an increase in ductility leads to an increase in the chip compression ratio. The experiment conducted yielded the same result. Therefore, the chip thickness (x_1 in Figure 2.23a and in Equation (2.25)) increases with the ductility of workpiece material. A very important conclusion follows from Equation (2.25) — an increase in chip thickness x_1 leads to a reduction of distance y_1 between two successive sliding planes, increasing in the number of the sliding planes per unit chip length.

To support this conclusion, Figure 2.23b shows a micrograph of the structure of the chip formed in cutting of the second specimen. Here, a series of slip planes, one following another, is observed. A great number of the sliding planes, one closely followed by another, make this chip quite similar in its appearance to the continuous chip. However, the considered conditions of chip formation show that this chip is a continuous fragmentary chip with a very small distance between two successive fragments. As the number of the sliding planes here is much higher than in the cutting of the first specimen, the heat generation is also much higher and explains the reduced tool life in the cutting of highly ductile materials.

The difficulties observed in the cutting of highly ductile materials may be readily explained by a lack of a bending moment due to insufficient rigidity of the chip. We may assume that for chip formation to start, a certain ratio of bending stress/shear stress should be achieved. As a result, the chip thickness should increase with ductility to transmit the same bending moment into the deformation zone. The same explanation may be given to "negative" (the chip compression ratio is less than one) deformation in cutting of some titanium alloys.[33] Here, the chip of a relatively small thickness is rigid enough to transmit the necessary bending moment into the deformation zone to keep the required ratio of the stresses to cause chip formation.

2.3.1.6 *Model of chip formation when seizure occurs at tool-chip interface*

Seizure as a phenomenon in metal cutting was probably first introduced by Zorev[11] and Trent.[12] Zorev did not study this phenomenon thoroughly, as in

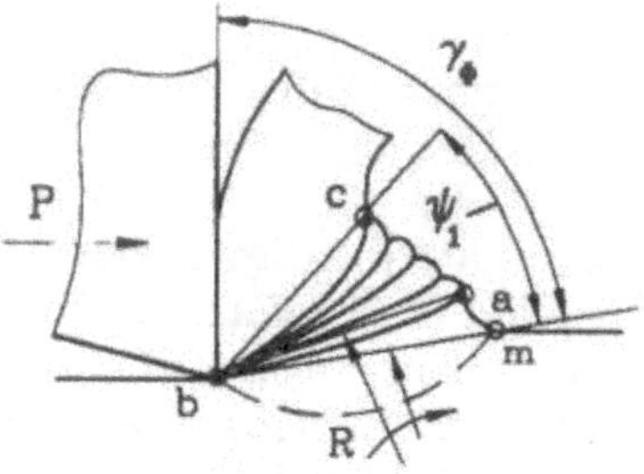

Figure 2.24 A model for chip formation when seizure occurs at the tool/chip interface. (From Astakhov, V.P. et al., *J. Mater. Process. Technol.*, 71(2), 247–257, 1997. With permission.)

the 1960s there were not many workpiece materials having properties which lead to seizure. Trent has presented evidence for seizure in his papers and books for more than 20 years. However, in one paper,[34] he has complained that though appreciation of the condition of seizure at the tool/chip interface is fundamental to understanding the whole process of metal cutting, it has not been recognized by most engineers or by tribologists. In the excellent *Wear – A Celebration Volume* (1984), the hypothesis of seizure in metal cutting was discussed in two paragraphs by J.M. Challen and P.L.B. Oxley as being a phenomenon which "is not ruled out for certain conditions," while the tribology of seizure is not discussed anywhere in the book. Recent books on metal cutting do not even mention this term.[13–16]

This situation may be explained by the fact that Trent has been concerned only with the metallographic aspect of the problem. Even in his book,[12] Trent first discusses the known model of chip formation and then presents evidence of seizure which cannot fit the existing model. It appears that the lack of a model for chip formation under conditions of seizure has caused the situation discussed. Therefore, we have found it necessary to clarify the place of seizure in chip formation.

Seizure, having an adhesion nature, ceases a normal chip sliding over the tool face because of the continuous contact in both the hills and valleys of the tool surface. For seizure to occur, a high contact temperature is the prime condition. As explained above, in cutting of highly ductile materials, the energy consumption per unit volume of the removed material is high and, when it is combined with a low thermoconductivity of workpiece material, high temperatures occur at the tool/chip interface which leads to seizure.

Figure 2.24 presents a model of chip formation when seizure occurs at the tool/chip interface.[6,7] It is assumed in this picture that the temperature at the tool/chip interface has reached the level when seizure occurs so that the chip is atomically bounded at the tool/chip interface. Surface *bc* of the motionless chip and surface *ab* (the boundary of plastic and elastic zones in the work material) are located by angle ψ_1 relative to each other. As the tool progresses in the cut, the material in the wedge-shaped zone *abc* is compressed and squeezed in the direction of boundary *ac*. When the load from the cutting tool increases further, the plastic zone in front of the tool expands — that is, the angle ψ_1 increases. This increase can take place only due to rotation of boundary *bm*, as boundary *ba* separates the plastic zone and the chip which is already heavily deformed compared to the rest of the workpiece material. Finally, boundary *ba* takes its limiting position, *bm*, and region *bmc* is formed. Here, gradually increasing the angle γ_ϕ up to 90° leads to a change in the mode of deformation from the compression-type to the shear-type. When γ_ϕ = 90°, fracture of the layer to be removed takes place along boundary *bm*. At the instant of fracture, resistance to the tool penetration is maximum, and the direction of the reaction *R* (from the elastically deformed part of the layer being removed) becomes parallel to the tool rake face. Therefore, the condition for shearing the entire layer along the rake face is created. The region in the vicinity of the cutting edge becomes free, and a new layer to be sheared starts to form.

The chip formed under these conditions may be referred to as the continuous humpbacked chip.[8] An emergency situation can occur in this type of cutting when the adhesion forces at the tool/chip interface are so high that the chip/tool contact cannot be separated by the growing cutting force. As a result, the chip cannot slide over the tool face, and the chip formation process becomes impossible. Moreover, when the growing cutting force reaches a certain limit, tool breakage is unavoidable. As such, the weakest components of the tool will be deformed and/or fractured.

Figure 2.25a presents system considerations of the model shown in Figure 2.24. In this figure, Phase 1 shows the instant when the sliding plane forms in the direction of the maximum combined stress (which corresponds to Phase 5 in Figure 2.11). When the chip slides over the rake face, the contact length is heated up mainly due to chip deformation and friction. When the contact temperature reaches a certain level, the chip and the tool are interlocked and atomically bonded to such an extent that "normal" sliding cannot occur. The interlocked part of the tool/chip contact changes the "normal" chip formation process, causing an increase in resistance to the tool penetration. As a result, the cutting force increases, leading to an increase in the dimensions of the plastic zone in front of the tool (Phase 2). In turn, the increase in size of the plastic zone leads to the formation of a growing deposit on the interlocked part (Phase 3). Here, zone *abc* is a current plastic zone formed after the chip sliding over the tool face ceased due to seizure. Its boundary *bc* separates the plastic zone and the motionless chip, while its boundary *ba* separates this zone from the undeformed workpiece. As the tool progresses in the cut, the material in the plastic zone *abc* is compressed and squeezed in the direction of the boundary *ac*, causing the plastic zone *abc* to expand. As the chip is motionless and has already been severely deformed, the plastic zone expands into the workpiece. During this transformation, the angle γ_ϕ gradually increases up to 90° which, in turn, leads to a change in the deformation mode of the deposit from the compression-type to the shear-type.

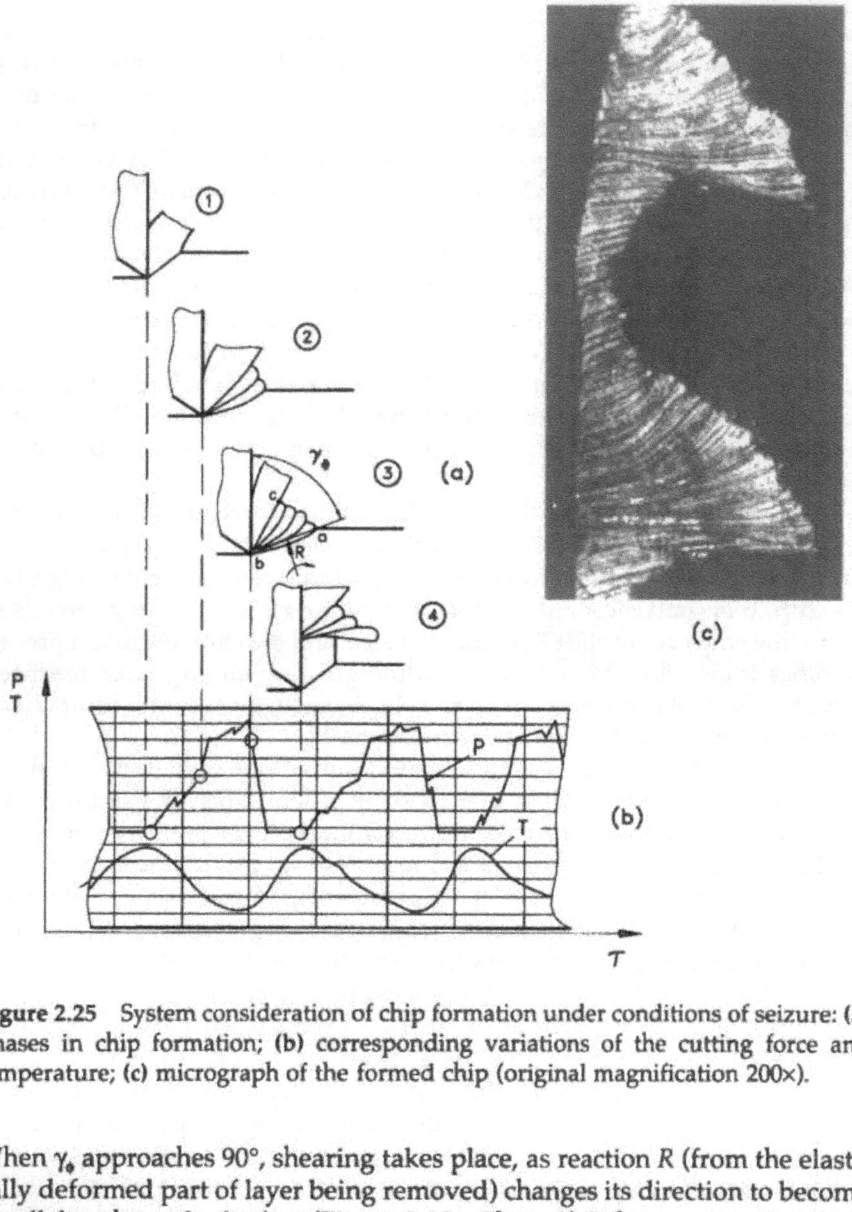

Figure 2.25 System consideration of chip formation under conditions of seizure: **(a)** phases in chip formation; **(b)** corresponding variations of the cutting force and temperature; **(c)** micrograph of the formed chip (original magnification 200×).

When γ_ϕ approaches 90°, shearing takes place, as reaction *R* (from the elastically deformed part of layer being removed) changes its direction to become parallel to the tool rake face (Figure 2.25a, Phase 3). When penetration force *P* reaches a certain limit, reaction *R* becomes big enough to shear the whole deposit along the tool face. Once the deposit has been sheared off, the "normal" chip formation takes place until the contact temperature becomes high enough to start the formation of a new deposit.

To verify the discussed model, a cutting experiment was carried out. A low-carbon, low-alloy steel (0.12% C, 1% Cr, 1% Mn, 1% T) was machined at a cutting speed of 70 m/min, feed of 0.14 mm/rev, depth of cut of 2.0 mm, with an M30 tool having a rake angle of –8°. The experimental setup is discussed in Chapter 6. Penetration force *P* and the temperature at the tool/chip interface were measured.

Corresponding variations in the tool/chip contact temperature and in the penetration force were found by recording the signals from the tool-work thermocouple (*T* in Figure 2.25b) and from the dynamometer (*P* in Figure 2.25b) simultaneously.[35] As seen from the results, within the period of time necessary for deposit formation, the penetration force increases. There is no sliding within this period, thus the contact temperature decreases. As described, this chip formation process has a periodic or almost periodic nature. As a result, the thickness of the chip formed almost shows periodic variations.

The experimental results show that the chip produced is the continuous humpbacked chip (Figure 2.25c) with a shape very similar to that predicted by the discussed model.

2.3.1.7 *Generalized model of chip formation and structure classification*

Understanding chip formation is the first step to good chip control, a necessity for automated machining. Moreover, lack of chip control often leads to coarse surface finish, poor machining accuracy, and problems with chip removal from the machining zone. Using chip-breakers on inserts has proven to be an effective way to curl chips, but their design is a matter of experimental findings and users' experience rather than the result of a thorough understanding of the mechanics of chip formation. Even chip flow direction and causes of chip curling are not yet clear.[27]

In metal cutting, the term "chip formation" has been in use since the last century. Its initial meaning is the formation of the chip in the primary and secondary deformation zones. Primary attention was devoted to the cutting force and contact process at the tool/chip interface. Later on, the chip-breaking problem became more and more important with increasing cutting speed and the development of new aerospace and stainless materials. Even though the term "chip formation" is still in use, its original meaning has been transformed. The modern sense of this term implies the chip which just left the tool/chip interface and has yet to be broken.[36]

As for geometry of chip formation, it has been pointed out by Merchant[2] that there are three basic types of chips found in the cutting of metal: discontinuous, continuous, and continuous with a built-up edge interposed between the chip and the tool in the vicinity of the cutting edge. This classification is generally accepted, at least in the theory of metal cutting, and will be referred to as the "known classification".

The above-mentioned classification cannot satisfy growing practical requirements. As a result, the national industries of developed countries have accepted some more practical chip classifications. For example, in Japan the Subcommittee "Chip Disposal" of the Japan Society for Precision Engineering

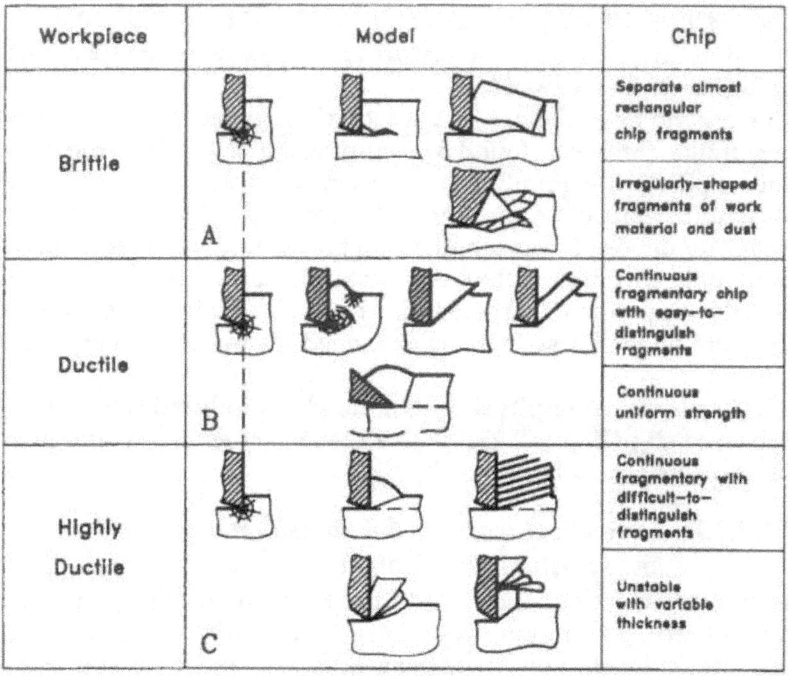
Workpiece
Model
Chip
Brittle
Separate almost rectangular chip fragments
Irregularly-shaped fragments of work material and dust
A
Ductile
Continuous fragmentary chip with easy-to-distinguish fragments
Continuous uniform strength
B
Highly Ductile
Continuous fragmentary with difficult-to-distinguish fragments
Unstable with variable thickness
C

Figure 2.26 The generalized model of chip formation. (From Astakhov, V.P. et al., *J. Mater. Process. Technol.*, 71(2), 247–257, 1997. With permission.)

(JSPE) formulated a revised system of chip form classification which included nine chip types basically according to their length. ISO 3685-1977 gives a comprehensive chip form classification based on the sizes and shapes of the chips that are generally obtained in metal machining. Other available classifications are well discussed in Jawahir and van Luttervelt.[26]

Unfortunately, the known classifications originate only from differences in the chip appearance but pay no attention to the physical state of the chip, including its state of stress and strain, hardness, texture, etc. Moreover, neither the tool geometry nor the cutting regime are taken into consideration. Thus, the known classifications have a post-process nature rather than being of a help in making preprocess decisions about chip-breaking. As a result, chip-breaking in metal cutting remains one of the fundamental problems that has to be solved for further advance in automated manufacturing.[27]

The purpose of this section is to present a new classification of the structure of chips which relates chip types to their formation mechanisms. Such knowledge enables an engineer to design a chip-breaker that is capable of breaking the chip with minimum energy consumption.

Analysis and generalization of results obtained in system studies of metal cutting enable the proposal of a generalized model for chip formation shown in Figure 2.26. The model includes three basic regions, *A*, *B*, and *C*, which correspond to the properties of workpiece materials and contact conditions.

Region *A* in this model generalizes models shown in Figures 2.7 and 2.9. An analysis of these models shows that even in the cutting of brittle workpieces, the chip shape is the controllable parameter. It is understood that when compression and bending act together, much less energy has to be supplied to the machining zone, and better working conditions (at least the absence of the dust) may be achieved. The tool geometry plays an important role here. As shown in Figure 2.7, the formed chip has an appearance of separate, almost rectangular elements and, therefore, is referred to as the regular broken chip. As one might argue, a positive rake angle is not very practical in cutting cast irons due to the presence of a significant amount of hard inclusions. In such a case, tungsten carbide, as a tool material, cannot withstand peak bending loads. As a result, practically all recommendations for tool geometry are the same, suggesting a high negative rake angle that unavoidably leads to the second model in Region *A* (Figure 2.26). As shown in Figure 2.9, the chip formed consists of irregularly shaped fragments of workpiece material and dust and, therefore, is referred to as irregular broken. To overcome this barrier and to shift from irregular broken to regular broken chip type, a high inclination angle combined with a positive rake angle is recommended as a compromise.

Region *B* of the model shown in Figure 2.25 represents a common case in the cutting of most engineering materials. The first model in this region has been already discussed (Figure 2.11), and it was shown that the chip formed according to the model is referred to as the continuous fragmentary chip. Because it already has non-uniform strength along its length, only the application of a relatively small shear stress is necessary to break this type of chip into separate pieces. This can be accomplished by placing a chip-breaking step above the rake face. The optimum size of the broken chips depends, under a given depth of cut and cutting feed, on the length, depth, and radius of the step.[37]

The second model in Region *B* (Figure 2.26) is a model of chip formation when cutting with a high rake angle. The details of this model are shown in Figure 2.27a. Here, the rake angle may reach 30 to 45°. As such, the reduction in the rate of plastic deformation with increasing rake angle leads to the condition where the angle, Θ, between the plane of maximum shear stress and the direction of compressive force *R*, approaches 90°. Such a representation allows one to compare the compression of the work material by the tool face with the pressing of a wedge-shaped workpiece between a pair of flat plates inclined with a small angle relative to each other. In cutting, the tool face plays the role of one of the plates, while the layer of metal being removed where the plastic deformation has not yet occurred (the conditional boundary between the plastic and elastic zone is shown in Figure 2.27 as line *ML*) plays the role of another plate. Under such conditions, the main part of the work material flows in the direction of the "thick" part of the wedge-shaped

(a)

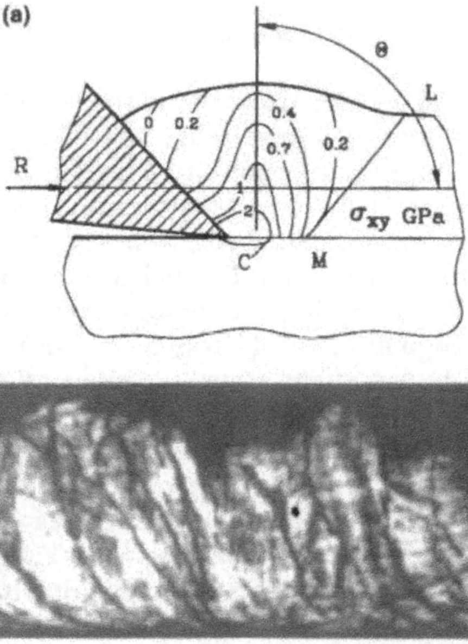

(b)

Figure 2.27 **(a)** A model for cutting with a high rake angle, and **(b)** the chip structure obtained in this process. (From Astakhov, V.P. et al., *J. Mater. Process. Technol.*, 71(2), 247–257, 1997. With permission.)

plastic deformation zone; the internal layers flow much more intensively than the external layers; and the deformation rate in the "thin" part of the wedge-shaped plastic-deformation zone is much higher than that in the "thick" part. The results of finite element simulation (discussed in Chapter 5), shown in Figure 2.27a, support the above qualitative analysis.

The foregoing considerations of cutting with high rake angles leads to the conclusion that if the interaction of metal-cutting-system components is as shown in Figure 2.27a, the work material fractures only along the line separating the workpiece and the layer being removed so there are no other sliding planes formed in the chip. As a result, a special chip type referred to as a continuous uniform-strength chip with wedge-shaped texture is formed. The wedge-shaped grains in the chip texture can be clearly seen by microphotography (Figure 2.27b) where a higher strain is observed at the thin part of grains. A chip-breaking step ground on the rake face would not be helpful in chip-breaking, while a radius groove may be beneficial. Such a groove should only increase the strain in the thin part of grain to the level of the chip fracture.

Region C of the model shown in Figure 2.26 represents two basic cases in cutting of highly ductile materials. The first model is basically the same as the first model of Region *B*. However, as discussed in Section 2.3.1.2., difficulties due to an insufficient rigidity of the partially formed chip are commonly observed in cutting of highly ductile materials. This insufficient rigidity is due not only to high ductility of the partially formed chip but also to the high temperature in the deformation zone which, in turn, reduces the elasticity of workpiece materials. Moreover, as these materials are usually much tougher, relatively low feed rates are common and make the uncut chip thickness smaller so that the cross-section of the partially formed chip becomes smaller, lowering its rigidity. A radius groove or even a two-radius groove chip-breaker is the best choice here. To deal with such a chip, Jawahir and Zhang[39] considered a four-stage chip-breaking cycle. The most important feature of this model is the necessity of increasing the bending moment in the deformation zone. According to the model, the bending moment is created when the free end of the chip comes into contact with the rotating workpiece.

When the workpiece material properties are such that the formed chip is not rigid enough to transmit the bending moment to its root, a groove and ridge type of chip-breaker must be used to increase the chip rigidity by changing its shape in the cross-section.[36,40] Such grooves are very helpful in machining with low cutting feeds when the chip thickness is small.

The second model of Region C of Figure 2.26 represents unstable chip formation. The instability occurs primarily due to seizure at the tool/chip interface. The chip formed under these conditions is referred to as the continuous humpbacked chip (Figure 2.25c). This type of chip is common in machining aerospace materials such as chromium- and/or nickel-based high alloys. Even though this chip looks easy to break, a special obstruction type of chip-breakers must be used. Grooves of special shapes help to reduce seizure due to reduction of actual tool/chip contact area.[26,37]

2.3.2 *Second step: controllability of the cutting system*

Control of the cutting system is carried out by the machining system. Because of insufficient information about the cutting system, this system is not considered as a system so that it is usually included in the machining system's hierarchy as an indivisible bottom component.

Considered as a system, the cutting system is characterized by the set of its output states, P; the set of its transition functions, M, and the set of their states, σ; the set of states of the cutting system, S; the set of output function, F; and the time scale of the cutting system, T. In terms of control, the number of transition functions of the cutting system defines the system order. The output and transition functions constitute an n-dimension vector of system states. The set of permissible states of this vector forms the vectorial space of system states. Considering the first approximation as a linear system, the cutting system may be represented by the normal form of the system-state equation:

$$\begin{aligned} \{\dot{S}\} &= [A]\cdot\{S\} + [B]\cdot\{P\} \\ \{F\} &= [C]\cdot\{S\} \end{aligned} \tag{2.26}$$

where $[A]$ is the system parametric matrix; $[B]$ is the system input matrix; $[C]$ is the system output matrix.

The choice of dimensions of the system vectors $\{S\}$, $\{F\}$, and $\{P\}$ is the matter of a particular problem:

$$\{S\} = \left[x_1, x_2, x_3, x_4, \ldots, x_{n-1}, x_n\right] \tag{2.27}$$

where x_1 is the tool wear; x_2 is the cutting force; x_3 is the cutting temperature; x_4 is the chip compression ratio; ... , x_{n-1} is the stress in the chip-formation zone; x_n is the strength of the work material.

$$\{F\} = \left[y_1, y_2, y_3, \ldots, y_m\right] \tag{2.28}$$

where y_1 is the residual stress in the machined surface; y_2 is the tool life; y_3 is the roughness of the machined surface; ... , y_m is the economy index.

$$\{P\} = \left[g_1, \ldots, g_i, g_{i+1}, \ldots, g_j\right] \tag{2.29}$$

where i is the number of regulatable input parameters; j–i is the number of non-regulatable input parameters; g_1 is the cutting speed; ... , g_i is the tool geometry; g_{i+1} is the variation of properties of the work material; ... , g_j is the temperature in the machining zone.

To the first approximation, matrixes $[A]$, $[B]$, and $[C]$ may be considered as having the same dimensions that yields:

$$A = \begin{bmatrix} a_{11} & \cdots & a_{1n} \\ & \cdots\cdots & \\ a_{nl} & \cdots & a_{nn} \end{bmatrix}; \quad B = \begin{bmatrix} b_{11} & \cdots & b_{1n} \\ & \cdots\cdots & \\ b_{nl} & \cdots & b_{nn} \end{bmatrix}; \quad C = \begin{bmatrix} c_{11} & \cdots & c_{1n} \\ & \cdots\cdots & \\ c_{nl} & \cdots & c_{nn} \end{bmatrix} \tag{2.30}$$

Insufficient knowledge about the cutting system results in the lack of the functional relationships between the input and output parameters. As a result, in practice, the set of transition functions M and the set of their states σ are ignored, and the set of states of the cutting system S are treated as a constant so that $\{S\} = 0$. Therefore, Equation (2.26) can be represented as:

$$\{F\} = [D] \cdot \{P\} \tag{2.31}$$

where

$$[D] = -\frac{[C] \cdot [B]}{[A]} \tag{2.32}$$

Therefore, the cutting process designer excludes the unknown set of transition functions M and the set of their states σ so that only the input and output parameters are used in the system optimization procedure. The problem here is that the matrix $[D]$ is obtained empirically under certain assumptions and agreements. The entries of this matrix are the regression coefficients of polynomial equations obtained experimentally. For example, the linear regression equation widely used as a mathematical model in metal cutting has the following form:

$$y = b_0 + b_1 x_1 + \ldots + b_n x_n \tag{2.33}$$

In this equation $\{F\} = y$; $[D] = [b_0, b_1, \ldots, b_n]$; $\{P\} = [x_0, x_1, \ldots, x_n]$.

For optimization, a model of the cutting system based on the technical, technological, and process economy restrictions is also often used. According to this approach, selection of the maximum values of the system parameters is carried out to justify the restrictions. Commonly, this type of system control aims to reduce the machining time and, as a result, minimize the other parameters dependent upon this time. The mathematical description of the discussed model may be presented in the following form:

$$\begin{aligned} x_1 + y_1 x_2 &= g_1 \\ x_1 + y_2 x_2 &= g_2 \\ y_3 x_2 &= g_3 \\ y_4 x_2 &= g_4 \\ x_2 &= g_5 \\ x_2 &= g_5 \\ x_2 &= b_6 \\ x_1 &= b_7 \\ x_1 &= b_8 \end{aligned} \tag{2.34}$$

Equation (2.34) is a system of linear equations obtained by taking logarithms of the standard power polynomial representation of the experimental data in metal cutting studies.[41] As such, if the parameters of the cutting system state may be treated as $x_1 = \ln n_s$, $x_2 = \ln s_f$ (n_s is rpm of the spindle; s_f is the cutting speed, mm/rev), the first equation in Equation (2.34) relates the output parameters with the cutting speed; the second, with the power requirement; the third, with the surface finish, etc. In the matrix form, Equation (2.34) is

$$[x_1\ x_2]\cdot\begin{bmatrix}1 & 1 & 0 & 0 & 0 & 0 & 1 & 1\\ y_1 & y_2 & y_3 & y_4 & 1 & 1 & 0 & 0\end{bmatrix}=[b_1,b_2,b_3,b_4,b_5,b_6,b_7,b_8] \tag{2.35}$$

or simply:

$$[b]=\{F\}\cdot\{P\} \tag{2.36}$$

This form is similar to Equation (2.31) so that there is no additional controllability achieved.

Control of a system is carried out by changing the system input parameters rather than changing the system transition functions and/or their state. The optimization control is often used and such a procedure includes minimization of a defined function correlating the input parameters and system-state vectors. Depending on a method of the representation of this functional, the system control includes the Lagrange, Mayer, or Bolza optimization problems.[42]

In cases when the Lagrange optimization problem is used, the functional has a form:[37]

$$J\equiv J(x)=\int_{t_a}^{t_b} f\,(x,\dot{x},t)\,dt \tag{2.37}$$

where t_a, t_b, $x(t_a)=c_a$, and $x(t_b)$ are fixed; f is a real-valued function of class C^2 with respect to all of its arguments; t_b-t_a is the time necessary to machine one part. In such a procedure, the tool life is to be given but can also be considered as corresponding to the given trajectory of the system's movement. In the latter case, the tool life is considered to be an additional optimization parameter.

In cases when the Mayer optimization problem is used, the functional has a form:

$$J_\mu=f_\mu\left(x(t),t\right)\Big|_{t_a}^{t_b}=f_\mu\left[x(t_b),t_b\right]-f_\mu\left[x(t_a),t_a\right] \tag{2.38}$$

and the functional for the Bolza problem is

$$J_\beta=f_{\beta 1}\left(x(t),t\right)\Big|_{t_a}^{t_b}+\int_{t_a}^{t_b} f_{\beta 2}\,(x,\dot{x},t)\,dt \tag{2.39}$$

In optimization of metal cutting operations, the machining time often has to be minimized so that the Mayer problem is mainly used. As such, the corresponding functional is $J_0 = L/S_{min}$, where L is the length of cut, and s_{min} is the feed rate.

The foregoing considerations indicate that in terms of control, the cutting system is treated as an assemblage, as the state vector, the set of transition functions, and the state of transition functions are ignored. Therefore, it is very important to prove that this assemblage which includes the cutting tool, the chip, and the workpiece is a system.

The simplest proof may be represented as follows. For the assemblage, $S = P = T = R$ is one of the admissible states. Set M may be considered here as the set of mappings α_r of S into itself, defined for every $r \in R$ and $x \in S$ by the equation $\alpha_r = xe^r$. Because the input and output parameters in metal cutting are real values and correlated during the time period of machining, we may define the set of output functions F as the set of all functions f defined on R with values in P, such that the Rienmann integral form 0 to t of $f(\tau)d\tau$ exists for all finite $t \in R$. Now, to verify that the assemblage is the cutting system, we have to verify that for each $(f, t) \in F \times T$, the set of the transition functions $\sigma(f, t)$ is the mapping α_r where:

$$r = \int_0^1 f(\tau)\, d\tau \tag{2.40}$$

First of all, it has to be verified that the set including the cutting tool, the chip, and the workpiece constitutes an assemblage according to the definition given by Equation (2.4). Specified are a set S (not empty), a set P (not empty), and a set F (not empty) of functions defined on R with values in P under transition and segmentation:

$$\int_0^1 (f \rightarrow r)(\tau)\, d\tau = \int_0^{r+1} f(\tau)\, d\tau - \int_0^r f(\tau)\, d\tau \tag{2.41}$$

Therefore, $f \rightarrow r \in F$ if $f \in F$, and:

$$\begin{aligned} \int_0^1 (f \mid g)(\tau)\, d\tau &= \\ &= \int_0^1 f(\tau)\, d\tau \text{ if } t < 0 \\ &= \int_0^1 g(\tau)\, d\tau \text{ if } t \geq 0 \end{aligned} \tag{2.42}$$

thus, $(f \mid g) \in F$ if f and $g \in F$. Furthermore, the specification includes a set M of mappings of S into itself and a mapping σ from $F \times R$ to M. Therefore, the set including the cutting tool, the chip, and the workpiece constitutes an assemblage.

The second step is to prove that this assemblage is a system. It is clear that the assemblage is a system as follows:

$$\sigma(f,0)(x) = x\left(\exp\left(\int_0^0 f(\tau)d\tau\right)\right) = x \tag{2.43}$$

therefore, $\sigma(f, 0) = 4$ for every $f \in F$; given $f, g \in F$, and $s, t \in R$, let $h = (f \to s \mid g \to s) \to -s$. Then,

$$\sigma(h, s+t)(x) = x \times \exp\left(\int_0^{s+t} h(\tau)d\tau\right) =$$
$$x \times \exp\left(\int_0^s h(\tau)d\tau + \int_s^{s+t} h(\tau)d\tau\right) = x \times \exp\left(\int_0^s h(\tau)d\tau + \int_0^t (h \to s)(\tau)dt\right) \tag{2.44}$$

Therefore:

$$\begin{aligned}
\sigma(h, s+t)\,(x) = \\
&= x \times \exp\left(\int_0^s f(\tau)d\tau + \int_0^t (g \to s)(\tau)d\tau\right) \text{ if } s \geq 0, t \geq 0 \\
&= x \times \exp\left(\int_0^s g(\tau)d\tau + \int_0^t (g \to s)(\tau)d\tau\right) \text{ if } s \geq 0, t < 0 \\
&= x \times \exp\left(\int_0^s f(\tau)d\tau + \int_0^t (g \to s)(\tau)d\tau\right) \text{ if } s < 0, t \geq 0 \\
&= x \times \exp\left(\int_0^s g(\tau)d\tau + \int_0^t (f \to s)(\tau)d\tau\right) \text{ if } s < 0, t < 0 \\
&= \sigma(g \to s, t)(\sigma(f, s)(x)) \text{ if } s \geq 0, t \geq 0 \\
&= \sigma(f \to s, t)(\sigma(f, s)(x)) \text{ if } s \geq 0, t < 0 \\
&= \sigma(g \to s, t)(\sigma(g, s)(x)) \text{ if } s < 0, t \geq 0 \\
&= \sigma(f \to s, t)(\sigma(g, s)(x)) \text{ if } s \geq 0, t \geq 0
\end{aligned} \tag{2.45}$$

Equations (2.45) constitute the system properties according to the considerations given in Section 2.2.3.

The preceding proof shows, at least, what has to be done to prove that an assemblage is a system and gives an example on which to try new concepts.

2.3.3 Principle of control of the cutting system

As considered in the previous section, the measure of performance in the cutting system can be expressed in terms of a functional, which here assumes the form:

$$J_m(m) = f_m(x, m)\Big|_{t_a}^{t_b} \tag{2.46}$$

where $x = x(t)$ is the $n \times 1$ state vector of the system; $m = m(t)$ is the $r \times 1$ control vector of the system referred to as a control function.[42]

It is often desirable, however, that m should be found as an explicit function of the state vector x and the external inputs. In this case, m is called a control law. Therefore, x and m are related by the vector state equation

$$\dot{x} = q(x, m) \tag{2.47}$$

The functions $q_1, q_2, \ldots, q_n$ and f are assumed to be of class C^1 with respect to the x_1's, and of class C^0 with respect to the m_i's. All functions are assumed to be real valued with respect to x's and m's of interest. Various conditions may be imposed upon t_a, t_b, $x(t_a)$, and $x(t_b)$; if all of these are not specified numerically, transversality conditions[42] are of consequence in solution.

As follows from Equation (2.47), for the cutting system to be controllable, the relationships among major system variables should be defined in form of Equation (2.47) that may be represented as:

$$\frac{dx^i}{dt} = q_i(x_1, \ldots, x_n, m_1, \ldots, m_r), \quad i = 1, \ldots n \tag{2.48}$$

where $x_1, \ldots, x_n$ are the phase coordinates of the process at time instant t; $m_1, \ldots, m_r$ are the parameters of the control law.

In terms of system control, a set of differential equations (Equation (2.48)) forms the system state vector. The state of the cutting system is considered to be defined over a time period $t_0 \le t \le t_1$ if the time-dependent control parameters:

$$u_j = u_j(t), \quad j = 1, \ldots, r \tag{2.49}$$

are defined over this time interval. Then, for the given initial conditions:

$$x^i(t_0) = x_0^i, \quad i = 1, \ldots, n \tag{2.50}$$

a solution of the system defined by Equation (2.48) can be found that defines the initial phase state of the cutting process. The course of the process is defined by the optimization functional (Equations (2.37) to (2.39)) so that the next phase state of the cutting system is defined by:

$$x^i(t_1) = x_1^i, \quad i = 1, \ldots, n \tag{2.51}$$

The problem here is to define the parameters of the control law (Equation (2.49)) which governs this transition.

To analyze the state of stress and deformation in the cutting system, the finite element method (FEM) seems to be the most suitable. In the application of FEM, distributions of the normal and shear stress at the tool/chip interface are assumed to be known. In other words, the components of the cutting system are considered separately, and their dynamic interactions are substituted by the static stress distributions. According to the introduced system approach, the components of the cutting system should be considered together so that the mentioned distribution will be the result of the dynamic interactions of the components.

According to FEM, the continuous media are represented by a system of elements with the finite degrees of freedom. Because it is assumed that the stress and strain in each particular element are uniformly distributed, the accuracy of calculations increases with increasing the number of the elements. In the context of the application of FEM, the components of the cutting system can be considered as subsystems (of the cutting system) consisting of a large number of elements with different physical and mechanical properties. The states of stress and strain of these elements correlate in a way to yield the compatibility condition for the deformations and equilibrium condition. This consideration leads one to conclude that there is a close similarity between the system approach and FEM. This similarity may be traced in the sequence used in both methods: decomposition, analysis, and then synthesis, so that finite-element mapping of the cutting system can be introduced. Therefore, FEM representation of the cutting system is found to be suitable for the proposed system approach in metal cutting.

To support the above qualitative discussion, the mathematical fundamentals of FEM have been analyzed. Consider a typical finite element *e* defined by nodes *i, j, m,* etc. and having straight-line boundaries (Figure 2.28). Let the displacements at any point within the element be defined as a column vector $f\{x,y\}$:[38]

$$\{f\} = [N]\delta^e = [N_i, N_j, N_m, \ldots] \begin{Bmatrix} \delta_i \\ \delta_j \\ \delta_m \\ \vdots \end{Bmatrix} \tag{2.52}$$

in which the components of $[N]$ are in general functions of position and $\{\delta\}^e$ represents a listing of nodal displacements for a particular element.

In the case of plane stress, for instance:

$$\{f\} = \begin{Bmatrix} u(x, y) \\ v(x, y) \end{Bmatrix} \tag{2.53}$$

represents horizontal and vertical movements of a typical point within the element and

$$\{\delta_i\} = \begin{Bmatrix} u_i \\ v_i \end{Bmatrix} \tag{2.54}$$

the corresponding displacements of a node I.

The functions N_i, N_j, and N_m have to be selected to give appropriate nodal displacements when the coordinates of the appropriate nodes are substituted in Equation (2.52).

With displacements known at all points within the element, the strains, which in this context are generally internal distortions, at any point can be determined. These will always result in the following matrix relationship:

$$\{\varepsilon\} = [C]\{\delta\}^e \tag{2.55}$$

For the plane stress case, the relevant strains of interest are those occurring in the plane and are defined according to the results of Chapter 3 as:

$$\{\varepsilon\} = \begin{Bmatrix} \varepsilon_x \\ \varepsilon_y \\ \gamma_{xy} \end{Bmatrix} = \begin{Bmatrix} \dfrac{\partial u}{\partial x} \\ \dfrac{\partial v}{\partial y} \\ \dfrac{\partial u}{\partial y} + \dfrac{\partial v}{\partial x} \end{Bmatrix} \tag{2.56}$$

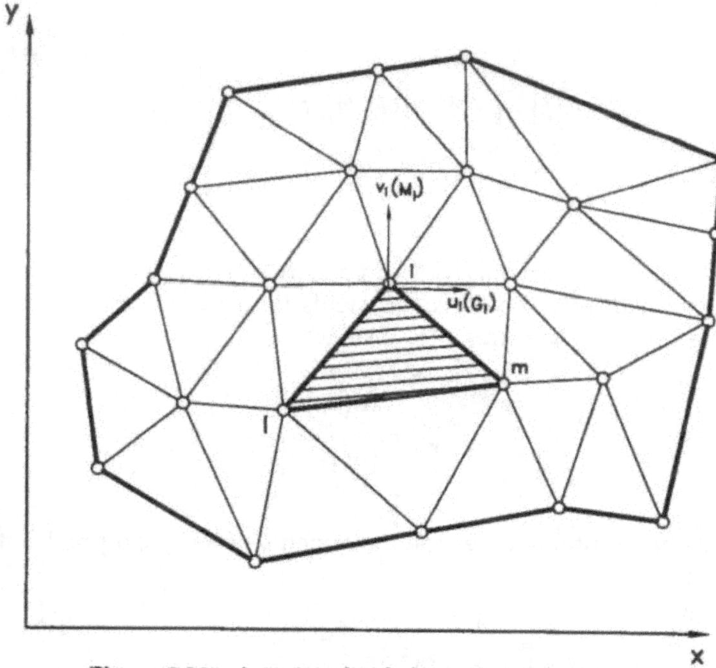

Figure 2.28 A region divided into finite elements.

It is seen that matrix [C] is formed by the coordinates of the deformed element. For the triangle element shown in Figure 2.28, this matrix becomes:

$$C = \frac{1}{\Delta}\begin{bmatrix} y_2 - y_3 & 0 & y_3 - y_1 & 0 & y_1 - y_2 & 0 \\ 0 & x_3 - x_2 & 0 & x_1 - x_3 & 0 & x_2 - x_1 \\ x_3 - x_2 & y_2 - y_3 & x_1 - x_3 & y_3 - y_1 & x_2 - x_1 & y_1 - y_2 \end{bmatrix} \quad (2.57)$$

where:

$$\Delta = \begin{bmatrix} x_1 & y_1 & 1 \\ x_2 & y_2 & 1 \\ x_3 & y_3 & 1 \end{bmatrix} \quad (2.58)$$

is the doubled area of the triangle.

From Equation (2.52), with the functions N_i, N_j, and N_m already determined, the matrix [C] will be obtained. If the linear form of these functions is adopted, then, in fact, the strains will be constant throughout the element.

In general, the material within the element boundaries may be subjected to initial strains such as those due to temperature changes, shrinkage, crystal growth, and so on. If such strain is denoted by $\{\varepsilon_0\}$, then the stress will be caused by the difference between the actual and initial strains. Assuming general elastic behavior, the relationship between stresses and strains will be linear and of the form:

$$\{\sigma\} = [D](\{\varepsilon\} - \{\varepsilon_0\}) \tag{2.59}$$

where $[D]$ is an elasticity matrix containing the appropriate material properties.

For the particular case of plane stress, three components of stress corresponding to the strains already defined have to be considered. In the notations used so far:

$$\{\sigma\} = \begin{Bmatrix} \sigma_x \\ \sigma_y \\ \sigma_{xy} \end{Bmatrix} \tag{2.60}$$

and the $[D]$ matrix is simply obtained from the usual isotropic stress-strain relationship discussed in Chapter 3:

$$\begin{aligned} \sigma_x &= \frac{E}{1-\mu^2}\left(\varepsilon_x - (\varepsilon_x)_0\right) + \frac{E-\mu}{1-v^2}\left(\varepsilon_y - (\varepsilon_y)_0\right) \\ \sigma_y &= \frac{E-v}{1-v^2}\left(\varepsilon_x - (\varepsilon_x)_0\right) + \frac{E}{1-v^2}\left(\varepsilon_y - (\varepsilon_y)_0\right) \\ \sigma_{xy} &= \frac{E}{2\left(1-v^2\right)}\left(\gamma_{xy} - (\gamma_{xy})_0\right) \end{aligned} \tag{2.61}$$

or in matrix notation:

$$\{\sigma\} = [D]\{\varepsilon\} = \frac{E}{1-v^2}\begin{bmatrix} 1 & v & 0 \\ v & 1 & 0 \\ 0 & 0 & \frac{1-v}{2} \end{bmatrix}\{\varepsilon\} \tag{2.62}$$

The nodal forces which are equivalent to the boundary stress and distributed loads on the element are

$$\{F\}^e = \begin{Bmatrix} F_i \\ F_j \\ F_m \\ \vdots \end{Bmatrix} \tag{2.63}$$

Each of the forces $\{F_k\}$ must contain the same number of components as the corresponding nodal displacement $\{\delta_k\}$ and be ordered in appropriate corresponding directions.

In the particular case, the nodal forces may be represented as:

$$\{F_k\} = \begin{Bmatrix} M_k \\ G_k \end{Bmatrix} \tag{2.64}$$

with components M and G corresponding to the directions of u and v displacements, and the distributed load is

$$\{p\} = \begin{Bmatrix} X \\ Y \end{Bmatrix} \tag{2.65}$$

in which X and Y are the "body force" components.

For linear elastic material the nodal and body forces are related as:

$$\sum_{i=1}^{n} F_i \delta_i = \sum_{i=1}^{m} N_j \, d\varepsilon_j \tag{2.66}$$

Here F_i, N_j are components of nodal and internal forces, respectively.

Equation (2.66) written in matrix notation becomes:

$$\{F\}^e = [B]\{p\} \tag{2.67}$$

Here, $[B]$ is the matrix of coefficients.

Because the internal forces are stress, Equation (2.67) may be written, accounting for the result given by Equation (2.62), as:

$$\{F\}^e = [B][D]\{\sigma\} \tag{2.68}$$

It is known[38] that:

$$[B] = [C]^T \tag{2.69}$$

and, therefore:

$$\{F\}^e = [C]^T [D][C]\{\delta\}^e \tag{2.70}$$

or, designating $[K] = [C]^T [D] [C]$, the final form is

$$\{F\}^e = [K]\{\delta\}^e \tag{2.71}$$

Here $[K]$ is the external stiffness matrix,

$$[K] = \begin{bmatrix} K_{11} & K_{12} & \dots & K_{1m} \\ K_{21} & K_{22} & \dots & K_{2m} \\ \dots & \dots & \dots & \dots \\ K_{n1} & K_{n2} & \dots & K_{nm} \end{bmatrix} \tag{2.72}$$

Each entire K_{ij} of matrix $[K]$ is a force applied along the direction i and causes unit displacement along direction j. From Equation (2.44), the displacement matrix $\{\delta\}$ can be defined as:

$$\{\delta\}^e = [K]^{-1}\{F\}^e \tag{2.73}$$

When the system of the finite elements instead of one element is considered, then the displacement of each node, δ_c, should be evaluated using the nodal system matrix:

$$\{\delta_c\} = [K_c]^{-1}\{R\}^e \tag{2.74}$$

where $[K_c]$ is the external stiffness matrix for the system of finite elements, and $\{R\}^e$ represents external concentrated forces.

Equation (2.74), solved under displacement restrictions of some nodes, gives the displacements δ. Using known δ, the stress and deformations for the system are defined using Equations (2.62) and (2.55).

It is known[43] that if the system of displacements is defined throughout the structure by the element displacement functions, with the nodal displacements acting as undetermined parameters, then the procedure discussed above is equivalent to that of minimizing the total potential energy of the system.

The total potential energy of the system is equal to the sum of potential energies of internal (W) and external (A) forces, so:

$$U = W + A \tag{2.75}$$

and the problem may be represented in the following form:

$$d[U] = 0 \tag{2.76}$$

This is simply a statement that the total potential energy must be a stationary quantity. As the strain energy W is always a positive quantity, and A can take up positive or negative values, the stationary position of equilibrium is that of a minimum, thus establishing the principle that, for a prescribed set of displacements and strains, the equilibrium conditions are satisfied when the total potential energy is at its minimum.

The continuum is considered again as divided by imaginary surfaces into "elements" (Figure 2.22). The nodal displacements listed at all nodes in the standard way:

$$\{\delta\} = \begin{Bmatrix} \delta_1 \\ \vdots \\ \delta_n \end{Bmatrix} \tag{2.77}$$

will define the state of displacement throughout the continuum, element by element.

To minimize total potential energy U with respect to the nodal displacements treated as variable parameters, it is necessary to establish a system equation of the type:[43]

$$dU = \frac{\partial U}{\partial \delta_1} d\delta_1 + \frac{\partial U}{\partial \delta_2} d\delta_2 + \cdots + \frac{\partial U}{\partial \delta_n} d\delta_n \tag{2.78}$$

According to the principle of minimum potential energy, if U for stationary conditions is minimum, then $dU = 0$ for any admissible set of infinitesimal displacements, thus:

$$\frac{\partial U}{\partial \delta_i} = 0, \quad i = 1, 2, \ldots, n \tag{2.79}$$

It follows from Equations (2.78) and (2.79) that:

$$dW + dA = 0 \tag{2.80}$$

or:

$$-dW + dA = 0 \tag{2.81}$$

Because the increments in the work done by the external forces are

$$-dW = \sum_{i=1}^{n} F_i d\delta_i \tag{2.82}$$

and the corresponding increments in the work done by the internal forces are

$$dA = \sum_{i=1}^{m} N_j d\varepsilon_j \tag{2.83}$$

Equations (2.78) and (2.79) are equivalent to Equation (2.66).

Because the structures of Equations (2.78) and (2.79) are equivalent to those of Equations (2.48) and (2.49), the cutting system may be analyzed using FEM.

References

1. Wymore, A.W., *A Mathematical Theory of System Engineering – The Elements*, John Wiley & Sons, New York, 1967.
2. Merchant, M.E., Mechanics of the metal cutting process, *J. Appl. Phys.*, 16, 267, 1945.
3. Vidosic, J.P., *Metal Machining and Forming Technology*, The Ronald Press Company, New York, 1964.
4. Johnson, W. and Mellon, P.B., *Engineering Plasticity*, Ellis Horwood, Oxford, 1983.
5. Madhavan, V. and Chandrasekar, S., Some observations on the uniqueness of machining, in *Manufacturing Science and Engineering*, Vol. 6(2), Proc. 1997 American Society of Mechanical Engineers International Mechanical Engineering Congress and Exposition, November 16–21, Dallas, TX, 1997, p. 99.
6. Shvets, S.V., Astakhov, V.P., and Osman, M.O.M., A system approach in metal cutting, in *Manufacturing Science and Engineering*, Proc. 1996 CSME Forum, 13th Symposium on Engineering Applications of Mechanics, Hamilton, OH, 1996, p. 656.
7. Astakhov, V.P. and Shvets, S.V., A system concept in metal cutting, *J. Mater. Proc. Technol.*, 82, 1–3, 289.
8. Astakhov, V.P., Shvets, S.V. and Osman, M.O.M., Chip structure classification based on mechanism of its formation, *J. Mater. Proc. Technol.*, 71/2, 247, 1996.
9. Astakhov, V.P., Shvets, S.V. and Osman, M.O.M., The bending moment as the cause of chip formation, in *Manufacturing Science and Engineering*, Vol. 6(2), Proc. 1997 American Society of Mechanical Engineers International Mechanical Engineering Congress and Exposition, November 16–21, Dallas, TX, 1997, 53.
10. Touret, R., *The Performance of Metal Cutting Tools*, American Society for Metals, Heston, 1957.
11. Zorev, N.N., *Metal Cutting Mechanics*, Pergamon Press, Oxford, 1966.
12. Trent, E.M., *Metal Cutting*, Butterworth-Heinemann, Oxford, 1991.

13. Stephenson, D.A. and Agapiou, J.S., *Metal Cutting Theory and Practice*, Marcel Dekker, New York, 1977.
14. Oxley, P.L.B., *Mechanics of Machining: An Analytical Approach To Assessing Machinability*, John Wiley & Sons, New York, 1989.
15. Shaw, M.C., *Metal Cutting Principles*, Clarendon Press, Oxford, 1984.
16. Boothroyd, G. and Knight, W.A., *Fundamentals of Machining and Machine Tools*, 2nd ed., Marcel Dekker, New York, 1989.
17. Standard ISO 3685-77.
18. Byars, E.F. and Snyder, R.D., *Engineering Mechanics of Deformable Bodies*, International Textbook Company, Scranton, PA, 1969.
19. Dieter, G.E., *Mechanical Metallurgy*, 3rd ed., McGraw-Hill, New York, 1986.
20. *Metal Handbook*, Vol. 8, *Mechanical Testing*, 9th ed., American Society for Metals, Metals Park, OH, 1985.
21. *Metals Handbook*, Vol. 9, *Fractography and Atlas of Fractography*, 8th ed., American Society for Metals, Metals Park, OH, 1974.
22. Weinmann, K.J., The use of hardness in the study of metal deformation processes with emphasis on metal cutting, in *Material Issues in Machining* and *The Physics of Machining Processes*, Proc. Symp., American Society of Mechanical Engineers, Metals Park, OH, 1992, p. 1.
23. Rosenberg, A.M. and Rosenberg, O.A., On the stressed-deformed state in metal cutting (in Russian), *Sverhtverdye Materaily*, 5, 41, 1988.
24. Beer, F.P. and Johnston, E.R., Jr., *Mechanics of Material*, McGraw-Hill Ryerson, Toronto, 1985.
25. Nakayama, K., Basic rules on the form of chip in metal cutting, *CIRP Ann.*, 27, 1, 97, 1978.
26. Jawahir, I.S. and van Luttervelt, C.A., Recent developments in chip control research and applications, *CIRP Ann.*, 42, 659, 1993.
27. Jawahir, I.S., Balaji, A.K., Stevenson, R., and van Luttervelt, C.A., Towards predictive modeling and optimization of machining operations, in *Manufacturing Science and Engineering*, Vol. 6(2), Proc. 1997 American Society of Mechanical Engineers International Mechanical Engineering Congress and Exposition, November 16–21, Dallas, TX, 1997, p. 3.
28. Astakhov, V.P. and Osman, M.O.M., An analytical evaluation of the cutting forces in self-piloting drilling using the model of shear zone with parallel boundaries. Part 2. Application, *Int. J. Machine Tools Manuf.*, 36(11), 1335, 1996.
29. Mills, B. and Redford, A.H., *Machinability of Engineering Materials*, Applied Science Publishers, London, 1983.
30. Komanduri, R., On the mechanism of chip segmentation in machining, *ASME J. Eng. Industry*, 103, 33,1981.
31. Lindberg, B. and Lindstrom, B., Measurements of the segmentation frequency in the chip formation process, *CIRP Ann.*, 32(1), 17, 1983.
32. Komaduri, R., Schroeder, T., Hazra, J., von Turkovich, B.F., and Flom, D.G., On the catastrophic shear instability in high-speed machining of an AISI 4340 steel, *ASME J. Eng. Industry*, 104, 121, 1982.
33. Kobayashi, S. and Thomsen, E.G., Metal-cutting analysis II: new parameters, *ASME J. Eng. Industry*, 84, 71, 1962.
34. Trent, E.M., Metal cutting and the tribology of seizure. Part 1. Seizure in metal cutting, *Wear*, 128, 29, 1988.

35. Hayajneh, M.T., Astakhov, V.P., and Osman, M.O.M., An analytical evaluation of the cutting forces in orthogonal cutting using the dynamic model of shear zone with parallel boundaries, *J. Mater. Proc. Technol.*, 82(1–3), 220, 1998.
36. Nakayama, K. and Arai, M., Comprehensive chip form classification based on the cutting mechanism, *CIRP Ann.*, 71, 1992.
37. Boyes, W.E., Dunlap, D.D., Hoffman, E.G., Jacobs, P., Orady, A.E., and Smith, D.A., *Fundamentals of Tool Design*, 3rd ed., Society of Manufacturing Engineers, 1991.
38. Jawahir, I.S. and Zhang, J.P., An analysis of chip curl development, chip deformation and chip-breaking in orthogonal machining, *Trans. NAMRI*, XXIII, 109, 1995.
39. Nakayama, K., Chip control in metal cutting, *Bull. Jpn. Soc. Pres. Eng.*, 18(2), 97, 1984.
40. Astakhov, V.P., Osman, M.O.M., and Al-Ata, M., Statistical design of experiments in metal cutting. Part 2. Applications, *J. Testing Evaluation, JTEVA*, 25(3), 328, 1997.
41. Pierre, D.A., *Optimization Theory with Applications*, John Wiley & Sons, New York, 1969.
42. Zienkiewicz, O.C. and Cheung, Y.K., *The Finite Element Method in Structural and Continuum Mechanics*, McGraw-Hill, London, 1968.

chapter three

Parallel-sided deformation zone theory

3.1 The role of engineering plasticity in metal cutting studies

The primary objective of engineering plasticity in metal cutting studies is to develop mathematical techniques for determining non-uniform stress and strain distributions. As a result, it is possible to predict the plastic deformation of the workpiece in various circumstances particular to a complex stress state. Although some of the laws that agree satisfactorily with experimental evidence have been established mainly for metals and their alloys, the non-linearity of the principal laws (and hence the basic equations) inevitably presents considerable mathematical difficulties.

The history of the science of the plasticity of metals began in 1854 when Tresca[1] presented an analysis of his experimental results on some forming processes. The analysis led to the formulation of a yield criterion which states that a metal yields plastically when the maximum shear stress attains a critical value. Using this criterion, Saint-Venant,[2] in 1870, published the first known mathematical theory of plasticity where a complex state of stress is considered, resulting in a system of stress-strain relations for plane-strain deformation. Following the ideas of Saint-Venant, the three-dimensional relations between stress and plastic strain were proposed by Lévy[3] in 1870, who also introduced the method of linearization of the plane-strain problem. At this stage, the mathematical theory of plasticity was beyond the experimental techniques available and as a result was poorly supported by experiments. Subsequent development of the theory of plasticity proceeded slowly. Due to great practical demand, most of the experiments performed at this stage were concerned with the establishment of a practical yield criterion for various states of stress. As a result, a great number of different yield criteria have been proposed. However, a new yield criterion found to be satisfactory for many metals was formulated by von Mises[4] in 1913 on the basis of purely mathematical considerations. Although the criterion is in wide use today, for our consideration here it is important to note that the von Mises criterion was interpreted later on by Hencky[5] as implying that yielding occurs when the shear strain energy attains a critical value.

The problem of indentation was studied by Prandtl,[6] who determined the pressure required to indent a half-space by a smooth flat punch using a solution to a hyperbolic problem. Hencky[7] produced the general theory applicable to the special solutions of Prandtl and in this way defined the geometrical properties of slip-line fields for plane-strain deformation; however, these solutions were not easy to implement in practice. The problem was solved by Geiringer,[8] who obtained the velocity compatibility equations for flow along slip lines.

Although a number of attempts have been made to implement the theory of plasticity in the study of forming processes,[9–11] it was not until 1926 that the Lévy-von Mises stress-strain relationships were validated to a first approximation when Lode[12] published the results of experiments studying the deformations of various metal tubes subjected to a combined state of stress. The theory of plasticity was first generalized to a reasonable degree by Reuss,[13] who introduced the elastic strain component into the stress-strain relations, and by Odquist,[14] who showed how the Lévy-von Mises equations could be modified to take into account the effect of strain hardening.

The theoretical and experimental advances made in the 1930s and 1940s were summarized in the classic textbook on the mathematical theory of plasticity by Hill.[15] In Hill's book, techniques for analyzing and predicting forming parameters and deformation were reasonably well defined for a number of metal forming processes; however, the solutions presented were too complicated to be used in practice. The introduction of the velocity diagram by Prager[16] in 1953 brought significant simplification of slip-line field solutions. An extensive bibliography of published work concerned with the plane-strain, slip-line theory is given in Johnson and Mellor.[17]

In 1951, Drucker et al.[18] stated three so-called limit theorems from which upper and lower bound estimates (that is, overestimates and underestimates) may be established for the forming parameters in metal-forming processes. However, it is recognized that these theorems were deducible from the work principles published in the textbook by Hill.[15] Numerous examples of the use of the upper and lower bound theorems in the field of metal-forming processes are presented in the textbook by Johnson and Mellor.[17] Upper bound solutions for plane-strain problems were obtained by Kudo,[19] who introduced the concept of the unit rectangular deforming region. Several types of admissible velocity fields in the deforming regions were considered, but it appeared that the rigid triangle velocity field, in which the deformation zones are bounded by plane surfaces, was the best.

The application of the mathematical theory of plasticity to the problem of metal cutting originates from Hill.[15,20–22] Using the results reported in his book, Hill[15] first hypothesized that the solution to the metal cutting process is not unique and that a steady-state mode of deformation would depend on conditions encountered in the initial transient period of deformation. For an assumed class of solutions having a shear plane, when machining a rigid, perfectly plastic material with a sharp tool, Hill was able to determine a range of admissible shear angles. The limits on the range were imposed by the requirement that the metal should not be overstressed at any point. Analyzing the hypotheses proposed by Merchant[23] that the shear plane in metal cutting would assume an inclination that would ensure that the work performed in the cutting process would be a minimum, Hill came to the conclusion that, because the geometry of deformation in machining is not known *a priori*, the principle of minimum work could not be applied. Later on, in 1983, Rubenstein[24] presented more evidence to support this fact. In the absence of a way to choose a particular solution uniquely from the possible range, Hill conjectured that the actual mode of deformation would probably be dependent on the exact conditions during the initial stage of cutting. Dewhurst[25,26] has also strongly argued for such non-uniqueness of machining.

At this point, it is worthwhile to discuss the application of the principle of minimum energy in metal cutting. On one hand, this is one of the fundamental principles of nature and thus should be applicable in the analysis of any real physical system. On the other hand, the application of this principle in metal cutting leads to conditions which cannot be fulfilled physically and which are at odds with theoretical and, more importantly, experimental results. In the author's opinion, this contradiction stems from the fact that the researchers have utilized an under-determined hypothetical system provided by Merchant[23] as the model for orthogonal cutting. If the non-uniqueness of machining is the case, the search for predictive models of machining could prove to be futile.

The ideas proposed by Hill have had a great influence on further work in the theoretical modeling of the metal cutting process. The slip-line field method has been used to find a solution to the metal cutting problem.[27,28]

Lee and Shaffer[29] proposed a slip-line solution to the cutting problem. Two essential aspects of their consideration are of importance. First, they considered the state of stress of the chip through which forces exerted by the tool were transmitted to the shear plane. It is worthwhile to note here that earlier and later researchers considered a laterally inverted force picture, which is convenient for modeling but is at odds with physical sense. Second, the shear plane was considered as a line in a two-dimensional cut along which the tangential component of velocity is discontinued and is in agreement with the mathematical theory of plasticity. Unfortunately, the later studies ignored these important findings. The proposed slip-line solution considers that the plastic deformation in metal cutting takes place in a plastic

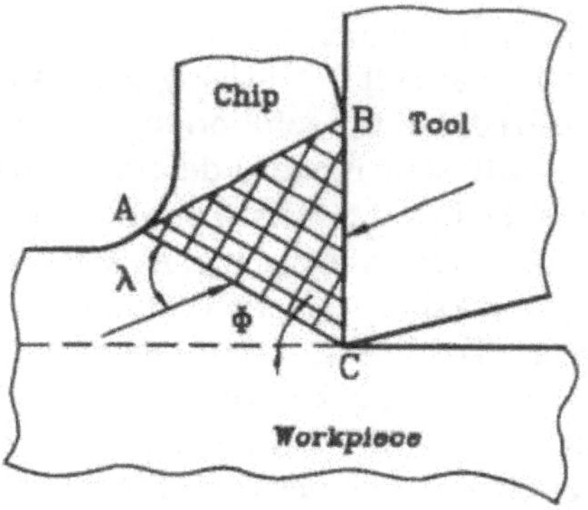

Figure 3.1 Lee and Shaffer (1951) model.

zone which exists within the chip (Figure 3.1). This zone is composed of two families of parallel straight slip lines. The stresses are assumed to be uniform throughout the plastic zone, which is bounded by the shear plane, tool face, and imaginary boundary across which no stresses are transmitted. Though the discussed solution was a huge methodological step ahead, the authors essentially used the known (Merchant[23]) model and velocity diagram that led to the uncommon results, which were criticized by Hill[22] and other researchers,[30] without noting the real reason for such results.

Shaw et al.[30] revised the Lee and Shafer model by drawing attention to the interrelationship between shear and friction processes in metal cutting. To bring this interrelationship into the theory and to fit the available experimental data, they had no choice but to assume that the shear plane need not be in the direction of the maximum shear stress. They even introduced a deviator angle of the shear plane from the maximum shear stress direction that is in obvious contradiction with material test practice, where the direction of deformation always coincides with the direction of the maximum stress.

The single shear plane model (discussed in Chapter 2 and shown in Figure 2.1) has been studied since the last century and, therefore, cannot be referred to as the Merchant model, as has become customary. The model has been constructed using simple observations of the metal cutting process. To the best of the author's knowledge, Usachev,[6] in 1883, was probably the first to propose the single shear plane model; he studied the chip structure and introduced the shear angle and texture angles as the model's basic parameters. As early as 1896, Bricks[33] justly criticized the single shear plane model for chip formation studied earlier by Zvorykin.[34] To solve the contradictions associated with this model, Bricks suggested that the plastic deformation takes place in a certain zone which is defined as consisting of a family of slip-lines (OA_1, OA_2, ..., OA_n) arranged fanwise, as shown in Figure 3.2. As such,

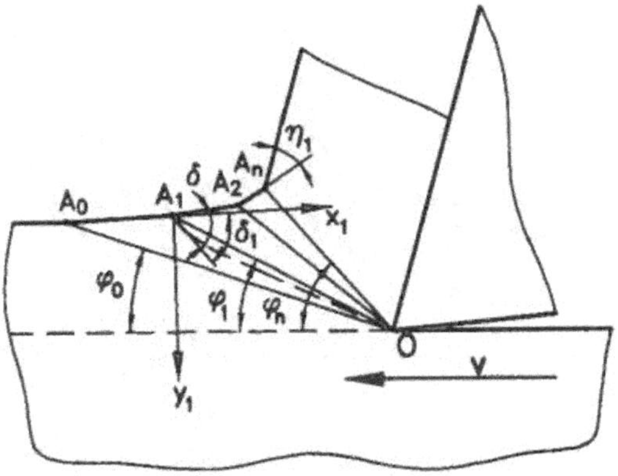

Figure 3.2 Briks (1896) model.

the outer surface of the workpiece and the chip-free surface are connected by a certain transition curve A_0A_n. Although the model proposed by Bricks solved the most severe contradictions associated with the single shear plane model, it also has a number of drawbacks which are well analyzed by Zorev.[27]

Analyzing Bricks' model, Zorev[27] attempted to construct the slip-line field in the deformation zone using the basic properties of slip lines. According to his consideration, the deformation process in metal cutting involves shear and, therefore, is characterized by the lines of maximum shear stress (i.e., by characteristic curves or slip lines). For the first and, unfortunately, last time since then, Zorev considered two zones of stress concentration — namely, zone *A*, which is immediately adjacent to the transition curve between the workpiece outer surface and the free surface of the chip, and zone *B*, which is located in front of the cutting edge (Figure 3.3). Applying the basic properties of slip lines, Zorev suggested that one family of the slip lines passes through the deformation zone from zone *B* to *A* (as shown in Figure 3.4 by solid lines). Portions of the second orthogonal family of slip lines are illustrated in Figure 3.4 by dashed lines. Zorev pointed out that, besides special cases, shearing along these lines is comparatively small.

According to Figure 3.4 the deformation zone *LOM* is limited by the slip line *OL*, along which the first plastic deformation in shear occurs; slip line *OM*, along which the last plastic deformation occurs; and line *LM*, which is the deformed section of the workpiece's outer surface. The particles of the work surface pass through the plastic region intersecting the shear lines and are in turn subjected to successively increasing deformations from zero to a certain maximum value peculiar to the final chip.

The application of the slip-line method has been pushed by Oxley for many years, and the results are well summarized in his book.[28] In his book,

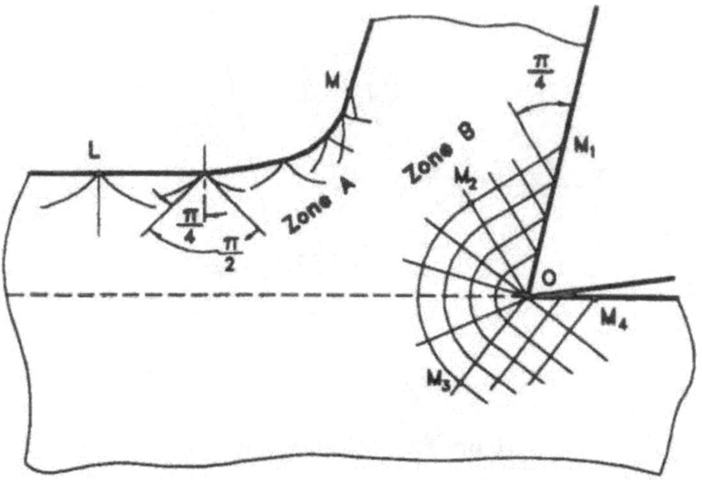

Figure 3.3 Slip-line fields in zones *A* and *B* when no friction on the tool/chip interface is considered.

Chapter 3 deals with slip-line field analysis of experimental flow fields which originates from the earlier work[35] and aims to estimate the boundaries of the deformation zone experimentally. Zorev,[27] considering the proposed slip-line field (shown in Figure 3.4 in Oxley's book[28] and in Figure 7b in Zorev's book[27]), commented that observations of the chip formation process under a microscope showed an instability in the process of plastic deformation in the chip formation zone that makes it impossible to determine the exact boundaries of the chip formation zone. Oxley, however, did not use the results obtained for further analysis. Instead, his next chapter (Chapter 4) deals with analysis of the parallel-sided shear zone model. The main problem associated with the derived slip-line solution is that the model was thought of as having a single shear plane which is assumed to open up so that the boundary between the shear zone and work and the boundary between the shear zone and chip are parallel to and equidistant from the shear plane. Even though these boundaries were considered as slip lines,[28] the corresponding velocity discontinuity[17] was not considered. Therefore, the overall impression is that the known consideration was taken to avoid some severe problems associated with a single shear plane model. To comment on how far the known slip-line solutions in metal cutting have gone, here is a quote from Rahman et al.:[36] "Jawahir and Oxley developed a chip formation model based on the theory of slip-line field. Chip backflow angle which is independent of BUE (built-up edge) was derived by its hodographs and Mohr's circle. It looks more reasonable because most of the work showed that the chip backflow effect occurs without a BUE. Unfortunately, the experimental verification was not satisfactory, especially for the case of cutting with negative rake angle." Eggelestone et al.[37] have concluded that, "Neither the Ernest and

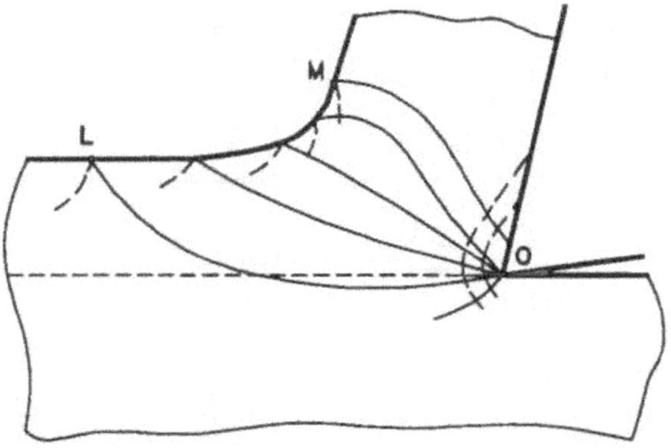

Figure 3.4 Schematic representation of the slip-line structure of the deformation zone. (Adapted from Zorev, N.N., *Metal Cutting Mechanics, Pergamon Press,* Oxford, 1966.)

Merchant minimum energy criterion, nor the ideal plastic-solid solution of Lee and Shafer, nor that of Hill is in agreement with all the experimental observations."

The objective of this chapter is to explain why the known models of chip formation based on the existing velocity diagrams have failed to match the known experimental results and, based on such an understanding, suggest new principal relationships among the parameters of the cutting process.

3.2 Velocity diagram: what seems to be the problem?

The model of chip formation constitutes the very basis of any metal cutting theory. Although a number of models are known to specialists in the field, the single shear plane model, originally proposed in the 19th century, survives all of them and is still the first choice for student textbooks. A simple explanation for this fact is that the model is easy to teach and learn, and simple numerical examples for calculating the cutting force and energy required for cutting can be provided for students' assignments. It is true that it is usually pointed out that the model represents an idealized cutting process; however, it is never mentioned how far this idealization deviates from reality.

The single shear plane model was discussed in Chapter 2 (Figure 2.1) and is shown again here in Figure 3.5a. A real problem begins when one tries to compare the relationships for velocities, strain, shear angle, etc. obtained on the basis of the velocity diagram and the actual velocity diagram shown in Figure 3.5b. Using this diagram, Merchant[23] derived the following:

$$v_s = v_c \frac{\cos\gamma}{\cos(\varphi-\gamma)} \qquad (3.1)$$

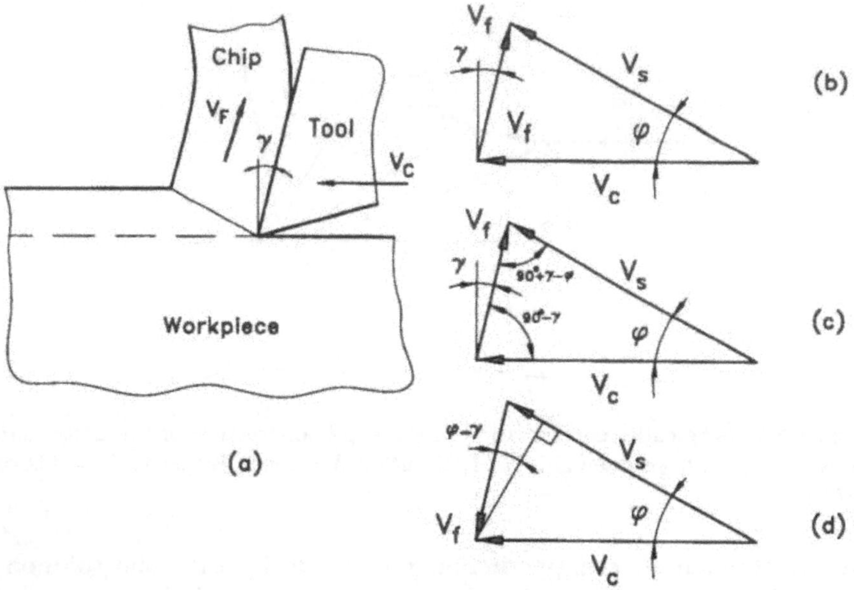

Figure 3.5 The single shear plane model (a) and three known velocity diagrams (b), (c), and (d).

$$v_f = v_c \frac{\sin\varphi}{\cos(\varphi - \gamma)} \tag{3.2}$$

Combining Equations (3.1) and (3.2), Merchant[23] obtained the well-known formula for the shear angle:

$$\tan\varphi = \frac{\xi \cos\gamma}{(1 - \xi \sin\gamma)} \tag{3.3}$$

Two basic problems with the discussed velocity diagram are

1. Equation (3.2) does not follow from the velocity diagram (Figure 3.5b), as the sign "–" is missing in this equation. As a result, Equation (3.3) cannot be derived. Nevertheless, the same velocity diagram has been used in subsequent publications on metal cutting.
2. The velocity of deformation (the shear velocity) may be greater than that of the applied load, which is physically impossible because the model is one of multi-degrees-of-freedom.

Black[38] "silently" corrected the velocity diagram shown in Figure 3.5b, offering a version shown in Figure 3.5c. Although this new diagram solved the "sign" problem and thus made the derivation of Equation (3.3) possible, the magnitude of the shearing velocity is still the same as in the diagram shown in Figure 3.5b so tht Problem 2 remains unsolved. Moreover, it is well known that the layer to be removed on passing through the shear plane deforms in the direction opposite to that assumed in Figure 3.5c. This is physically impossible, as the direction of the velocity of deformation must coincide with the direction of deformation.

Stephenson and Agapiou[39] went even further. Trying to solve the discussed "sign" problem, they proposed the velocity diagram shown in Figure 3.5d, where the direction of the chip velocity is assumed to be opposite to the direction of its motion. It is true that the chip shrinks in cutting (i.e., its length is commonly less than that of the removed layer). However, when the single shear plane model is considered, this process is assumed to occur "instantly" as the chip crosses the shear plane having zero thickness. Then the chip moves over the tool face with the velocity of chip that, within the tool/chip interface, is parallel to this face and directed outward toward the shear plane.

3.3 Role of the velocity diagram in the theory of metal cutting

Having read the above discussion, one may ask logically if the velocity diagram plays any significant role in the development of the theory of metal cutting or is just of passing interest. We may answer that the velocity diagram, being invariant to any particular model of chip formation, is the fundamental issue in this theory. This statement can be supported by the following.

Knowing the velocity diagram, one can calculate the total energy consumed per unit time that fundamentally defines all other cutting parameters. As discussed by Zorev[27] and Shaw,[40] this energy u can be thought of as:

$$u = u_s + u_F \tag{3.4}$$

Here, u_s and u_F are the shear and friction energy per unit volume, calculated as:

$$u_s = \tau_f \left(\frac{v_s}{v \sin \varphi} \right) \tag{3.5}$$

$$u_F = \frac{F_c v_c}{v b t_1} \tag{3.6}$$

In these equations, τ_f is the shear flow stress, F_c is the force along the shear plane, b is the width of cut.

Knowing the chip velocity, one can define the velocities at the tool/chip and tool/workpiece interfaces, which, in turn, define the physical nature of these contacts, chip shape, tool wear, etc.

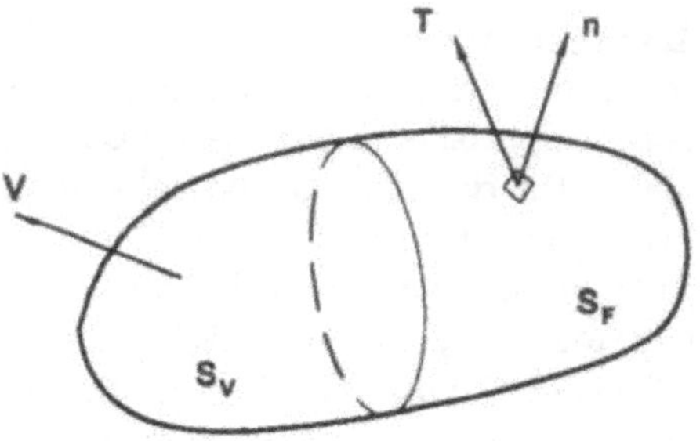

Figure 3.6 A body, considered to comprise two parts, S_F and S_V.

3.4 The discontinuity of the tangential velocity

As is discussed in Chapter 2, chip formation is a cyclic process. Each cycle basically consists of two successive phases: (1) compression of the workpiece material ahead of the tool, and (2) deformation and then sliding of a chip fragment along a certain sliding plane which appears to be the plane of the maximum combined stress. The mechanics of the first phase will be considered using the finite-element method of modeling discussed in Chapter 5. Then, the second phase occupying 85 to 95% of the cycle time and involving the discussed velocity diagram is considered.

When one applies the basics of engineering plasticity to analyze the chip formation process, one inevitably starts such an analysis with the known differential equation of force equilibrium,[17] which can be written in tensor notation as:

$$\frac{\partial \sigma_{ij}}{\partial x_i} + X_i = 0 \tag{3.7}$$

where σ_{ij} is the stress tensor and X_i is the components of the body force per unit volume.

Many problems in plasticity theory are regarded as quasi-static, as it is assumed that the inertia forces due to plastic flow may be neglected. It is true for metal cutting, as the dimensionless parameter $\rho_w v^2/\sigma_Y \ll 1$, where ρ_w is the density of workpiece material, v is a characteristic velocity of deformation (can be regarded as the shear velocity), and σ_Y is the uniaxial yield stress.

Consider a body which occupies a volume, V, and is bounded by the surface, S, as shown in Figure 3.6. The surface, S, is considered to comprise two parts, S_F and S_V. Assume σ_{ij} to be any stress field which satisfies the differential equation of equilibrium (Equation (3.7)) which is consistent with the prescribed tractions T_i on the surface S_F of the body and body forces X_i. Then,

$$T_i = \sigma_{ij} n_j \tag{3.8}$$

where n_j are directional cosines of the outward unit normal n.

On the other part, S_V, of the surface, let the velocity v be prescribed and its components be denoted as v_i. Corresponding to this field, the strain-rate components are

$$\dot{\varepsilon}_{ij} = \frac{d\varepsilon_{ij}}{dt} = \frac{1}{2}\left(\frac{\partial v_i}{\partial x_j} + \frac{\partial v_j}{\partial x_i}\right) \tag{3.9}$$

where t denotes time.

If it is assumed that the configuration of the considered body after deformation is negligibly different from its initial state defined by S and V, then the rate of the external work done over the complete body can be stated as:

$$\dot{W} = \int_S T_i v_i dS + \int_V X_i v_i dV \tag{3.10}$$

where the first integration extends over the whole surface S and the other integration over the whole volume of the body V. As the surface integral in Equation (3.10) can be expressed using Equation (3.8):

$$\int_S T_i v_i dS = \int_S \sigma_{ij} n_j v_i ds \tag{3.11}$$

then Equation (3.10) can be represented in the following form:

$$\int_S T_i v_i dS + \int_V X_i v_i dV = \int_V \sigma_{ij} \dot{\varepsilon}_{ij} dV \tag{3.12}$$

which is known as the virtual work equation.

Equations (3.7) and (3.12) have been used in multiple attempts to apply the various aspects of plasticity theory (slip-line and upper-bound solutions) in the metal cutting mechanics.[17,22,25–29,41] However, the results obtained using the single shear plane model are found to be closer to those obtained experimentally than those obtained using the theory of engineering plasticity. This paradox should be explained.

A detailed analysis of the known solutions shows that the main problem is associated with the use of incorrect velocity diagrams shown in Figure 3.5b–d. As such, the researchers had to employ Equations (3.7) and (3.12) in their analyses. It will be shown here that these equations are simply inapplicable to the study of the chip formation process.

Equations (3.7) and (3.12) assume continuous stress and velocity fields in the deformation zone. As such, the lower boundary of the deformation zone in the known solutions is defined as the line where the first deformation occurs and the upper boundary is the line where the deformation is complete.

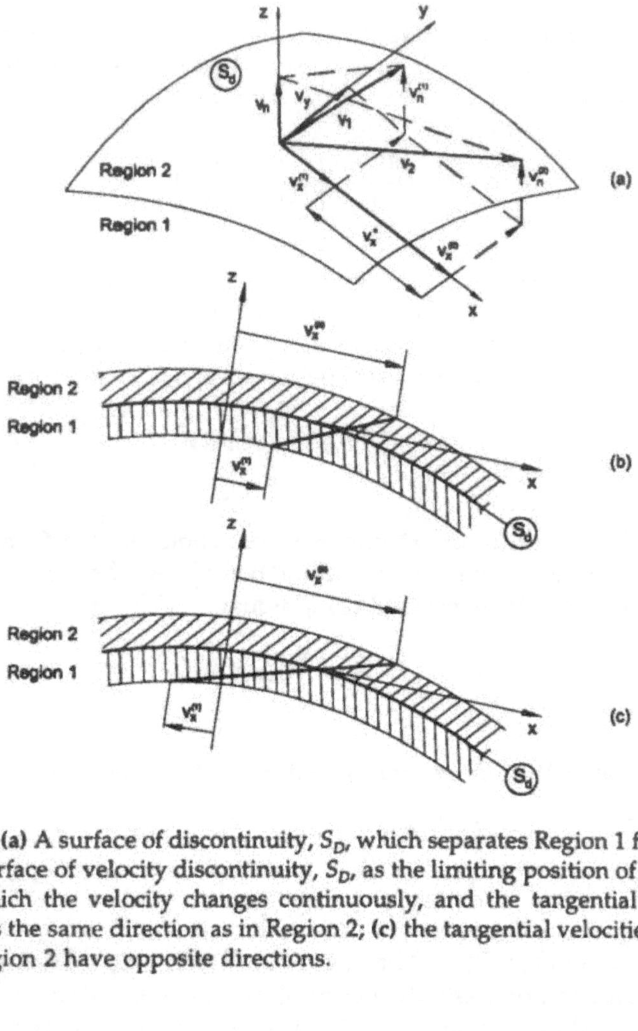

Figure 3.7 (a) A surface of discontinuity, S_D, which separates Region 1 from Region 2; (b) the surface of velocity discontinuity, S_D, as the limiting position of a thin layer through which the velocity changes continuously, and the tangential velocity in Region I has the same direction as in Region 2; (c) the tangential velocities in Region I and in Region 2 have opposite directions.

Although these are recognized and defined as slip lines, they cannot be regarded as sliding lines. To the contrary, the shear plane attributed to the single shear plane model is the sliding plane, although this fact has never been admitted. In the considered sense, the latter model is much more close to the cyclic model of chip formation discussed in Chapter 2, particularly to its second phase.

The mechanism of sliding which leads to fracture is explained in Chapter 2 and in this chapter. Excellent experimental verification of this fact is presented by Sidjanin and Kovac.[42] As such, discontinuities in stresses and velocities are inevitable in plastic deformation. It is therefore necessary to consider the generalization of Equation (3.12).

Consider the case where the stress is discontinuous at a certain surface S_D as shown in Figure 3.7a. The surface divides the deformation zone into two parts — namely, Region 1 and Region 2. Let the traction $T_i^{(1)} = \sigma_{ij} n_j^{(1)}$ act on one side of the surface S_D and $T_i^{(2)} = \sigma_{ij} n_j^{(2)}$ act on the other.

The equilibrium requirement for an element at a surface of stress discontinuity is

$$\sigma_{ij} n_j^{(1)} - \sigma_{ij} n_j^{(2)} = 0 \tag{3.13}$$

Therefore, the integrals over the surface of stress discontinuity cancel each other. This means that the normal components $\sigma_{ij}\ n_j$ must be continuous across this surface (i.e., the existence of stress discontinuities does not affect the virtual work equation). In other words, Equation (3.12) is still valid when the equilibrium stress field contains surface(s) of stress discontinuity.

Next we assume that the body under consideration is divided into two regions — namely, Region 1 and Region 2 — by a surface of velocity discontinuity, S_D, where the velocity field v_i is otherwise continuous as shown in Figure 3.7a. Because incompressibility is assumed during plastic deformation, the condition of the permanent contact between Region 1 and 2 may be represented by the continuity condition in the following form:[17]

$$\frac{\partial v_x}{\partial x} + \frac{\partial v_z}{\partial z} = 0 \tag{3.14}$$

For a continuous solid media, the sliding velocity is constant: $v_x = Const$. This leads to:

$$\frac{\partial v_x}{\partial x} = 0 \tag{3.15}$$

and Equation (3.14) becomes:

$$\frac{\partial v_z}{\partial z} = 0 \Rightarrow v_z = Const \Rightarrow proj._z v^{(1)} = proj._z v^{(2)} \Rightarrow v_n^{(1)} = v_n^{(2)} \tag{3.16}$$

Therefore, the component of velocity normal to the surface of velocity discontinuity S_D must be continuous across the surface. Hence,

$$v_i^{(1)} n_i = v_i^{(2)} n_i \tag{3.17}$$

and discontinuities can only occur in the velocity components which have a direction tangential to the surface S_D.

Consider some point O on the surface of velocity discontinuity S_D as shown in Figure 3.7a. In this figure, a local rectangular Cartesian coordinate system with the origin at O is set so that:

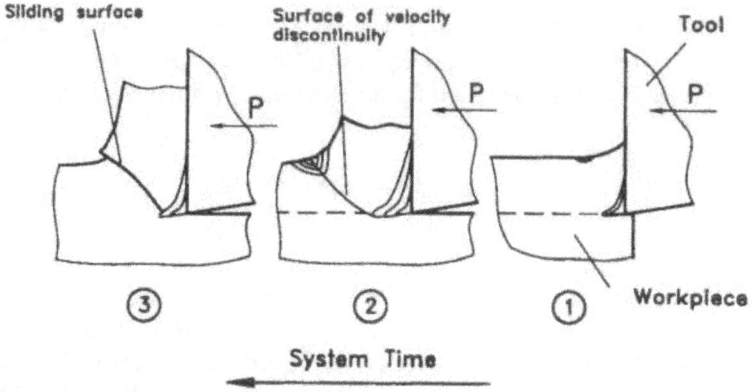

Figure 3.8 Three distinctive stages in chip formation cycle.

- The *z*-axis coincides with the normal to the surface at point *O*.
- The *x*-axis is tangent to the surface at point *O*.

The velocities as the surface of discontinuity is approached from Region 1 and Region 2 are $v^{(1)}$ and $v^{(2)}$, and the tangential velocity components on the two sides of S_D are $v_x^{(1)}$ and $v_x^{(2)}$, respectively. Therefore, the tangential velocity discontinuity is

$$v_x^{\bullet} = v_x^{(2)} - v_x^{(1)} \tag{3.18}$$

If the surface of velocity discontinuity has zero thickness as shown in Figure 3.7a, then the shear strain rate component $\gamma'_{zx} \to \infty$, and this is physically impossible.

To solve the contradiction, we should consider a simplified version of the model of chip formation in cutting of ductile materials, originally shown in Figure 2.11. This simplified model, shown in Figure 3.8, illustrates three distinctive stages of a chip formation cycle.

At the first stage, stress grows in the workpiece and, as a result, the deformation of the layer to be removed takes place locally, originating from two distinctive zones. The stresses and deformations in the deformation zone change with the applied penetration force *P* so that this deformation zone has ever-changing boundaries. At this stage, the model of chip formation corresponds to that proposed by Lee and Shaffer[29] and modified later by Zorev.[27]

Studying the quick-stop samples of the deformation zone obtained at this stage, Zorev arrived at the conclusion that the deformation zone is "a living body" rather than having a defined shape that can be drawn as a chip formation model. There is no unique shear plane; the initial and final boundaries of this zone have ever-changing positions due to instability of the process of plastic deformation in the zone. This fact has never been acknowledged by subsequent researchers who continue to represent a model of chip formation by a single picture (i.e., assuming the existence of its steady-state configuration). Zorev concluded further that a complex state of stress exists in the deformation zone. This stress is not steady state and is non-uniform throughout this zone. As a result, the process of plastic deformation during chip formation is not a case of simple loading. Due to these phenomena, a precise solution to the chip formation problem cannot be obtained using the known methods of engineering plasticity.

Hill[22] and Dewhurst[25,26] arrived at similar conclusions, arguing for the non-uniqueness of the machined process. As both researchers used the engineering plasticity principles in their considerations, it is obvious that the non-uniqueness of the instant parameters of the metal cutting process is meant. In other words, no steady-state solution exists at the first stage of a chip formation cycle.

Subsequent researchers argued for the uniqueness of the machined process. Instead of proving that the solutions of Zorev, Hill, and Dewhurst are wrong, they conducted research using a wide range of machining parameters and concluded that the discussed non-uniqueness was not observed in their experimental results.[43] It should be clear, however, that in their experiments, the average (per cutting cycle) values of machining parameters were measured, while the instant values were discussed by Zorev, Hill, and Dewhurst, so the theoretical and experimental results cannot be compared in this way.

Some studies went further, arguing against the non-uniqueness of the machining process on the basis of results from their finite element modeling (FEM).[45] However, such modeling, being of a steady-state nature with a stationary input force being assumed, inherently cannot distinguish the instant values in the dynamic process. Unfortunately, more and more researchers do not seem to be aware of the simple fact that mathematics cannot create physics, physical bodies, or phenomena, just as it cannot create biology, life, love, money, or the economy of a country, though it may be very helpful in all these endeavors. Mathematical models, including FEM models, cannot output more physics than what was put into them. If a mathematical model or treatment is not based on the correct physical model of a physical event, then the results of FEM are useless.

A complete analysis of the first stage of a chip formation cycle is presented in Chapter 5 using FEM modeling. It is demonstrated that the growing resistance of workpiece material to the tool penetration results in the dynamic (time-dependent) stress fields in the deformation zone and in the region adjacent to the tool/chip interface. At this stage neither mentioned fields has a steady-state mode (i.e., one that can be defined uniquely only for a given combination of the cutting parameters and at a given time). This is in perfect agreement with the results of Hill,[22] Zorov,[27] and Dewhurst.[25,26]

At the second stage of a chip formation cycle, the deformation becomes global, which leads to the formation of a surface of velocity discontinuity as shown in Figure 3.8. At this stage the surface of velocity discontinuity may be considered as the limiting position of a thin layer through which the velocity changes continuously and rapidly from $v^{(1)}$ and $v^{(2)}$. For the situation considered in Figures 3.7b and c, only the component in the x-direction experiences a rapid change, while the components in the other two directions are essentially constant through the layer. It follows that the shear strain-rate component $\dot{\gamma}_{zx}$ is very much greater than the other strain-rate components. Figures 3.7b and c illustrate that, depending on the geometry of the deformation zone, the tangential velocity components on the two sides of S_D ($v_x^{(1)}$ and $v_x^{(2)}$) may have either the same or opposite directions. At this stage, the model of chip formation corresponds to that considered by Kececioglu[45] and then by Oxley,[28,46] although they both used the velocity diagram shown in Figure 3.5b — the existence of velocity discontinuity was not admitted.

The third stage of a chip formation cycle (Figure 3.8) begins when the combined stress in the deformation zone reaches a certain limiting value as discussed in Chapter 2. As such, a sliding surface forms in the direction of the maximum combined stress. As soon as the sliding surface forms, all the chip-cantilever material starts to slide along this surface and thus along the tool face. Because no further chip deformation (in the direction of the sliding surface) occurs on sliding (i.e., the chip thickness has completely been formed at the second stage), the shear strain-rate component $\dot{\gamma}_{zx}$ is equal to zero rather than infinity, as is thought presently. As a result, the penetration forces reduce on sliding.

At this stage, the model of chip formation corresponds to that known as the single shear plane model considered by Usachev,[32] Merchant,[23] and many others.[31,33,34,38–44] Although the sliding plane was termed as the shear plane, the known schematic representation of the model consisting of card-like elements displaced by the cutting tool (i.e., Figure 4 in Merchant[23]) supports the author's point.

The foregoing analysis reveals that there are no contradictions among the known concepts of the chip formation process. Using the proposed system approach, they simply describe the different stages of this process, and in this sense they cannot be compared to each other as is commonly done in many works on metal cutting.

3.5 Real velocity diagram in metal cutting

3.5.1 Velocity diagram at the second stage

It follows from the foregoing discussion that the velocity diagram makes sense only at the second and third distinctive stages of a chip formation cycle, because the deformation of workpiece material takes place locally at the first stage. As argued in Chapter 2, a surface of velocity discontinuity, which eventually turns to be a sliding surface, forms in each cycle of chip formation and can be well approximated by a plane. Therefore, the deformation zone

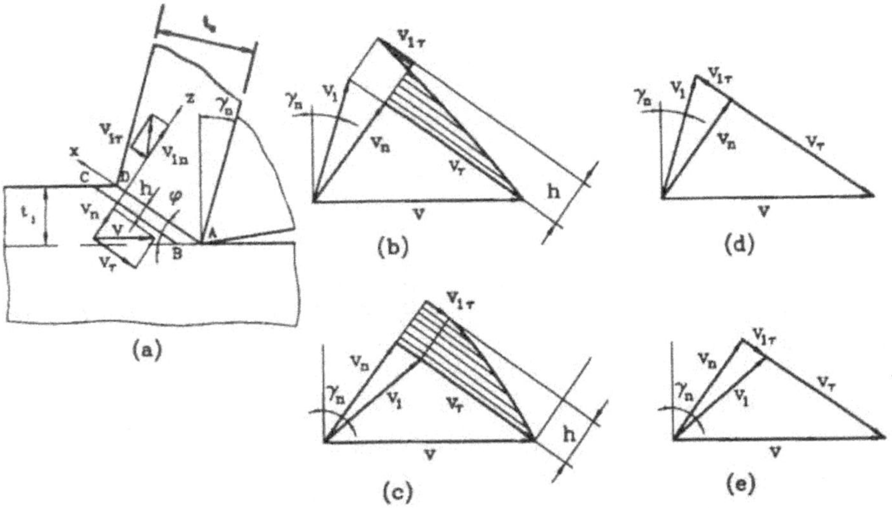

Figure 3.9 Model for orthogonal cutting: (a) model of shear zone with parallel boundaries; (b) velocity diagram at the second stage when the tool rake angle is positive; (c) velocity diagram at the second stage when the tool rake angle is negative; (d) velocity diagram at the third stage when the tool rake angle is positive; (e) velocity diagram at the third stage when the tool rake angle is negative.

at the second stage can be thought of as having parallel boundaries (Figures 2.12 and 2.23).

Although the parallel-sided deformation zone theory and corresponding model of chip formation had been originally proposed by Kececioglu[45] and then further developed for many years by Oxley,[28] it appears to be fruitless compared to the single shear plane model. In the author's opinion, the cause of it is in the use of the velocity diagram shown in Figure 3.5b which forced Oxley to make a number of physically unjustifiable assumptions about stress and strain in this zone as well as about deformation mode. For example, simple shearing was assumed to be the deformation mode that never leads to the third stage, as this mode does not lead to fracture.

Astakhov and Osman[47] suggested a new avenue in the analysis of the problem, and their approach will be used in further considerations. A simple model of two-dimensional machining (orthogonal cutting) is shown in Figure 3.9a. The tool is a single-point tool characterized by the rake angle γ_n. The forces imposed on the tool create intense shearing action on the metal ahead of the tool. The metal in the chip is severely deformed on changing from an undeformed chip of thickness $t_1 = a_1$ to a deformed chip with thickness $t_c = a_c$ as a result of shearing in the shear zone *ABCD* with parallel boundaries, occurring at a shear angle φ.

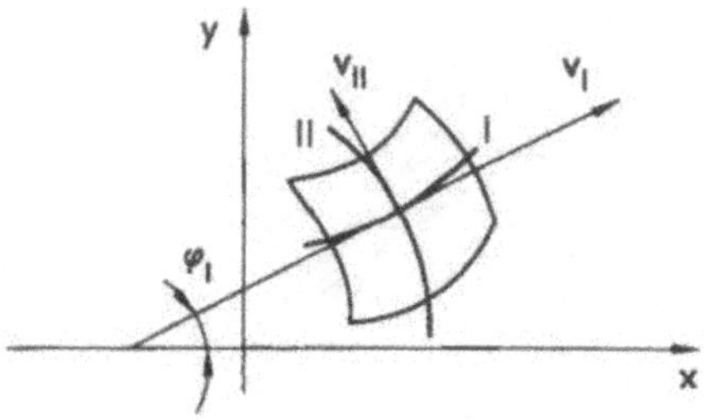

Figure 3.10 Location of the tangent velocities relative to the slip lines. (From Astakhov, V.P. and Osman, M.O.M., *Int. J. Machine Tools Manuf.*, 36(11), 1996, pp. 1187–1200. With permission.)

The coordinate system, illustrated in Figure 3.9a, is set as:

- The x-axis coincides with the upper boundary of the shear zone $ABCD$.
- The z-axis is perpendicular to the x-axis, as shown in Figure 3.9a.

According to the plasticity theory[17] and above considerations, the following should not be forgotten:

- The boundaries BC and AD of the deformation zone are slip lines.
- Discontinuity of the tangential velocities on the boundaries should take place.

As it is assumed that the upper boundary of the deformation zone is a straight line, the continuity condition, Equation (3.14), appears to be the condition of the permanent contact between the deformation zone and the chip. This condition can be simplified using the fact that the upper boundary of the shear zone is a slip line. It is known that for velocities v_I and v_{II}, tangent to the mutual perpendicular slip lines I and II (Figure 3.10), the Geiringer equations are valid:[17]

$$dv_I - v_{II} d\varphi_I = 0 \tag{3.19}$$

$$dv_{II} + v_I d\varphi_{II} = 0 \tag{3.20}$$

Because the upper boundary of the shear zone is a straight line, then $d\varphi_I = 0$. Thus, from Equation (3.19):

$$dv_I = 0 \Rightarrow v_I = Const \tag{3.21}$$

Hence, the velocity along a straight slip line is constant. This leads to the conclusion obtained earlier — Equation (3.15), $\partial v_z/\partial x = 0$ — and, hence, Equation (3.16) can now be written in the new notations as:

$$\frac{\partial v_z}{\partial z} = 0 \Rightarrow v_z = Const \Rightarrow proj._z v^{(1)} = proj._z v^{(2)} \Rightarrow v_n = v_{1n} \tag{3.22}$$

Referring to Figure 3.9a, Equation (3.22) shows that the velocity in the z-direction in the deformation zone is equal to that of the chip in the same direction.

Because $v_z = Const$, it follows from Equation (3.20) that $dv_{II} = 0$. This, in turn, results in the following:

$$d\varphi_{II} = 0 \Rightarrow \varphi_{II} = Const \tag{3.23}$$

Therefore, the second family of the slip lines is a set of straight lines. Because the slip lines must be orthogonal,[17] the second family of lines must cut the first family at right angles. Because φ_{II} is constant for the entire region, then from the well-known Hencky equations:[17]

$$p + 2k\varphi_I = constant\ along\ the\ \mathrm{I}\ slip\ line \tag{3.24}$$

$$p - 2k\varphi_{II} = constant\ along\ the\ \mathrm{II}\ slip\ line \tag{3.25}$$

it follows that the normal stress p must be constant over the entire region.

Referring to Figure 3.9a, Equation (3.22) may be written as follows:

$$v \sin \varphi = v_1 \cos(\varphi - \gamma_n) \tag{3.26}$$

The ratio of velocities v and v_1, or the ratio of the length of cut L_p to the chip length L_c is known as the chip compression ratio[27] (its reciprocal is also used and known as the chip ratio[40]):

$$\frac{v}{v_1} = \frac{v\,\Delta t}{v_1\,\Delta t} = \frac{L_p}{L_c} = \zeta = \frac{\cos(\varphi - \gamma)}{\sin \varphi} \tag{3.27}$$

As might be expected, this well-known equation is another form of the continuity condition, Equation (3.16).

According to the continuity condition, Equation (3.16), the boundary conditions for the tangential and normal velocity components (Figure 3.9a) are

$$v_x = \begin{Bmatrix} v_\tau = v \cos\varphi & when\ z = -h \\ v_{1\tau} = -v_1 \sin(\varphi - \gamma_n) & when\ z = 0 \end{Bmatrix}$$
$$v_y = v_n = v \sin\varphi \tag{3.28}$$

The component of the workpiece velocity v_τ and the component of the chip velocity $v_{1\tau}$ with respect to the deformation zone can have either the same (Figure 3.9b) or opposite direction (Figure 3.9c), depending on a particular combination of the angles φ and γ_n. As seen in Figure 3.9, the velocity components v_τ and $v_{1\tau}$ have the same direction if and only if the tool rake angle γ_n is greater than the shear angle φ.

The difference between the velocities v_τ and $v_{1\tau}$ defines the discontinuity v_x^* of the tangential velocity v_x:

$$v_x^* = v_\tau = v_{1\tau}$$
$$v_x^* = v\cos\varphi + v_1 \sin(\varphi - \gamma_n) = v\frac{\cos\gamma_n}{\cos(\varphi - \gamma_n)} \tag{3.29}$$

The continuous transformation of the velocity v_x from v_τ to $v_{1\tau}$ takes place in the deformation zone with parallel boundaries. Thus, the change may be approximated by the suitable continuous transition curve justifying the boundary conditions defined in Equation (3.28) (see Figures 3.9b,c). Because the slip lines are straight lines, the velocity v_x depends on the z-coordinate only. The law governing this dependence is unknown.

A detailed analysis of the transformation of the tangential velocity in the deformation zone had been carried out by Kushnir.[48] Analyzing the multiple experimental data, he has shown that the curve $v_x(y)$ can be approximated by the following analytical function:

$$v_x(z) = v_\tau - v_x^* \left|\frac{z}{h}\right|^n \tag{3.30}$$

Here, n is a parameter (power) characterizing the non-uniform distribution of the tangential velocity in the deformation zone. Kushnir[48] demonstrated that in the machining of ductile materials at low cutting speeds $n = 4$ and at high cutting speed $n = 8$.

Figure 3.11 illustrates that the greater the value of n the more non-uniform change of velocity ratio v_x/v_n takes place in the deformation zone. This zone can be thought of as being divided into two regions. The first part can be called the wide region, where the change of velocities ratio v_x/v_n takes place at a low rate. The second is the narrow region where this change takes place at a high rate.

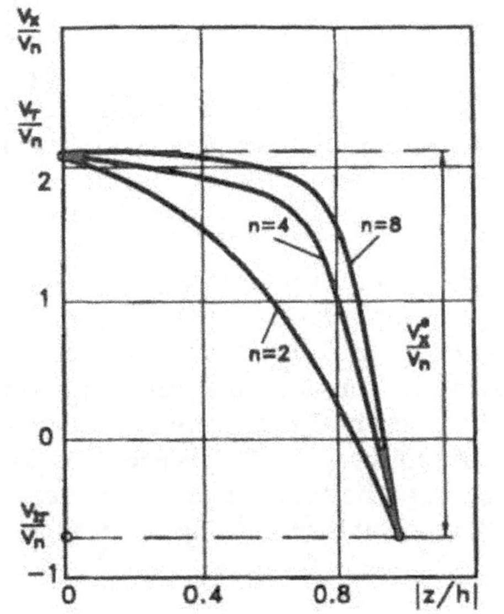

Figure 3.11 Transformation of the tangential velocity within the deformation zone under different n.

It is important for further considerations to obtain the equation for displacements in the deformation zone. Consider a microvolume of workpiece material which is entering into the deformation zone. Its displacement as a function of time $t = z/v_n$ can be expressed as:

$$D_x(y) = \int_0^t v_x(\tau)d\tau = \frac{v_\tau}{v_n}y - \frac{\dot{v}_x}{v_n}\frac{h}{n+1}\left(\frac{z}{h}\right)^n \tag{3.31}$$

It follows from Equation (3.31) that the displacement is maximum within the deformation zone if the velocities $v_{1\tau}$ and v_τ have opposite directions ($\varphi > \gamma_n$). When n increases, this maximum approaches to the upper boundary of the deformation zone.

The results of this theoretical finding can be verified experimentally by analysis of the chip structure. It has to be pointed out that the experimental verification is rather simple and has been laying on the surface for many decades. For example, Figure 2.15 (Chapter 2) shows a micrograph of a chip fragment. Here, two distinctive regions of deformation can be clearly distinguished. As seen, the deformation of grains is quite moderate in the first region, while it becomes severe in the second narrow region. As was discussed in Chapter 2, when the initial structure of the workpiece material contains very fine grains, a change in the grain structure after cutting cannot be distinguished readily with an optical microscope. In this case, a microhardness scanning test is a solution to the problem. Figures 2.16 and 2.17 (Chapter 2) show the test results which fully support the theoretical conclusions. Similar experiments can be performed by anyone, and their results will be identical to those discussed, regardless of the cutting conditions used.

3.5.2 Velocity diagram at the third stage

At the third stage of a chip formation cycle, a sliding surface forms in the direction of the maximum combined stress (Chapter 2). As such, the entire chip fragment slides over the sliding surface which, for the time of sliding, mechanically separates the chip from the rest of the workpiece. As pointed out above, the chip thickness has been formed completely at the second stage so that the shear strain-rate component γ'_{zr} is equal to zero rather than infinity at the third stage. Because the sliding surface has been approximated by a plane, the velocity diagram at the third stage corresponds to those shown in Figures 3.9d,e.

3.6 Real virtual work equation

To derive the virtual work equation that corresponds to the real model of chip formation discussed above, consider an infinitesimal element dS_D of the surface of velocity discontinuity, S_D, shown in Figure 3.7a. In Region I, the rate of work done by the stresses is

$$-\left(\sigma_n v_n + \tau_y v_y + \tau_x v_x^{(1)}\right) dS_D \tag{3.32}$$

where $\sigma_n = \sigma_z$ and $v_n = v_z$ are the components of normal stress and velocity (in the z-direction); τ_y and v_y, τ_x, and $v_x^{(1)}$ are the components of shear stress and tangential velocity in the y- and x-direction, respectively.

The corresponding rate of work done by the stresses in Region 2 is

$$+\left(\sigma_n v_n + \tau_y v_y + \tau_x v_x^{(2)}\right) dS_D \tag{3.33}$$

and the algebraic sum of the rates of work done by the stress acting on the element dS_D is

$$u_{dS_D} = -\tau\left(v_x^{(1)} - v_x^{(2)}\right) dS_D = \tau_x v_x^{*} dS_D \tag{3.34}$$

Consequently, the rate of work done by the stress acting on the surface of velocity discontinuity, S_D, is given by:

$$u_{S_D} = \int_{S_D} \tau_x v_x^{*} dS \tag{3.35}$$

Here, τ_x may be thought of as the shear flow stress of the workpiece material — $\tau_x = \tau_f$.

Therefore, the virtual work equation, Equation (3.12), then becomes:

$$\int_S T_i v_i dS + \int_V X_i v_i dV = \int_V \sigma_{ij} \dot{\varepsilon}_{ij} dV + \int_{S_D} \tau_y v_x^* dS \qquad (3.36)$$

and thus Equation (3.5) should be modified to become:

$$u_s = \tau_f \left(\frac{v_x^*}{v \sin \varphi} \right) \qquad (3.37)$$

Referring to Figure 3.9, it is seen that $v_x^* = v_\tau - v_{1\tau}$.

3.7 Analysis of plastic deformation in metal cutting

If the relative position of any two points in a continuous body is changed, then the body is said to be deformed or strained. The analysis of strain is the study of displacements of points in a body relative to one another when the body is deformed and is, therefore, essentially a geometrical problem which is unrelated to material properties.

As it is recognized[23,27,28,40] that the deformation process in orthogonal cutting may be considered as a plane-strain problem, our attention will be concentrated on a two-dimensional strain consideration.

3.7.1 Chip compression ratio and strain — what is a real measure of plastic deformation in metal cutting?

As might be expected, the well-known Equation (3.27) is another form of the continuity condition which is, in fact, the condition of contact between the chip and the deformation zone along its straight upper boundary. Due to the relative simplicity of its experimental determination, the chip (compression) ratio has been widely used in metal cutting studies as a quantitative measure of the total plastic deformation.[27] Numerous attempts have been made to establish analytically a relationship to predicting the chip compression ratio in terms of fundamental variables of the cutting process. However, none of these attempts has produced results matching experimentally obtained data for a reasonable variety of input conditions. Later research has abandoned this route in favor of the "modern" metal cutting approach in which this parameter is expected to be defined experimentally. Because the chip compression ratio competes for the role of a measure of plastic deformation encountered in metal cutting, it seems only logical to check up on the justification of its usage as the measure.

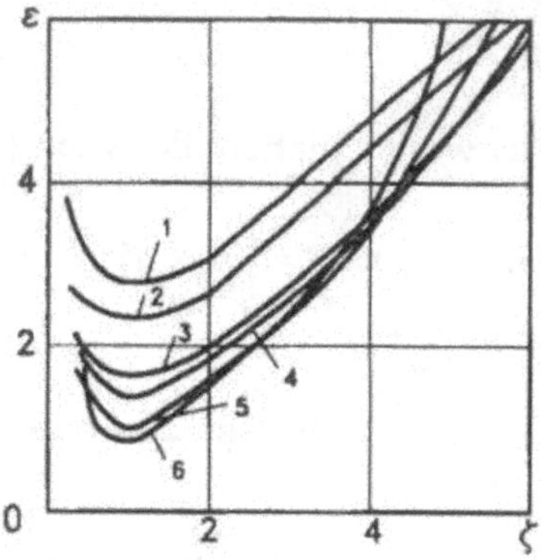

Figure 3.12 Shear strain vs. the chip compression ratio for different rake angles: (1) –15°, (2) 0°, (3) 15°, (4) 45°, (5) 60°.

Merchant[23,49] proposed the following equation for shear strain ε:

$$\varepsilon = \frac{v_2}{v_n} = \frac{\cos\gamma}{\cos(\varphi - \gamma_n)\sin\varphi} = \frac{\zeta^2 - 2\zeta\sin\gamma_n + 1}{\zeta\cos\gamma_n} \tag{3.38}$$

To better visualize the correlation between the final shear strain and the chip compression ratio, the results of calculations using Equation (3.38) are illustrated in Figure 3.12 for different rake angles. As might be expected, the shear strain depends to a large extent on the rake angle γ and decreases rapidly when the chip compression ratio tends to 1.

It is known in orthogonal cutting that the chip width is equal to the width of cut[27] (i.e., no side spread occurs), so the plastic deformation may be thought of as the plane deformation. When $\zeta = 1$, the chip thickness is equal to the uncut chip thickness. This reveals a contradiction, as the chip compression ratio, considered to be a measure of the plastic deformation, indicates that no plastic deformation occurs in the chip, while the final shear strain, defined using the same geometry of the deformed body, remains significant (Figure 3.12). At this point, it is worthwhile to mention that the final strain defined in Equation (3.38) does not consider the change of internal energy of the chip due to the changed chip density, the increased dislocation concentration, or the imposed stresses on the boundaries of the grains (which may be considered as the residual stresses), even though all of these increase the shear strength of the chip compared with the original workpiece material. Shear strain, according to Equation (3.38), is defined only by the changes in the dimensions of a deformed body as compared with the original dimensions. As the chip compression ratio indicates that there is no change in the dimensions, there is no "geometrical" deformation, so strain should be equal to zero. However, it does not follow from the equation for strain.

Proponents of the Merchant theory often try to defend the validity of Equation (3.38) by saying that it is physically impossible for the chip compression ratio to be equal to one. In other words, the chip thickness cannot be equal to the uncut chip thickness. To address that issue, we would like to refer to the experimental works by Kobayashi and Thomsen[50] and Zorev,[27] in which the experimentally obtained chip compression ratios for different workpiece materials are presented; their values vary in a wide range including one and even less. In particular, it is interesting to consider the work by Kobayashi and Thomsen,[50] especially their discussion section and Tables 2 and 3. In the body of the paper, the authors state, "This suggests that the actual chip ratio will never be larger than unity." (In other words, the chip compression ratio will never be less than one). In the discussion section, Kronenberg presented the experimentally obtained chip compression ratios having values greater than, equal to, and less than unity. Interesting to note is that the data were obtained under real rather than exotic cutting conditions. Kobayashi and Thomsen in their conclusions had no choice but to recognize the fact. As an excuse, they suggested that the experimental data "apparently indicate that the deformation mechanism (in metal cutting) is not one of the simple shear plane type, even though continuous chips were formed." Unfortunately, nobody has been paying any attention to this discussion, which is one of the best in the field.

Although the chip compression ratio may not be a perfect measure of plastic deformation in metal cutting, it does directly reflect (when properly measured) the final plastic deformation. In contrast, Equation (3.38), the expression for strain, has nothing to do with plastic deformation because ε can reflect the displacement of a series of elements even without their plastic deformation. In Merchant's example,[23,40] Equation (3.38) is valid for the case of displacement of a deck of cards without any plastic deformation.

Few attempts to question the validity of Equation (3.38) are known in the history of metal cutting. Okusima and Hitomi[51] studied the grid deformation in orthogonal cutting with a lead specimen in a shaper under various cutting conditions. They concluded that the theoretical results derived using Equation (3.38) are not in complete accord with the experimentally observed values of strain, especially for negative rake angles. Spaans[52] used improper equations for the local shear strain increment (see his Equation 10) and obtained the equation for strain in the differential form (Equation 13 in Spaans[52]), for which he did not comment. A significant advantage of this work is that the author properly set coordinate axes which give meaning to shear strain in metal cutting. It is worthwhile to mention here that Oxley did not define a Cartesian coordinate system in his book;[28] however, in his earlier paper[46] he defined a coordinate system which employs the use of the complete strain tensor to analyze the deformation in metal cutting. Only shear strain, though, has been analyzed. Thé and Scrutton[53] attempted to study deformation in metal cutting using strain tensor analysis. No coordinate systems have been set, however, in their model, and the analysis repeats the same inaccuracy seen in Equation 5 in Spaans.[52] von Turkovich,[54] in his attempt to improve Equation (3.38), suggested that the chip contraction in cutting is due to friction at the tool/chip interface. This suggestion allowed him to represent Equation (3.38) as a sum of the frictionless part and the redundant strain due to friction. However, if a zero rake angle and a unit chip compression ratio are substituted into the obtained equation, the calculated strain is equal to 2. This result cannot be explained based on the data provided in this work.

The results of the above analysis show:

- Equation (3.38) has been derived based on the assumption that the deformation process in metal cutting corresponds to simple shearing, although the known studies of grid deformations[27,51] do not conform to this assumption
- Because Equation (3.38) has been derived using qualitative analysis, the kind of shear strain it represents is not understood. Is it direct engineering shear strain, strain in a given point or strain of a line element, finite strain, principal strain, maximum shear strain, octahedral strain, equivalent strain, spherical strain, deviator strain, natural strain? This question has to be answered if the theory of plasticity is to be applied to metal cutting studies.

Because there are a number of questions to be answered, it has been found useful to analyze strain using the mathematical theory of plasticity, concentrating on its practical side.

3.7.2 *Infinitesimal strain*

Consider a small rectangular element, *ABCD* (Figure 3.13a). After shearing, the element is both displaced and deformed to $A_1B_1C_1D_1$. Consider two arbitrary points *A* and *C* in an unstrained body where *C* is very close to *A*; after shearing, the points move to A_1 and C_1, respectively. The distance AA_1 is the displacement of *A*, and CC_1 is the displacement of *C*. If the distance A_1C_1 is exactly equal to *AC*, then the displacement is one of translation, as might occur in the case of a rigid body. If, however, the distances are not equal then there is a displacement of *C* relative to *A* and a state of strain exists in the body.

In general, the strain will not be homogeneous but will be different at different points. However, if a sufficiently small element of the body is considered, the strain may be regarded as essentially homogeneous. In this case, it may be assumed that parallel straight lines remain straight and parallel, plane surfaces remain plane, and, for all straight lines having the same direction, the ratio of alteration in length to original length will be the same. Two parallel straight lines of the same length will then be equally elongated or contracted. With reference to the Cartesian coordinate system shown in Figure 3.13, let the coordinates of *A* in the illustrated state be (x,y) and of A_1 after shearing $(x + u,$

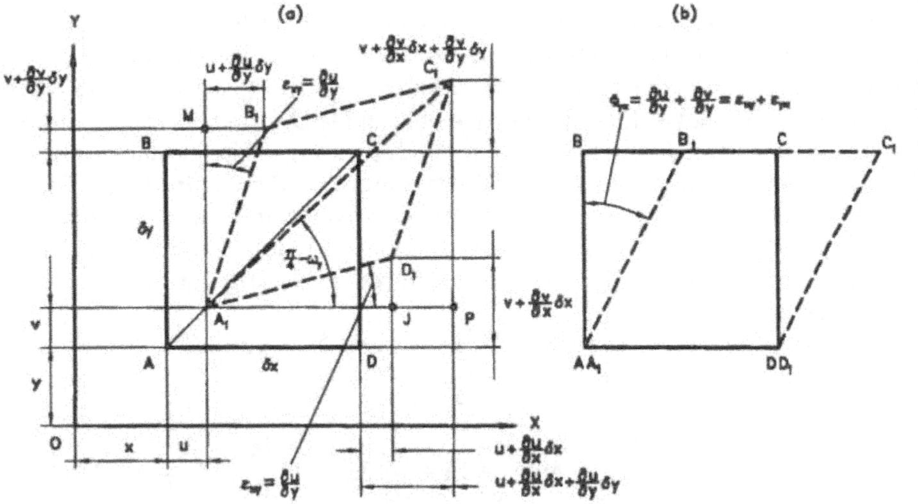

Figure 3.13 Components of displacement at a point, direct strains, engineering shear strain (a), and rotation in a plane strain condition (b).

$y + v$), where u and v are the projections of displacement of A — that is, AA_1 on the x- and y-coordinate axes, respectively. It is assumed that these quantities are infinitesimal and that they are continuous functions of the coordinates x and y.

When point A moves to A_1, the side AD elongates to A_1D_1 (as shown in Figure 3.13a) and has an angular movement JA_1D_1. Identically, side AB elongates to A_1B_1 and has an angular movement MA_1D_1. Therefore, by referring to Figure 3.13a, it will be appreciated that the coordinates of C are ($x + \delta x$, $y + \delta y$) and after straining become (($x + \delta x$) + ($u + \delta u$), ($y + \delta y$) + ($v + \delta v$)) where δu and δv are evidently projections of the displacement of C relative to A on the corresponding coordinate axes.

As u is assumed to be a continuous function of x and y, ($u + \delta u$) will be the same function of ($x + \delta x$), ($y + \delta y$). Thus, if $u = f(x,y)$, $(u + \delta u) = f\{(x + \delta x), (y + \delta y)\}$, and this latter expression is expanded by employing Taylor's theorem:

$$(u+\delta u) = f(x,y) + \frac{\partial f}{\partial x} + \frac{\partial f}{\partial y} + (\textit{terms in higher powers of } \delta x \textit{ and } \delta y) \tag{3.39}$$

As $u = f(x,y)$ is assumed to be a very small quantity, these latter terms can be neglected. Then,

$$\delta u = \frac{\partial u}{\partial x}\delta x + \frac{\partial u}{\partial y}\delta y \tag{3.40}$$

It is evident that $(\partial u/\partial x)\delta x$ is the component of ∂u independent of ∂y and is the projection of D relative to A on the OX axis. Hence, $\partial u/\partial x$ is the direct strain at A in the direction OX which is denoted e_{xx}. This nomenclature may be interpreted as the rate of movement in the OX direction of a point on a line parallel to OX at A. The part of δu depending on δy alone — that is, $(\partial u/\partial y)\,\delta y$ — is that part of the displacement of C measured parallel to the OX axis resulting from the angular movement of AB as shown in Figure 3.13a, and $\partial u/\partial y$ is the rate of shear of sides parallel to OX and perpendicular to OY and can be considered as the angular strain of AB denoted by e_{xy}.

The equation for the relative displacement δv is similar to that for δu. Then,

$$\begin{aligned}\delta u &= \frac{\partial u}{\partial x}\delta x + \frac{\partial u}{\partial y}\delta y \\ \delta v &= \frac{\partial v}{\partial x}\delta x + \frac{\partial v}{\partial y}\delta y\end{aligned} \tag{3.41}$$

or, in tensor notation:

$$\delta u_i = e_{ij}\,\delta x_j \tag{3.42}$$

The tensor:

$$e_{ij} = \begin{bmatrix} \dfrac{\partial u}{\partial x} & \dfrac{\partial u}{\partial y} \\ \dfrac{\partial v}{\partial x} & \dfrac{\partial v}{\partial y} \end{bmatrix} \tag{3.43}$$

is called the relative displacement tensor and, as can be seen, is not generally symmetric about its main diagonal. The reason for this is that it is subject to rigid rotational motion about the coordinate axes.

The change in the right angle BAD to the angle $B_1A_1D_1$ is shown in Figure 3.13a. The line AB moves to A_1B_1, and the line AD to A_1D_1. In moving from AD to A_1D_1, the line AD moves through an angle $JA_1D_1 = \alpha_{yx}$, the projection of which in the plane XOY is $\partial v/\partial x$. Because the strain is considered to be sensibly homogeneous, the line A_1B_1 will be parallel to the line D_1C_1; therefore, the angular movement of AB will be equal to $MA_1B_1 = \alpha_{xy}$, the projection of which in the plane XOY is $\partial u/\partial x$.

As α_{yx} is assumed to be small, then:

$$\alpha_{yx} \approx \tan\alpha_{yx} = \frac{D_1J}{A_1J} = \frac{\frac{\partial v}{\partial x}\delta x}{\delta x + \frac{\partial u}{\partial x}\delta x} = \frac{\frac{\partial v}{\partial x}}{1+\frac{\partial u}{\partial x}} \tag{3.44}$$

Therefore,

$$\alpha_{yx} \approx \frac{\partial v}{\partial x} = e_{yx} \tag{3.45}$$

As α_{xy} is assumed to be small, then:

$$\alpha_{xy} \approx \tan\alpha_{xy} = \frac{B_1M}{A_1M} = \frac{\frac{\partial u}{\partial y}\partial y}{\delta y + \frac{\partial v}{\partial y}\delta y} = \frac{\frac{\partial u}{\partial y}}{1+\frac{\partial v}{\partial y}} \tag{3.46}$$

Therefore,

$$\alpha_{xy} \approx \frac{\partial u}{\partial y} = e_{xy} \tag{3.47}$$

The difference in angles BAD and $B_1A_1D_1$ (shown in Figure 3.13b) is given by $(\partial v/\partial x + \partial v/\partial y)$, which is referred to as the engineering shear strain and is designated φ_{xy}. This expression is composed of two angular strains, $e_{xy} = \partial v/\partial x$ and $e_{yx} = \partial u/\partial y$, so that:

$$\varphi_{xy} = \left(\frac{\partial v}{\partial x} + \frac{\partial u}{\partial y}\right) = e_{yx} + e_{xy} \tag{3.48}$$

The infinitesimal strain of the rectangular element at point A is thus defined by two direct strains and one engineering shear strain, and, for simplicity, the direct strains are designated by using a single subscript:

$$\begin{aligned} e_x &= \frac{\partial u}{\partial x} \\ e_y &= \frac{\partial v}{\partial y} \\ \varphi_{xy} &= \frac{\partial v}{\partial x} + \frac{\partial u}{\partial y} \end{aligned} \tag{3.49}$$

3.7.3 The rotational element

It should be noted that it is impossible to express the four components of Equation (3.43) in terms of three components of strain; therefore, Equation (3.49) is inadequate, as the geometrical representation of the deformations at a given point is incomplete.

Let the small rectangular element considered in Figure 3.13a be a square ($\delta x = \delta y$), and the direct strains $\varepsilon_{xy} = \varepsilon_{yx} = 0$. Referring to Figure 3.13a, the line AC, where DAC is 45°, is rotated through an angle (say, $+\omega_z$) — that is, clockwise about the OZ axis when viewed from origin O along the OZ axis in the positive direction to take up its strained position, A_1C_1. Then:

$$\tan\left(\frac{\pi}{4} - \omega_z\right) = \frac{C_1P}{A_1P} = \frac{\delta y + \dfrac{\partial v}{\partial x}\delta x + \dfrac{\partial v}{\partial y}\delta y}{\delta x + \dfrac{\partial u}{\partial x}\delta x + \dfrac{\partial u}{\partial y}\delta y} = \frac{1 + \dfrac{\partial v}{\partial x} + \dfrac{\partial v}{\partial y}}{1 + \dfrac{\partial u}{\partial x} + \dfrac{\partial u}{\partial y}} = \frac{1 + \dfrac{\partial v}{\partial x}}{1 + \dfrac{\partial u}{\partial y}} \tag{3.50}$$

when $\partial u/\partial x = \partial v/\partial y = 0$ and $\partial x = \partial y$.

Therefore,

$$\tan\left(\frac{\pi}{4} - \omega_z\right) \approx \left(1 + \frac{\partial v}{\partial x}\right)\left(1 - \frac{\partial u}{\partial y}\right) \approx 1 + \frac{\partial v}{\partial x} - \frac{\partial u}{\partial y} \tag{3.51}$$

if the product of partial derivatives is neglected.

Alternatively,

$$\tan\left(\frac{\pi}{4} - \omega_z\right) = \frac{\tan\dfrac{\pi}{4} - \tan\omega_z}{1 + \tan\dfrac{\pi}{4}\tan\omega_z} = \frac{1 - \tan\omega_z}{1 + \tan\omega_z} \approx \frac{1 - \omega_z}{1 + \omega_z} \tag{3.52}$$

Using the result of Equation (3.52), one may obtain:

$$\begin{aligned} 1 - \omega_z &\approx (1 + \omega_z)\left(1 + \frac{\partial v}{\partial x} - \frac{\partial u}{\partial y}\right) \approx \\ & 1 + \frac{\partial v}{\partial x} - \frac{\partial u}{\partial y} + \omega_z\left(1 + \frac{\partial v}{\partial x} - \frac{\partial u}{\partial y}\right) \end{aligned} \tag{3.53}$$

and:

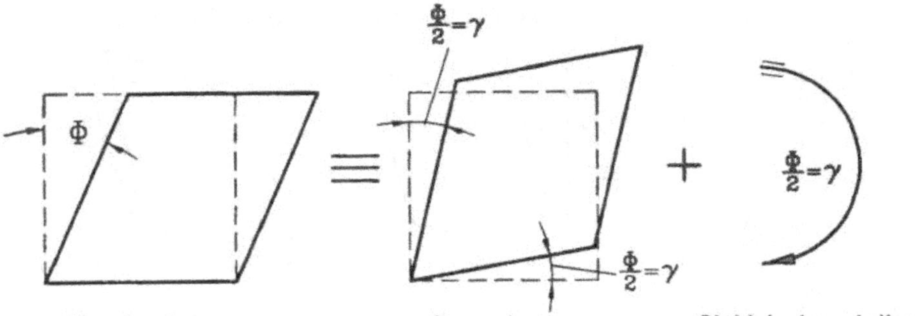

Figure 3.14 Representation of the general case of deformation as consisting of a shear strain and a rigid body rotation.

$$\omega_z\left(2+\frac{\partial v}{\partial x}-\frac{\partial u}{\partial y}\right)\approx\frac{\partial u}{\partial y}-\frac{\partial v}{\partial x} \tag{3.54}$$

from where:

$$\omega_z\approx\frac{\dfrac{\partial u}{\partial y}-\dfrac{\partial v}{\partial x}}{2+\dfrac{\partial v}{\partial x}-\dfrac{\partial u}{\partial y}} \tag{3.55}$$

Therefore,

$$\omega_z\approx 1/2\left(\frac{\partial u}{\partial y}-\frac{\partial v}{\partial x}\right)=1/2(e_{yx}-e_{xy}) \tag{3.56}$$

The deformation can be considered diagrammatically[55] to consist of a shear strain and a rigid body rotation, as shown in Figure 3.14. The distinction between simple shear and pure shear is demonstrated in Figure 3.15.

It is known[15,55] that every second order tensor can be resolved into a symmetric tensor and a skew-symmetric tensor. Therefore, if tensor e_{ij} is resolved into symmetric and skew-symmetric tensors, the former will represent pure deformation, while the latter will represent rigid body rotations without deformation.

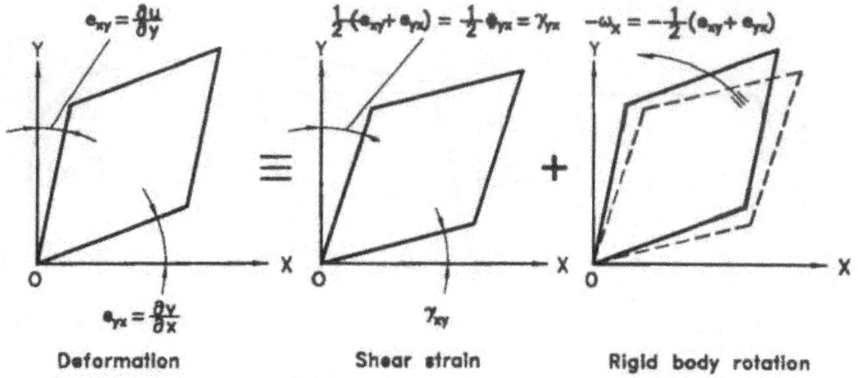

Figure 3.15 Distinction between simple and pure shear.

Tensor e_{ij} may formally be represented as:

$$e_{ij} = 1/2(e_{ij} + e_{ji}) + 1/2(e_{ij} - e_{ji}) \tag{3.57}$$

or:

$$e_{ij} = \varepsilon_{ij} + \omega_{ij} \tag{3.58}$$

Thus,

$$e_{ij} = \begin{bmatrix} \frac{\partial u}{\partial x} & \frac{\partial u}{\partial y} \\ \frac{\partial v}{\partial x} & \frac{\partial v}{\partial y} \end{bmatrix} = \varepsilon_{ij} = \begin{bmatrix} \frac{\partial u}{\partial x} & 1/2\left(\frac{\partial u}{\partial y} + \frac{\partial v}{\partial x}\right) \\ 1/2\left(\frac{\partial u}{\partial y} + \frac{\partial v}{\partial x}\right) & \frac{\partial v}{\partial y} \end{bmatrix} + \omega_{ij} =$$

$$= \begin{bmatrix} 0 & 1/2\left(\frac{\partial u}{\partial y} - \frac{\partial v}{\partial x}\right) \\ 1/2\left(\frac{\partial v}{\partial x} - \frac{\partial u}{\partial x}\right) & 0 \end{bmatrix} \tag{3.59}$$

where ε_{ij} is called the irrational or pure strain tensor:

$$\varepsilon_{ij} = \begin{bmatrix} \varepsilon_x & 1/2\varphi_{xy} = \gamma_{xy} \\ 1/2\varphi_{yx} = \gamma_{yx} & \varepsilon_y \end{bmatrix} \tag{3.60}$$

and ω_{ij} is the rotation tensor:

$$\omega_{ij} = \begin{bmatrix} 0 & -\omega_z \\ \omega_z & 0 \end{bmatrix} \tag{3.61}$$

3.7.4 *Infinitesimal strain of a line element*

Consider diagonal AC of rectangular $ABCD$ as shown in Figure 3.13a. Denoting its length as r, after deformation, this diagonal becomes A_1C_1 having length $r + \delta r$. Then:

$$\begin{aligned} &r^2 = \delta x^2 + \delta y^2 \\ &(r+\delta r)^2 = (\delta x + \delta u)^2 + (\delta y + \delta v)^2 = \\ &\delta x^2 + \delta y^2 + \delta u^2 + \delta v^2 + 2\,(\delta x \delta u + \delta y \delta v) \end{aligned} \tag{3.62}$$

Therefore,

$$(r+\delta r)^2 - r^2 = \delta u^2 + \delta v^2 + 2\,(\delta x \delta u + \delta y \delta v) \tag{3.63}$$

or:

$$r\delta r = \delta x \delta u + \delta y \delta v \tag{3.64}$$

if small quantities of higher order are neglected.

Substituting for δu and δv from Equation (3.41) gives:

$$\begin{aligned} r\delta r &= \delta x \left[\frac{\partial u}{\partial x}\delta x + \frac{\partial u}{\partial y}\delta y\right] + \delta y \left[\frac{\partial v}{\partial x}\delta x + \frac{\partial v}{\partial y}\delta y\right] = \\ &= \delta x^2 \frac{\partial u}{\partial x} + \delta y^2 \frac{\partial v}{\partial y} + \delta x \delta y \left(\frac{\partial v}{\partial x} + \frac{\partial u}{\partial y}\right) \end{aligned} \tag{3.65}$$

If l and m are the direction cosines of AC, then $\delta x = rl$ and $\delta y = rm$, then:

$$\delta r = r\left[l^2 \frac{\partial u}{\partial x} + m^2 \frac{\partial v}{\partial y} + ml\left(\frac{\partial v}{\partial x} + \frac{\partial u}{\partial y}\right)\right] \tag{3.66}$$

The strain in the direction AC is then given as:

$$e_r = \frac{\delta r}{r} = e_x\, l^2 + e_y\, m^2 + \varphi_{xy}\, lm \tag{3.67}$$

As Equation (3.67) is valid for all values of l and m, a necessary condition for Equation (3.41) to represent rigid body rotation is $e_r = 0$ if l and $m \neq 0$. By this is meant that:

$$\frac{\partial u}{\partial x} = \frac{\partial v}{\partial y} = 0 \tag{3.68}$$

and:

$$\left(\frac{\partial v}{\partial x} + \frac{\partial u}{\partial y}\right) = 0 \rightarrow \frac{\partial v}{\partial x} = -\frac{\partial u}{\partial y} \tag{3.69}$$

or:

$$e_{xy} = e_{yx} \tag{3.70}$$

From Equation (3.70) it follows that for rigid body motion the tensor of Equation (3.43) is skew-symmetric.

3.7.5 Strain compatibility equations

The displacement of a given point in a deforming body for a plane strain condition is determined by the two components u and v as continuous functions of x and y, and the deformation at the point is defined by the three components — e_x, e_y, φ_{xy} — of Equation (3.49). Therefore, if the two displacement components are specified, all three strain components can be determined uniquely, being expressed in terms of the first derivatives of the displacement components. It may, however, be anticipated that the three strain components cannot be defined arbitrarily and that there must be a relationship between them.

A problem is encountered in calculating displacement from the strains. Equation (3.49) provides three equations for the two unknowns, u and v. It is evident, therefore, that these equations will not have a unique solution for arbitrary chosen strains and that some restrictions must be imposed in order for Equation (3.49) to have a solution. The interrelationships that must exist between the strains for a body to remain continuous after straining (that is, the displacements are continuous functions of the coordinates) are established by the strain compatibility equations.[55]

From Equation (3.48):

$$\varphi_{xy} = \left(\frac{\partial v}{\partial x} + \frac{\partial u}{\partial y}\right) \tag{3.71}$$

and from Equation (3.56):

$$2\omega_z = \left(\frac{\partial u}{\partial x} - \frac{\partial u}{\partial x}\right) \tag{3.72}$$

Subtracting Equation (3.72) from Equation (3.71) gives:

$$2\frac{\partial u}{\partial y} = \varphi_{xy} = 2\omega_z \tag{3.73}$$

and adding Equations (3.71) and (3.72) produces:

$$2\frac{\partial v}{\partial x} = \varphi_{xy} + 2\omega_z \tag{3.74}$$

and because $e_x = \partial u/\partial x$ and $e_y = \partial v/\partial y$, it follows that:

$$\begin{aligned} \frac{\partial}{\partial x}\left(\varphi_{xy} - 2\omega_z\right) &= 2\frac{\partial^2 u}{\partial x \partial y} = 2\frac{\partial e_x}{\partial y} \\ \frac{\partial}{\partial y}\left(\varphi_{xy} + 2\omega_z\right) &= 2\frac{\partial^2 v}{\partial x \partial y} = 2\frac{\partial e_y}{\partial x} \end{aligned} \tag{3.75}$$

Hence:

$$\frac{\partial \varphi_{xy}}{\partial x} - 2\frac{\partial \omega_z}{\partial x} = 2\frac{\partial e_x}{\partial y} \tag{3.76}$$

and:

$$\frac{\partial \varphi_{xy}}{\partial y} + 2\frac{\partial \omega_z}{\partial y} = 2\frac{\partial e_y}{\partial x} \tag{3.77}$$

The compatibility equations may be derived using Equations (3.76) and (3.77):

$$\frac{\partial}{\partial y}\left(\frac{\partial \varphi_{xy}}{\partial x} - 2\frac{\partial e_{xy}}{\partial y}\right) = 2\frac{\partial^2 \omega_z}{\partial x \partial y} = \frac{\partial}{\partial x}\left(\frac{2\partial w_z}{\partial y}\right) = \frac{\partial}{\partial x}\left(\frac{2\partial e_y}{\partial x} - \frac{\partial \varphi_{xy}}{\partial y}\right) \tag{3.78}$$

or, finally:

$$\frac{\partial^2 e_x}{\partial y^2} + \frac{\partial^2 e_y}{\partial x^2} = \frac{\partial^2 \varphi_{xy}}{\partial x \partial y} \tag{3.79}$$

3.7.6 *Finite displacement of a point in a continuous body*

Though the previously derived results have been used in the known attempts to employ the results of the theory of plasticity in metal cutting,[15,28,53] the displacements and their derivatives have been previously assumed to be extremely small so that infinitesimal deformation was considered. As the plastic deformation in metal cutting is finite and strains are large, it is obvious that the relationships previously developed for infinitesimal strains became increasingly obsolete. In this case, the strains will no longer be linearly related to derivatives of the displacement. Furthermore, the equilibrium equations must be satisfied in the deformed body and should therefore be considered in terms of the deformed coordinates if these are considerably different from the undeformed coordinates.

There are two commonly used methods of describing the deformation of a continuous body when the deformations are finite—namely, the Lagrangian and the Eulerian.[55] The former method uses the initial coordinates of each particle to describe the deformation, while the latter method uses the coordinates of the particles in the deformed state to describe the deformation. However, the practice of metal-forming modeling requires more exact relations than that provided by Equation (3.49), which corresponds to the finite deformation developed by Slater.[55]

The projections of the deformed line element A_1C_1 shown in Figure 3.13a on the coordinate axes are obtained by subtracting the coordinates A_1 from C_1. Let the projections of A_1C_1 on the OX and OY axes be ξ and η, respectively, then:

$$\xi = x + \delta x + u + \delta u - (x + u) = \delta x + \delta u \tag{3.80}$$

Similarly,

$$\eta = \delta y + \delta v \tag{3.81}$$

Equation (3.41) can be rewritten as:

$$\begin{aligned} \xi &= \left(1 + \frac{\partial u}{\partial x}\delta x\right) + \frac{\partial u}{\partial y}\delta y \\ \eta &= \frac{\partial v}{\partial x}\delta x + \left(1 + \frac{\partial v}{\partial y}\delta y\right) \end{aligned} \tag{3.82}$$

3.7.7 Finite strain

Dividing Equation (3.82) by the deformed length of the line element $A_1C_1 = r + \delta r$ gives:

$$\frac{\xi}{r+\delta r} = l_1 = \left[\left(1+\frac{\partial u}{\partial x}l\right)+\frac{\partial u}{\partial y}m\right]\frac{r}{r+\delta r}$$
$$\frac{\eta}{r+\delta r} = m_1 = \left[\frac{\partial v}{\partial x}l+\left(1+\frac{\partial v}{\partial y}m\right)\right]\frac{r}{r+\delta r} \tag{3.83}$$

where l_1 and m_1 are direction cosines after deformation, and:

$$l = \frac{\delta x}{r}; \quad m = \frac{\delta y}{r} \tag{3.84}$$

were direction cosines of the line element before deformation.

Equation (3.83) can be rewritten as follows:

$$l_1 = A\frac{r}{r+\delta r}$$
$$m_1 = B\frac{r}{r+\delta r} \tag{3.85}$$

The direction cosines are not known in advance of the deformation process and can be eliminated from Equation (3.83) by squaring these two equations and adding:

$$l_1^2 + m_1^2 = 1 = \left(A^2+B^2\right)\left(\frac{r}{r+\delta r}\right)^2 \tag{3.86}$$

which gives:

$$\left(\frac{r}{r+\delta r}\right)^2 = A^2 + B^2 \tag{3.87}$$

It can be shown that:

$$\left(\frac{r}{r+\delta r}\right)^2 = A^2 + B^2 =$$
$$l^2\left[1+2\frac{\partial u}{\partial x}+\left(\frac{\partial u}{\partial x}\right)^2+\left(\frac{\partial v}{\partial x}\right)^2\right]+$$
$$m^2\left[1+2\frac{\partial v}{\partial y}+\left(\frac{\partial u}{\partial y}\right)^2+\left(\frac{\partial v}{\partial y}\right)^2\right]+ \tag{3.88}$$
$$2ml\left[\left(\frac{\partial u}{\partial y}+\frac{\partial u}{\partial x}\frac{\partial u}{\partial y}\right)+\left(\frac{\partial v}{\partial x}+\frac{\partial v}{\partial x}\frac{\partial u}{\partial y}\right)\right]$$

The term $((x + \delta r)/r)^2$ can then be written as equal to $(G + 1)$, say, because in the first two terms of the right-hand side of this equations it will be seen that there is the sum $l^2 + m^2 = 1$. Then:

$$\left(\frac{r+\delta r}{r}\right)^2 - 1 = G \tag{3.89}$$

or:

$$\left[\left(\frac{(r+\delta r)-r}{r}\right)+1\right]^2 - 1 = (e_r+1)^2 - 1 = e_r^2 + 2e_r = G \tag{3.90}$$

where e_r is the conventional or engineering strain of the line element AC.

Then:

$$e_r + 2e_r = 2\left(\varepsilon_x l^2 + \varepsilon_y m^2 + \varepsilon_{xy}\, lm\right) \tag{3.91}$$

where:

$$\varepsilon_x = \frac{\partial u}{\partial x}+\frac{1}{2}\left[\left(\frac{\partial u}{\partial x}\right)^2+\left(\frac{\partial v}{\partial x}\right)^2\right]$$
$$\varepsilon_y = \frac{\partial v}{\partial y}+\frac{1}{2}\left[\left(\frac{\partial u}{\partial y}\right)^2+\left(\frac{\partial v}{\partial y}\right)^2\right] \tag{3.92}$$
$$\varepsilon_{xy} = \frac{\partial v}{\partial x}+\frac{\partial u}{\partial x}+\frac{\partial u}{\partial x}\frac{\partial u}{\partial y}+\frac{\partial v}{\partial x}\frac{\partial v}{\partial y}$$

Comparing Equations (3.91) and (3.67), it can be seen that the coefficients of Equation (3.92) are related to the components of the tensor:

$$\varepsilon_{ij} = \begin{bmatrix} \varepsilon_x & \frac{1}{2}\varepsilon_{yx} \\ \frac{1}{2}\varepsilon_{yx} & \varepsilon_y \end{bmatrix} = \begin{bmatrix} \varepsilon_x & \gamma_{xy} \\ \gamma_{xy} & \varepsilon_y \end{bmatrix} \tag{3.93}$$

which is known as the finite or small strain tensor, distinguishing it from the infinitesimal pure strain tensor of Equation (3.60).

3.7.8 *Principal strains and strain invariants*

In the study of the stress at a point, it was found that two orthogonal planes exist on which there are no shear stresses, and these are known as the principal planes.[55] As a result, planes exist on which there are no shear strains. The normals to these planes will not change orientation when the body is deformed. Thus, a line element normal to such a plane will either elongate or contract but will not change direction. The normal directions to these planes are the principal directions, and the corresponding strains are known as principal strains.[55]

For a state of plane strain, consider a line element r normal to the plane AB which is normal to XOY plane (Figure 3.16). Upon straining, it is assumed that the line element changes length by an amount δr but its direction remains the same if AB is a principal plane. The components r and δr in the directions OX and OY are then proportional, that is,

$$\varepsilon = \frac{\delta r}{r} \rightarrow \delta u = \varepsilon \delta x; \quad \delta v = \varepsilon \delta y \tag{3.94}$$

Therefore, Equation (3.41) becomes:

$$\begin{aligned} \delta u &= \varepsilon_x \delta x + \varepsilon_{xy} \delta y \\ \delta v &= \varepsilon_{yx} \delta x + \varepsilon_y \delta y \end{aligned} \tag{3.95}$$

Substituting for δu and δv from Equation (3.94) into Equation (3.95) gives:

$$\begin{aligned} (\varepsilon_x - \varepsilon)\,\delta x + \varepsilon_{xy} \delta y &= 0 \\ \varepsilon_{yx} \delta x + (\varepsilon_y - \varepsilon)\delta y &= 0 \end{aligned} \tag{3.96}$$

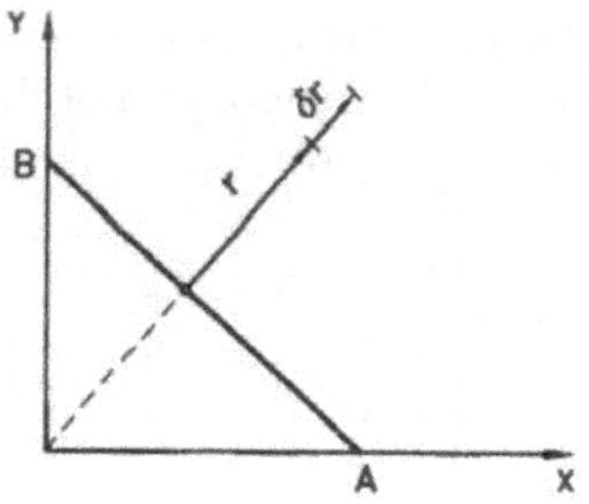

Figure 3.16 Principal strain vector.

According to Cramer's theorem, Equation (3.96) will have a nontrivial solution only if the determinant of the coefficients is equal to zero, that is,

$$\begin{bmatrix} (\varepsilon_x - \varepsilon) & \varepsilon_{xy} \\ \varepsilon_{yx} & (\varepsilon_y - \varepsilon) \end{bmatrix} = 0 \tag{3.97}$$

If this determinant is expanded, a square equation is obtained:

$$\varepsilon^2 - \varepsilon(\varepsilon_x + \varepsilon_y) - (-\varepsilon_x \varepsilon_y + \varepsilon_{xy}) = 0 \tag{3.98}$$

or:

$$\varepsilon^2 - I_1 \varepsilon - I_2 = 0 \tag{3.99}$$

where I_1 and I_2 are called the strain invariants:

$$\begin{aligned} I_1 &= \varepsilon_x + \varepsilon_y \\ I_2 &= -\varepsilon_x \varepsilon_y + \varepsilon_{xy}^2 \end{aligned} \tag{3.100}$$

If the two roots of Equation (3.99) are the principal strains and the plane *AB* of Figure 3.16 is a principal plane, then $\varepsilon_{xy} = 0$ and $(\varepsilon - \varepsilon_1)(\varepsilon - \varepsilon_2) = 0$, so that ε_1 and ε_2 are the principal strains.

3.7.9 *Increments in the strain components*

The plastic deformation in metal cutting involves very large strains. To study this deformation, the equations for finite strains can be used successfully in the incremental form.

The mechanical properties of metals, in conditions of relatively slow plastic deformation at not too high temperatures, are known to be practically independent of the rate of deformation. In this case, the main interest then lies in the small strain increment $\dot{\varepsilon}dt$ denoted $d\varepsilon_{ij}$. However, keep in mind that in general these quantities are not differentials of the strain components.

These strain increments are determined in accordance with:

$$d\varepsilon_{ij} = \frac{1}{2}\left(\frac{\partial}{\partial x_i}du_i + \frac{\partial}{\partial x_j}du_j\right) \tag{3.101}$$

and they also generate a tensor. Moreover, the components of the tensor have a physical meaning.

Equation (3.101) is useful for describing large strains which may be obtained by integrating the small changes. The increments in the strain components are evaluated with respect to the instantaneous state.

In the general case, the integrals $\int d\varepsilon_{ij}$ cannot be evaluated and do not have a physical meaning. These integrals can be found only if the strain path is known — that is, if the components ε_{ij} are known as functions of some parameter such as, for example, the deforming force. This limits the range of application of natural strains to the case of fixed principal directions.

If the principal axes do not rotate under deformation, then the integrals $\int d\varepsilon_{ij}$ have a simple physical meaning, being equal to the corresponding logarithmic or natural strains. In this case, the strains are additive — that is, the sum of successive natural strains is equal to the resultant natural strain. Therefore, the choice of the initial coordinate system in a strain analysis is of prime importance. Unfortunately, many known studies of the metal cutting mechanics do not show the coordinate system used in the analyses of the known model of orthogonal cutting, making it impossible to relate the metal cutting mechanics to the plasticity theory.

3.7.10 *Strain in metal cutting*

It follows from the foregoing analysis that the proper choice of a coordinate system in the analysis of deformations during chip formation might lead to significant simplifications. Bearing in mind the results of Section 3.5.1, we may set a Cartesian coordinate system as illustrated in Figure 3.17. If elementary transforms are carried out, it is not difficult to be certain that strain tensor is represented by two strain increments which are equal due to tensor symmetry:

$$d\varepsilon_{xy} = d\varepsilon_{yx} \tag{3.102}$$

It is easy to show that two other tensor components, $d\varepsilon_x$ and $d\varepsilon_y$, are equal to zero, as:

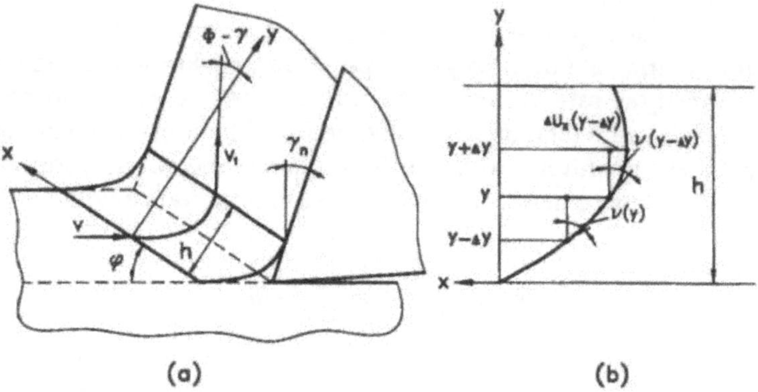

Figure 3.17 Trajectory of a particle of the layer to be removed through the deformation zone.

$$
\begin{aligned}
d\varepsilon_x &= \frac{\partial}{\partial x}\left[dD_x\,(y)\right] = 0 \\
d\varepsilon_y &= \frac{\partial}{\partial y}\left[dD_y(x)\right] = 0
\end{aligned}
\qquad (3.103)
$$

The former stems from the fact that increments along the x-axis depend on the y-coordinate alone and are independent on the x-coordinate, as the slip lines are straight lines, so that the Geiringer equations — Equations (3.19) and (3.20) — are valid. The latter follows from the continuity conditions, that is, from the constancy of the velocity v_n. Therefore, the plastic deformation in metal cutting can be characterized by only one component, namely, by shear strain.

In the coordinate system illustrated in Figure 3.17, Equation (3.31), the expression for the displacement, becomes:

$$
D_x\,(y) = -\frac{v_\tau}{v_n}y + \frac{v_2}{v_n}\frac{h}{n+1}\left(\frac{y}{h}\right)^{n+1} \qquad (3.104)
$$

According to Equation (3.101), an increment of shear strain can be thought of as the sum of particular increments along the coordinate axes. Using the obtained results for velocities and displacements, the shear strain increments are determined in accordance with:

$$
de_{xy} = \frac{\partial}{\partial y}\left[dD_x\,(y)\right] = \frac{d}{dy}\left[dD_x\,(y)\right] = d\left[\frac{dD_x\,(y)}{dy}\right] = \frac{n}{h}\frac{v_2}{v_n}\left(\frac{y}{n}\right)^{n-1} dy \qquad (3.105)
$$

Shear strain, which under accepted conditions coincides with natural strain, can then be calculated by integrating Equation (3.105):

$$\varepsilon_T = \int_0^y \frac{n}{h}\frac{v_2}{v_n}\left(\frac{y}{h}\right)^{n-1} dy = \frac{v_2}{v_n}\left(\frac{y}{h}\right)^n \tag{3.106}$$

The final true strain $\varepsilon_T = \varepsilon_T(h)$ when $y = h$ coincides with the commonly used expression (Equation (3.38)) known as shear strain:

$$\varepsilon_\tau(h) = \frac{v_2}{v_n} = \frac{\cos\gamma_n}{\cos(\varphi - \gamma_n)\sin\varphi} \tag{3.107}$$

As was expected, the final true strain ε_T does not depend on the width of the shear zone or the distribution of the tangential velocity in the deformation zone (parameter n). The considered strain is the pure strain, as it does not account for rotation.

Although the final true strain ε_T can be calculated using Equation (3.107), it provides no information on the deformation process in metal cutting. In contrast, the dynamics of strain in the deformation zone can be studied using Equation (3.106). It follows from Equation (3.106) that the deformation process in the deformation zone does not correspond to simple shearing.

The latter conclusion is of great importance for the mechanics of metal cutting. It shows that proper application of engineering plasticity to study even highly idealized models of the deformation zone yields to the proper velocity diagram and is more sophisticated than a simple shearing deformation process. This conclusion fully supports the results on the deformation mode in metal cutting obtained in Chapter 2.

Merchant illustrated[23] that the grains of workpiece material are found to be elongated as the material passes the deformation zone, and if the initial grains are approximated by circles then the elongated grain can be approximated by ellipses due to that fact that it is generally accepted[23,27,28,39,40,52] that the deformation process in metal cutting corresponds to simple homogeneous shear. As such, it follows from Figure 3.14 and was shown in details by Zorev[27] that circles should be transformed into ellipses; however, it does not coincide with Equation (3.106).

To understand the nature of plastic deformation, consider the transformation of the square 1234 in the layer to be removed in the deformation zone with parallel boundaries (Figure 3.18).[48] Let point A be a point selected on the side 23 of the square. To consider the transformation of A into A_1, Equation (3.31) may be used:

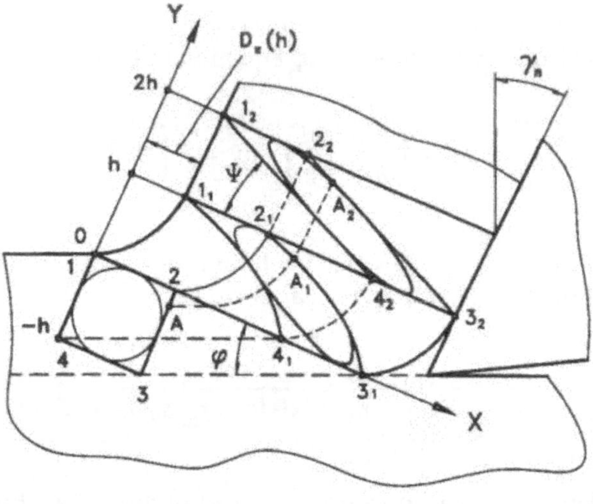

Figure 3.18 Transformation of the square in the layer to be removed in the deformation zone with parallel boundaries.

$$
\begin{aligned}
x_{A_1} &= x - \frac{v_2}{v_n}\frac{h}{n+1}\left(\frac{y+h}{h}\right)^{n+1} + \frac{v_\tau}{v_n}h \\
y_{A_1} &= 0 \cdot x + 1 \cdot y + h
\end{aligned}
\tag{3.108}
$$

As such, a square is transformed into a curvilinear rectangular and a circle transforms not into an ellipse but into a more complicated shape as illustrated in Figure 3.18. Because this transformation is not linear, the deformation does not correspond to the simple homogeneous shearing.

To understand why Equation (3.107) is still valid for the considered conditions, consider the next h-step in the displacement of the curvilinear rectangular $1_1 2_1 3_1 4_1$ in the y-direction, as shown in Figure 3.18. As seen, the square 1234 transforms into the parallelogram $1_2 2_2 3_2 4_2$, and the circle drawn within the square tangent to its sides transforms into an ellipse. As such, the coordinate transformation of A into A_2 is linear:

$$
\begin{aligned}
x_{A_2} &= x - \frac{v_2}{v_n}y + \frac{v_\tau + v_{1\tau} - \frac{1}{n+1}v_2}{v_n}h \\
y_{A_2} &= 0 \cdot x + 1 \cdot y + 2h
\end{aligned}
\tag{3.109}
$$

as the final transformation of A into A_2 coincides with simple homogeneous shearing, which makes Equation (3.107) applicable to calculate the final shear strain which appears to be the final true strain. Such a coincidence results from the fact that both strains depend only on the boundary conditions for the velocities.

3.8 Strain rate in metal cutting

The analysis of the strain rate in metal cutting is one of the most important yet not fully understood problems. The question concerning the order of strain rate in metal cutting is one of the oldest problems and has existed throughout the modern history of metal cutting. Because this problem is of great interest for both theory and practice, the next section addresses this topic.

3.8.1 Reported strain rates

The high strain rates which are believed to occur in metal cutting may affect the mechanical properties of workpiece material significantly by modifying the relevant mechanical properties of workpiece material. Moreover, if the strain rate encountered in the metal cutting process is high and can be adequately determined, then metal cutting may become a very important testing method for the determination of the dynamic physical properties of machinable materials. Such a possibility was suggested by Drucker[56] and Shaw.[40]

Drucker[56] assumed that the mean rate of shear strain $\dot{\varepsilon}_{xy}$ is equal to the mean shear strain ε_{xy} divided by the average time t_{av} required to traverse the parallel-sided shear zone of thickness $h \leq t_1/20$ (t_1 as before is the uncut chip thickness):

$$\dot{\varepsilon}_{xy} = \frac{\varepsilon_{xy}}{h} = \frac{0.2\,\varepsilon_{xy}\,v\sin\varphi}{h} \tag{3.110}$$

Assuming $\varepsilon_{xy} = 2$, $h = 0.010$ in, $v = 200$ fpm, and $\varphi = 15$ deg, Drucker calculated a $\varepsilon_{xy} = 40{,}000\ \text{s}^{-1}$ and stated that only for very low cutting speeds, v, and large uncut chip thicknesses, t_1, the values of $\dot{\varepsilon}_{xy}$ in metal cutting approach those obtained in high-speed impact experiments.

Freudenthal[57] assumed $v = 100$ fpm, $t_1 = 0.01$ in, $\varphi = 20$ deg, a thickness of glide lamellae h_1 in the order of magnitude of 4×10^{-5} in, and a ε_{xy} in the order of magnitude of 0.1. He then calculated the strain rate as:

$$\dot{\varepsilon}_{xy} = \frac{0.2\varepsilon_{xy}\sin\varphi}{h_1} = \frac{0.2 \cdot 0.1 \cdot 100 \cdot 0.34}{4 \cdot 10^{-5}} = 16{,}000\ s^{-1} \tag{3.111}$$

Freudenthal stated that the strain rate, obtained in this way, exceeds that imposed by impact loads applied at bullet velocities.

Chao and Bisacre[58] stated that the strain rate commonly found in machining operations is estimated to be of the order of 10^3 to 10^6 s^{-1}. Shaw[59] derived the following relationship for $\dot{\varepsilon}_{xy}$:

$$\dot{\varepsilon}_{xy} = \frac{\cos\gamma}{\cos(\varphi-\gamma)}\frac{0.2v}{\Delta_y} \tag{3.112}$$

where Δ_y is the spacing between successive shear planes. Assuming a reasonable cutting regime, Shaw calculated a mean shear strain rate of $\dot{\varepsilon}_{xy} = 213{,}000$ s^{-1}. Shaw stated that the maximum $\dot{\varepsilon}_{xy}$ might be many times this value and may reach the order of 10^6 s^{-1}, which is very high compared to the strain rate in an ordinary tensile test of about 10^{-3} s^{-1} and 10^3 s^{-1} in the most rapid impact tests.

Kececioglu[60] attempted to determine the strain rate in metal cutting experimentally. Studying different workpiece materials and using a variety of cutting regimes, he obtained a range for the mean shear strain rate from 2500 to 212,000 s^{-1}. Based on this result, Kececioglu concluded that under metal cutting conditions dislocations may be moving with the velocity of sound or faster. von Turkovich[54] concluded that the strain rate under common metal cutting conditions is of 2×10^4 c^{-1}, and at very high cutting speeds and very shallow feeds the strain rate can be very large, on the order of 10^7 c^{-1}.

Oxley studied the strain rate and its effect in metal cutting for many years[28,35,46] and concluded that the average shear strain rate in the shear zone lies in the range of 10^3 to 10^5 c^{-1} or even higher. Most of studies, published in recent years (see, for example, References 39 and 61) consider the same range for the rate of strain following his approach. The analysis above shows that there is a high consistency in the reported results. Any objection to the discussed results could not be located in the literature available to the author.

3.8.2 *Critical analysis of the reported strain rates*

From the foregoing, it is seen that, as believed, the strain rate in metal cutting is higher than that in most rapid impact tests and even exceeds that imposed by impact loads applied at bullet velocities. In the author's opinion, the strain rate in cutting is considered to be extremely high because it is thought of in terms of the known velocity diagram and the assumed model of a deformation zone with zero thickness. In this case, any strain rate, even an infinite, can be obtained. To understand this point, however, there is no need to consider even the speed of a bullet. As is well known, the working speeds encountered in different metal-forming operations are much higher than those in cutting (see Table 15.1 in Dieter[62]). It can be found on the very same page that the drawing of a fine wire at speed 120 ft/s (that is much higher than the cutting speed, and the deformation zone is very small) can result in a strain rate in excess of 10^5 s^{-1}. It is also shown in Dieter[62] that some

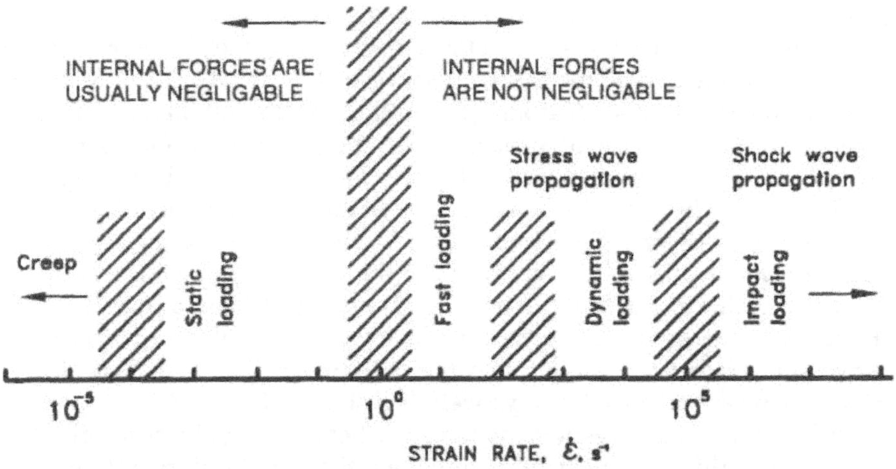

Figure 3.19 Classifications of various types of loading in terms of the strain rate.

modern metal-working processes utilize velocities as high as 700 ft/s to carry out forging, extrusion, sheet forming, etc. These processes are known as high-energy-rate forming. These high-velocity processes utilize the energy from explosion (gas or conventional explosive); however, the calculated strain rate in metal cutting is still higher than those in these processes. Moreover, contradictory to the common sense, the strain rate in cutting is even higher than that of a bullet hitting a wall.[63] In our opinion, the sense of order of the strain rate has been missing in metal cutting studies. We feel that a greater understanding of the mechanical properties of materials in metal cutting is necessary; therefore, the question about the real value of the strain rate in metal cutting has to be clarified.

Classification of various types of loading in terms of the strain rate is presented in Figure 3.19.[64] Upon examining the present state of research within the individual areas indicated in this figure, it is found that for strain rates of less that 10 s^{-1}, reliable, commercially available experimental apparatus and standard methods for the evaluation of results already exist. A rather different situation exists for strain rates, which are greater than about 10^4 s^{-1}. As can be seen in Figure 3.19, the role of material internal forces cannot be neglected in this region. Nevertheless, Oxley[28] and many other researchers neglected these forces while reporting high strain rates. Moreover, when the rate of strain falls in this region, a description of the corresponding force effects must include the propagation of the stress waves and/or stress pulse.[64] This approach has never been considered in metal cutting studies.

Additional support for the author's argument is the aspect of metal plastic deformation. There are two principle, different mechanisms of plastic deformation of metals: plastic deformation by slip and by twinning.[62,65]

The usual method of plastic deformation in metals is by the sliding of blocks of crystal over one another along definite crystallographic planes, called slip planes. Slip occurs when the shear stress exceeds a critical value. The atoms move an integral number of atomic distances along the slip plane, and a step is produced in the polished surface. When the polished surface is viewed from above with a microscope, the step shows up as a line called a slip line.

The second important mechanism by which metals deform is the process known as twinning. Twinning results when a portion of the crystal takes up an orientation that is related to the orientation of the rest of the untwined lattice in a definite, symmetrical way. The twinned portion of the crystal is a mirror image of the parent crystal. The plane of symmetry between the two portions is called the twinning plane. Twinning differs from slip in several specific ways well described in Dieter.[62]

It is known that, in general, when a steel experiences high strain rates, twinning should be a mechanism by which the metal deforms.[62] Furthermore, it is also known that increasing the strain rate by a factor of 10^3 greatly increases the number of observed twins.[66] Therefore, if the strain rate in cutting areas is as high as is believed, all the plastic deformation observed in cutting should be governed by twinning, and neither the slip-line theory nor other methods of the mathematical theory of plasticity are applicable to study the deformation process. However, this is not the case in metal cutting because:

1. The maximum contribution which the twining shear could make to the observed strain is about 17%, which is only a small fraction of the strain observed in cutting.
2. The multiple micrographs of the chip deformation zone presented in this book and those published in the literature[27,67] illustrate the result of typical plastic deformation by slip.
3. Microstructure of twins,[62,66] which is very different from that of slipping, has never been observed on the reported micrographs of the deformation zone and chip.

Therefore, this simple consideration shows that the calculated high strain rate in metal cutting does not conform to the available experimental data.

3.8.3 Strain rate tensor

From the finite strait tensor of Equation (3.93) and referring to Equation (3.67), the direct strain of a line element can be defined as:

$$\varepsilon_r = \varepsilon_x l^2 + \varepsilon_y m^2 + \varepsilon_{xy} lm \tag{3.113}$$

Then, considering a small time interval δt during which a line element is subjected to a small strain such that the coefficients of the finite strain tensor are small compared with unity when the products of derivatives in Equation (3.92) can be neglected, the rate of strain of the considered element may be represented as:

$$\lim\left(\frac{\delta\varepsilon_r}{\delta t}\right)=\frac{d\varepsilon_r}{dt}=\dot{\varepsilon}_r \tag{3.114}$$

Therefore,

$$\dot{\varepsilon}_r=\frac{\partial}{\partial x}\left(\frac{\partial u}{\partial t}\right)l^2+\frac{\partial}{\partial y}\left(\frac{\partial v}{\partial t}\right)m^2+\left[\frac{\partial}{\partial y}\left(\frac{\partial u}{\partial t}\right)+\frac{\partial}{\partial x}\left(\frac{\partial v}{\partial t}\right)\right] \tag{3.115}$$

or,

$$\dot{\varepsilon}_r=\dot{\varepsilon}_x\,l^2+\dot{\varepsilon}_y\,m^2+\dot{\varphi}_{xy}\,lm \tag{3.116}$$

where:

$$\dot{\varepsilon}_x=\frac{\partial\dot{u}}{\partial x}$$
$$\dot{\varepsilon}_y=\frac{\partial\dot{v}}{\partial y} \tag{3.117}$$

are direct strain rates, and

$$\dot{\varphi}_{xy}=\frac{\partial\dot{v}}{\partial y}+\frac{\partial\dot{u}}{\partial x} \tag{3.118}$$

is engineering shear strain rate.

The direct strain rates determine the rates of relative elongation or contraction in the direction of the *OX* and *OY* axes, whereas the shear strain rate determines the angular rate of change of initially right angles.

The rate of volumetric or dilatational strain is then given by:

$$\dot{\Delta}=\dot{\varepsilon}_x+\dot{\varepsilon}_y=\frac{\partial}{\partial t}\left(\frac{\partial u}{\partial x}\right)+\frac{\partial}{\partial t}\left(\frac{\partial v}{\partial y}\right)=div\,q \tag{3.119}$$

where q is the velocity vector. Hence, analogous to the infinitesimal pure strain tensor of Equation (3.58), a strain rate tensor can also be formed which is given by

$$\frac{\partial v_i}{\partial v_j} = \frac{1}{2}\left[\left(\frac{\partial v_i}{\partial x_j} + \frac{\partial v_j}{\partial x_i}\right) + \left(\frac{\partial v_i}{\partial x_j} - \frac{\partial v_j}{\partial x_i}\right)\right] = \dot{\varepsilon}_{ij} + \dot{\omega}_{ij} \qquad (3.120)$$

That is, it may be resolved into a symmetric tensor which actually represents the strain rate tensor and skew-symmetric part which corresponds to a rigid body rotation of the element considered.

As velocities are the total derivatives of the displacement with respect to time, it is evident that:

$$\dot{\varepsilon}_{ij} \neq \frac{d}{dt}\varepsilon_{ij} \qquad (3.121)$$

Therefore, in the analysis of large deformations, the strain rate cannot be thought of as simple time derivations of strain. Unfortunately, this fact has been ignored in the known studies of strain rates in metal cutting. For example, Oxley used the relationships for the strain rate components (see Equation A1.1 in Oxley[28] and Equation 1 in Oxley[46]) which are applicable only in the analysis of small strains. The identical approach has been used by Spaans[52] (see Equation 10), Thé,[67] and many others. It should be understood that only in the case of small deformations do simple relations exist between the strain components and the strain rate components, namely:

$$v_i = \frac{\partial}{\partial t} u_i \qquad (3.122)$$

and:

$$\dot{\varepsilon}_{ij} = \frac{\partial}{\partial t}\varepsilon_{ij} = \begin{bmatrix} \dot{\varepsilon}_x & \dot{\gamma}_{xy} \\ \dot{\gamma}_{xy} & \dot{\varepsilon}_y \end{bmatrix} \qquad (3.123)$$

It should be also stated that, in general,

$$\dot{\varepsilon}_i \neq \frac{\partial}{\partial t}\varepsilon_i \qquad (3.124)$$

as the principal axes of the strain tensor and the strain rate tensor do not coincide.[55]

References

1. Tresca, H., Sur l'ecoulement des corps solides soumis à de fortes pression, *C.R. Acad. Sci.*, Paris, 59, 754, 1864.
2. De Saint-Venant, B., Memoire sur l'establissement des équations différentielles des mouvements intérieurs opéres dans les corps solides ductiles au delà des limites où l'ésticité pourrait les ramener à leur premier état, *C.R. Acad. Sci.*, Paris, 70, 473, 1870.
3. Lévy, M., Memorie sur kes équations générales des mouvements intérieus des corps solids ductiles au delà des limites oú l'élasticité pourrait les ramener à leur premier état, *C.R. Acad. Sci.*, Paris, 70, 1323, 1870.
4. von Mises, R., Mechanik der festen körper im plastisch deformablen zustant, *Narch. Ges. Wiss. Göttingen*, 582, 1913.
5. Hencky, H.Z., Zur theorie plastischer deformationen und hierduch im material hervorgerufenen nebenspannunger, *Z. Angew. Mach. Mech.*, 4, 323, 1924.
6. Prandtl, L., Über die härte plastischer körper, *Nachr. Ges. Wiss. Göttingen*, 341, 1920.
7. Hencky, H.Z., Über einige statisch bestimmte fälle des gleichgewichts in prlastischen körpern, *Z. Angew. Mach. Mech.*, 3, 241, 1923
8. Geiringer, H., Beitrag zum vollständigen ebenen plastizitäts problem, *Proc. 3rd Int. Cong. Appl. Mech.*, 2, 185, 1930.
9. von Karman, T., Beitrag zur theorie des walzvorganges, *Z. Angew. Mach. Mech.*, 5, 130, 1925.
10. Sachs, G., Beitrag zur theorie des ziehvorganges, *Z. Angew. Mach. Mech.*, 7, 235, 1927.
11. Siebel, E., The plastic forming of metals, *Steel*, Oct. 16, 1933/May 7, 1934.
12. Lode, W., Versuch über den einfluss der mitteren hauptspannung auf das fliessen der metalle eisen, kupfer and nickel, *Z. Phys.*, 36, 913, 1926.
13. Reuss, A., Beruecksichtigung der elastischen formaendergen in der plastizitätstheorie, *Z. Angew. Mach. Mech.*, 10, 266, 1930.
14. Odquist, F.K.G., Die verfestigung von flussiesenähalichen körpern, *Z. Angew. Mach. Mech.*, 3, 215, 1933.
15. Hill, R., *The Mathematical Theory of Plasticity*, Oxford University Press, London, 1950.
16. Prager, W., A geometrical discussion of the slip-line field in plane plastic flow, *Trans. Roy. Inst. Technol., Stockholm*, 65, 27, 1953.
17. Johnson, W. and Mellor, P.B., *Engineering Plasticity*, John Wiley & Sons, New York, 1983.
18. Drucker, D.C., Greenberg, W., and Prager, W., The safety factor of an elastic plastic body in plane strain, *ASME J. Appl. Mech.*, 73, 371, 1951.
19. Kudo, H., Some analytical and experimental studies of axisymmetric cold forging and extrusion, *Int. J. Mech. Sci.*, 2, 102, 1961.
20. Hill, R., On the state of stress in a plastic-rigid body at yield point, *Phil. Mag.*, 42, 868, 1951.
21. Hill, R., On the limits set by plastic yielding to the intensity of singularities of stress, *J. Mech. Phys. Solids*, 2, 278, 1954.
22. Hill, R., The mechanics of machining: a new approach, *J. Mech. Phys. Solids*, 3, 47, 1954.

23. Merchant, M.E., Mechanics of the metal cutting process, *J. Appl. Phys.*, 16, 267, 1945.
24. Rubenstein, S., A note concerning the inadmissibility of applying the minimum work principle to metal cutting, *ASME J. Eng. Industry*, 105, 294, 1983.
25. Dewhurst, W., On the non-uniqueness of the machining process, *Proc. Roy. Soc. Lond. A*, 360, 587, 1978
26. Dewhurst, W., The effect of chip breaker constrains on the mechanics of machining process, *Ann. CIRP*, 28, 1, 1979.
27. Zorev, N.N., *Metal Cutting Mechanics*, Pergamon Press, Oxford, 1966.
28. Oxley, P.L.B., *Mechanics of Machining: An Analytical Approach To Assessing Machinability*, John Wiley & Sons, New York, 1989.
29. Lee, E.H. and Shaffer, B.W., The theory of plasticity applied to a problem of machining, *ASME J. Appl. Mech.*, 18, 405, 1951.
30. Kobayashi, S. and Tomsen, E.G., Metal-cutting analysis. 1. Re-evaluation and new method of presentation of theories, *ASME J. Eng. Industry*, 84, 63, 1962.
31. Shaw, M.C., Cook, N.H., and Finnie, I., The shear-angle relationship in metal cutting, *ASME J. Eng. Industry*, 75, 273, 1953.
32. Usachev, Ya. G. Phenomena occurring during the cutting of metals — review of the studies completed, (in Russian), *Izv. Petrogradskogo Politechnicheskogo Inst.*, XXIII(1), 321, 1915.
33. Bricks, A.A., *Metal Cutting* (in Russian), St. Petersbourg, Russia, 1896.
34. Zvorykin, K.A., On the force and energy necessary to separate the chip from the workpiece (in Russian), *Vestnik Promyslennostie*, 123, 1896.
35. Palmer, W.B. and Oxley, P.L. B., Mechanics of metal cutting, *Proc. Inst. Mech. Eng.*, 173, 623, 1959
36. Rahman, R, Seah, K.H.W., Li, X.P., and Zhang, X.D., A three-dimensional model of chip flow, chip curl and chip breaking under the concept of equivalent parameters, *Int. J. Mach. Tools Manuf.*, 35(7), 1015, 1995.
37. Eggeleston, D.M., Gerzog, R., and Tomsen, E.G., Observations on the angle relationships in metal cutting, *ASME J. Eng. Industry*, 81, 263, 1959.
38. Black, J.T. Mechanics of chip formation, in *Metals Handbook*, Ninth ed., 16, *Machining*, ASM, Metals Park, OH, 1985.
39. Stephenson, D.A. and Agapionu, J.S., *Metal Cutting Theory and Practice*, Marcel Dekker, New York, 1997.
40. Shaw, M.C., *Metal Cutting Principles*, Clarendon Press, Oxford, 1984.
41. Seethaler, R.J. and Yellowley, I., An upper-bound cutting model for oblique cutting tool with a nose radius, *Int. J. Mach. Tools Manufact.*, 37(2), 119, 1997.
42. Sidjanin, L. and Kovac, P., Fracture mechanisms in chip formation process, *Mater. Sci. Technol.*, 13, 439, 1997.
43. Stevenson, R. and Stephenson, D.A., The effect of prior cutting conditions on the shear mechanics of orthogonal cutting. *ASME J. Eng. Industry*, 120, 13, 1998.
44. Madhaven, V. and Chandrasekar, S., Some observations on the uniqueness of machining, in *Manufacturing Science and Engineering*, Vol. 6(2), Proc. 1997 American Society of Chemical Engineers International Mechanical Engineering Congress and Exposition, November 16–21, 1997, Dallas, TX, 1997, p. 99.
45. Kececioglu, D., Shear-zone size, compressive stress, and shear strain in metal-cutting and their effects on mean shear-flow stress, *ASME J. Eng. Industry*, 82, 79, 1960.

46. Oxley, P.L.B., Rate of strain effect in metal cutting, *ASME J. Eng. Industry*, 84, 335, 1963.
47. Astakhov, V.P. and Osman, M.O.M., An analytical evaluation of the cutting forces in self-piloting drilling using the model of shear zone with parallel boundaries. Part 1. Theory, *Int. J. Mach. Tools Manuf.*, 36(11), 1187, 1996.
48. Kushner, V.S., *Thermomechanical Approach in Metal Cutting* (in Russian), Irkutsk Univ. Pabl., Irkutsk, 1982.
49. Merchant, M.E., Basic mechanics of the metal-cutting process, *ASME J. Appl. Mech.*, A-168, 1944.
50. Kobayashi, S. and Tomsen, E.G., Metal-cutting analysis. II. New parameters, *ASME J. Eng. Industry*, 83, 71, 1962.
51. Ocusima, K. and Hitomi, K., An analysis of the mechanism of orthogonal cutting and its application.
52. Spaans, C., A treatise of the streamlines and the stress, strain, and strain rate distributions, and on stablility in the primary shear zone in metal cutting, *ASME J. Eng. Industry*, 94, 690, 1972.
53. Thé, J.H.L. and Scrutton, R.F., The stress-state in the shear zone during steady state machining, *ASME J. Eng. Industry*, 101, 211, 1979.
54. von Turkovich, B.F., Cutting theory and chip morphology, in King, R.I., Ed., *Handbook of High-Speed Machining Technology*, Chapman & Hall, New York, 1985.
55. Slater, R.A.S., *Engineering Plasticity: Theory and Application to Metal Forming Processes*, Macmillan, London, 1977.
56. Drucker, D., An analysis of the mechanics of metal cutting, *J. Appl. Phys.*, 20, 1013, 1949.
57. Freudenthal, A.M., *The Inelastic Behavior of Engineering Materials and Strictures*, John Wiley & Sons, New York, 1950.
58. Chao, B.T. and Bisacre, G.H., The effect of speed and feed on the mechanics of metal cutting, *Proc. Inst. Mech. Eng.*, 165, 1, 1951.
59. Shaw, M.C., *Metal Cutting Principles*, 3rd. ed., Massachusetts Institute of Technology, Cambridge, MA, 1954.
60. Kececioglu, D., Shear strain rate in metal cutting and its effect on shear flow stress, *ASME J. Eng. Industry*, 80, 158, 1958.
61. Lei, S., .Shin, Y.-C., and Incropera, F., Material constitutive modeling under high strain rates and temperatures through orthogonal machining tests, in *Manufacturing Science and Engineering*, Vol. 6(2), Proc. 1997 American Society of Chemical Engineers International Mechanical Engineering Congress and Exposition, November 16–21, 1997, Dallas, TX, 1997, p. 91.
62. Dieter, G.E., *Mechanical Metallurgy*, 3rd ed., McGraw-Hill, New York, 1986, p. 521.
63. Winter, R.E., Measurement of fracture strain at high strain rates, in *Mechanical Properties at High Rates of Strain*, Proc. Second Conf. on the Mechanical Properties of Materials at High Rates of Strain, Oxford, March 28–30, 1979, p. 81.
64. Buchar, J., Bilek, Z., and Dusek, F., *Mechanical Behavior of Metals at Extremely High Strain Rates*, Trans Tech Publications, Switzerland, 1986.
65. Hertzberg, R.W., *Deformation and Fracture Mechanics of Engineering Materials*, 3rd ed., Wiley, New York, 1989.
66. Reed-Hill, R.E., Role of deformation twinning in the plastic deformation of a polycrystalline anisotropic metals, *Deformation Twinning, AMS*, 25, 1, 1964.

67. Trent, E.M., *Metal Cutting*, 3rd. ed., Butterworth-Heinemann, Oxford, 1991.
68. Thé, J.H.L., The quadratic curve and the trajectory in the shear zone in metal cutting, *ASME J. Eng. Industry*, 96, 1105–1107, 1975.

chapter four

Work material considerations

The properties of metal have been studied ever since man discovered that he could change the hardness of steel by heating it to a bright cherry red and then quenching it in water or other suitable media. What happened to the metal was first theorized, but then, as instruments for study improved, facts replaced theory.[1] Mechanical metallurgy as a fact-based science has since developed. For a long time, the impression was that, with progress in the physics of solids, the development of a physically based theory of plastic deformation of metals was just a matter of time, but that has not yet happened. In spite of great advances in the investigation of the nature of deformation, strength, and fracture of solids, a lot of problems related to the physics of these phenomena remain unresolved. It seems that within the scope of the conventional ideas their solution is not possible.

One of the most impressive developments has undoubtedly been dislocation theory, which accounts for many of the characteristics of crystalline solids, particularly behavior during plastic deformation. The concept of dislocation was proposed independently by Taylor, Orowan, and Polanyi in 1934.[2] Although in the last 20 years extensive research has developed a variety of techniques for observing and studying dislocations in real materials, no practical way to apply the theory of dislocation behavior to study plastic deformation of polycrystalline aggregate has been suggested. Several comprehensive books have been written on this subject, but there seems to be a definite need for a text which systematically describes the actual behavior of metals and alloys during various types of deformation and attempts to explain this as far as it possible in terms of dislocation theory. Known attempts (see Honeycombe,[3] for example) are of academic rather than practical concern.

The theories of elasticity and plasticity describe the mechanics of deformation of most engineering solids. Both theories, as applied to metals and alloys, are based on the experimental studies of the relation between stress and strain in a polycrystalline aggregate under simple loading conditions.[1] Thus, they are of a phenomenological nature on the macroscopic scale and, as yet, pertain little to the structural knowledge of a metal.

Although it is recognized that during cutting processes many different phenomena occur — elastic and plastic deformations, external and internal friction, thermal phenomena, strain hardening and thermosoftening, phase transition, absorption, etc. — and the state of stress in the deformation zone is not uniaxial, the mechanical properties of workpiece material obtained in a simple tension test are used (sometimes with certain modifications accounting for high temperature, strain, and strain rate in the deformation zone). The book on plasticity by Johnson and Mellor[1] begins with the following quote (from the work by Orowan): "The tensile test is very easy and quick to perform but it is not possible to do much with its results, because one does not know what they really mean. They are the outcome of a number of very complicated physical processes. The extension of a piece of metal is in a sense more complicated than the working of a pocket watch and to hope to derive information about its mechanism from two or three data derived from measurement during tensile testing is perhaps as optimistic as would be an attempt to learn about the working of a pocket watch by determining its compressive strength." In other words, it is not clear how to correlate the properties obtained in the standard tensile test, where a uniaxial state of stress is the case, with those involved in deforming processes, where triaxial states of stress complicated by high strains and strain rates are common.

In the author's opinion, at the current stage of development, the predictability of metal cutting theory depends entirely on the accuracy with which it accounts for the properties of workpiece material, as the design and geometry of the cutting tool along with the properties of tool material are well known and the cutting regime can be set at any desirable level and/or varied according to any defined sequence. However, it is arguable that the mechanical properties of workpiece material seem to be also well known and tabulated in the corresponding reference books. Because the model for cutting is known, predicting metal cutting performance should be no problem. This is not the case, though, as it is unclear how to apply the known mechanical properties of workpiece materials in metal cutting studies and/or how to determine, if necessary, the addition properties which are not tabulated.

4.1 What has to be predicted according to existing theories?

Merchant's well-known force model,[4] is basically the only model accepted in practically all publications on metal cutting.[5-8] The core of the model includes the determination of the shear plane component F_S of the resultant cutting force as follows:

$$F_S = \pi_{fl} \frac{a_1 b_1}{\sin \varphi} \tag{4.1}$$

where τ_{fl} is the shear flow stress along the shear plane AB; a_1 is the uncut chip thickness; b_1 is the width of cut; φ is the shear angle.

Although many studies on metal cutting have attempted to theoretically derive or experimentally obtain the shear flow stress, this is still one of the most controversial issues in the field, as the results obtained using Equation (4.1) only occasionally match the experimental results. There are several principal questions concerning the shear flow stress of workpiece material in cutting: Could the stress-strain relations obtained in the standard tensile (compression) test be used in metal cutting? Would the stress in the deformation zone stop growing or even start to decrease (an idealistic idea which stands behind so-called high-speed machining[9]) in the case of large strains similar to those in metal cutting? Is the shear flow stress affected by high strain rates which are believed to be 10^4 to 10^8 higher in metal cutting than those in the standard tensile test and by the high temperatures occurred in cutting? Practically all serious studies done on metal cutting contain the direct or indirect answers to these questions.

4.2 Review of attempts made to predict shear flow stress

It is logical to assume that the theory of plasticity, known also as engineering plasticity, the main goal of which is to develop mathematical techniques for the prediction of the plastic deformation of the workpiece in various circum stances particularly in a complex stress state, has been used in attempts to predict the flow shear stress in metal cutting. As discussed in Chapter 3, the application of the mathematical theory of plasticity to the problem of metal cutting originates from Hill,[10] who first hypothesized that the solution to the metal cutting process is not unique and that a steady-state model of deformation would depend on the conditions encountered in the initial transient period of deformation. Later on, in 1983, Rubenstein[11] presented proof to support this fact. Dewhurst[12] has also strongly argued for such non-uniqueness of machining.

The main ideas proposed by Hill had a great influence on further works regarding theoretical modeling of the metal cutting process. There have been a number of attempts to develop models that can predict the resistance of workpiece material to cutting as a first step towards the prediction of cutting forces, tool wear, quality of the machined surface, etc. These models are now briefly considered.

Three major approaches to the determination of the thermomechanical behavior of metals in cutting may be distinguished, although they utilize the same or similar model of the cutting process proposed by Merchant[4] so that they consider the shear flow stress on the shear plane as the main mechanical property of workpiece material involved in cutting.

4.2.1 The first approach

This approach originates from Merchant himself[4] and is based on the assumption that the shear flow stress on the shear plane is equal to that obtained from the standard tensile test;[7,13–16] therefore, the stress-strain curves obtained in the standard tensile test can be used in metal cutting.

It is worthwhile to discuss here the results of works by Zorev[7] and von Turkovich.[17] Zorev has conducted a great number of cutting experiments to establish the shear flow stress. The essence of his work is the comparison of the resistance of workpiece material to plastic deformation during cutting and during mechanical testing, particularly during tension and compression. For the range of deformation studied in tension, the relation of the maximum shear stress (τ_{max}) (which is wrongly termed in the book as the tangential stress due to poor translation) to the true shear strain (ε_T) was represented in the form of the following equation:

$$\tau_{max} = A_\tau \varepsilon_T^m \tag{4.2}$$

On double logarithmic coordinates ($\log \varepsilon_T - \log \tau_{max}$), Equation (4.2) is represented as a straight line. By continuing these lines into the range of large plastic deformations, appropriate to the cutting process, it is possible to compare the values of the shear stresses during cutting (τ_{sp}) with extrapolated values of the maximum shear stress during tension (τ_{max}). The comparison was made using experimental data for a large quantity of different workpiece materials, cutting regimes, and tool geometries and shows that the extrapolation of the relationship defined by Equation (4.2) to deformation of $\varepsilon_T = 2.5$ produces approximately the same values of the maximum shear stress during tension (τ_{max}) as for the shear stresses during cutting (τ_{sp}). It is therefore possible to determine τ_{sp} approximately from the results of tensile tests by using the following formula:

$$\tau_{sp} \approx A_\tau \, 2.5^{m_\tau} \tag{4.3}$$

Studying the influence of the temperature on the shear flow stress in cutting, Zorev concluded that at high cutting speeds, the temperature has comparatively little influence on this stress. Moreover, the experimentally obtained results suggest that the assumption that thermal strain hardening of steels as the temperature in the chip formation zone rises to the blue shortness temperature leads to the reduction of deformation and the chip compression ratio has no ground.

von Turkovich compared the shear stress computed by means of equations attributable to Merchant with that calculated by means of the universal equation, which expresses the shear stress as a function of dislocation density in the following form:

$$\tau = A_1 G b_r \sqrt{\rho_d} \tag{4.4}$$

Here, A_1 is a constant of order of 1; G is the shear modulus; b_r is the Burgers vector magnitude; ρ_d is dislocation density.

The comparison, assuming that the process of deformation in metal cutting takes place at constant dislocation density, shows a fairly good agreement between the theoretical and experimental results for iron and copper when the value of the shear modulus in Equation (4.4) was selected according to the cutting temperature. Therefore, before proceeding further, it would be appropriate to note that the authors of these groups of works tried to reveal the influence of cutting temperature, high strain, and strain rate on the shear flow stress but were unsuccessful.

4.2.2 The second approach

According to the second approach, the shear flow stress in cutting appears to be much higher than that obtained in the standard material test. This is explained by the fact that the shear flow stress obtained in the standard tensile test is modified by high strain, strain rate, and/or their linear/nonlinear combination.[5,18–21] According to this approach, the stress-strain relationships obtained in the standard tensile test cannot be used in metal cutting. Instead, a number of new stress-strain relationships, mostly obtained from the work on material properties under high strain-rate conditions, have been proposed. As before, these studies also provide numerous theoretical and experimental results to supporting their conclusions.

It is worthwhile to discuss here the work done by Oxley.[19] Analyzing the results of the experimental studies of Kececioglu[22] and Nakayama[23] conducted at relatively high cutting speeds and of Palmer and Oxley[24] conducted at low cutting speeds and using Merchant's velocity diagram, Oxley concluded that for practical cutting speeds (in turning, for example), the average shear strain rate in the shear zone lies in a range from 10^3 to 10^5 s^{-1} or even higher. These values are much higher than the strain rates of 10^{-3} to 10^{-1} s^{-1} normally incorporated in conventional tension and compression tests. As it is known that the rate of strain has a marked effect on the strain-stress properties in standard material tests, Oxley assumed that, to be realistic, any machining theory should take account of strain rate effects. The temperatures generated in machining and their effect on flow stress would be expected to be equally important but this was not fully recognized in Oxley's study.[19]

4.2.3 The third approach

The essence of the third approach is consideration of the influence of the combined effects of strain hardening and thermal softening on shear flow stress. As before, a number of new stress-stain relations have been proposed.[9,25-29] Also, numerous theoretical and experimental proofs have been provided to support this approach.

The contributions made by Spaans[25] are also relevant. Spaans acknowledged the influence of temperature, strain, and strain rate on the shear flow stress and concluded that the effect of temperature is balanced by the strain rate effect in such a way that the shear flow stress remains equal to that in the standard tensile test.

The foregoing brief review of previous works reveals a broad scattering of descriptions of the thermomechanical behavior of metals in cutting. The important issue here is that the authors representing the second and third approach have never acknowledged (mentioned, disproved, discussed, etc.) the work of authors representing the first approach who studied the influence of high strain rates and temperatures on the thermomechanical behavior of materials in cutting but did not find it to be significant. Furthermore, Shaw[5] analyzed the shear flow stress in cutting and concluded that this stress cannot be predicted in terms of properties derived from ordinary material tests; therefore, it is next to impossible to predict metal cutting performance.

In the author's opinion, one reason for such a significant discrepancy in the reported results is consideration of the shear flow stress as a sufficient mechanical property completely representing the resistance of workpiece material to cutting.

4.3 What has to be predicted in reality?

Machining is a deforming process that forms and shapes metals and alloys.[2] However, it seems that no single study points out the principal difference that exists between machining and all other metal forming processes. In machining, the physical separation of the layer to be removed (in the form of chips) from the rest of the workpiece must occur. To achieve this, the stress in the deformation zone should exceed the ultimate stress of workpiece material, whereas other forming processes are performed by applying stress sufficient to achieve the well-known shear flow stress in the deformation zone. The objective of machining is to separate the layer to be removed with minimum possible plastic deformation, and, therefore, the energy spent on the plastic deformation in cutting may be considered as wasted. On the other hand, any metal deforming process, especially involving high strains (deep drawing, extrusion), uses plastic deformation to accomplish the process. Parts are formed into useful shapes such as tubes, rods, and sheets by displacing the metal from one location to another.[30] Therefore, a better material, from the viewpoint of metal forming, should exhibit a higher strain

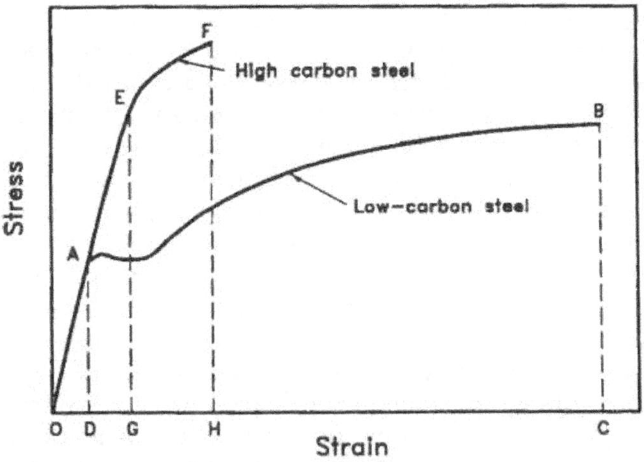

Figure 4.1 Comparison of stress-strain curves for high and medium carbon steels.

before fracture. It is understood that this is not the case for machining, as an optimal material should exhibit a fracture strain as small as possible. Everyday machining practice shows that the cutting forces, energy, tool wear, etc. are generally less in cutting brittle as opposed to ductile materials.

It follows, then, that among other important parameters characterizing the behavior of a material in cutting, toughness should be considered of prime importance. The toughness of a material is defined as its ability to absorb energy in the plastic range. One way of looking at toughness is to consider it as the total area under the stress-strain curve.[2] This area is an indication of the amount of work per unit volume which must be done on the material to cause its fracture. Figure 4.1 shows the strain-stress curves for materials of high and low toughness. High-carbon steel has a high-yield stress that makes the elastic energy necessary to reach the proportional limit (represented by triangle *OEG* in Figure 4.1) much higher than that for a low-carbon structural steel (represented by triangle *OAD* in Figure 4.1). However, the low-carbon steel is more ductile and thus has a greater total elongation. The total area under the stress-strain curve for this steel (denoted as *OABC* in Figure 4.1) is much greater than that for the high-carbon steel (*OFH* in Figure 4.1); therefore, the former is a tougher material. This illustrates that toughness is a parameter that comprises both strength and ductility.

Several mathematical approximations for the area under the stress-strain curve have been suggested. For ductile materials having stress-strain curves like that shown in Figure 4.1, it can be approximated by the following equation:[2]

$$U_T \approx \sigma_{UTS}\, \varepsilon_f \tag{4.5}$$

The terms of Equation (4.5) may be interpreted as follows: U_T is the amount of work per unit volume which must be done on the material to cause its fracture; σ_{UTS} is the ultimate tensile strength of a material; ε_f is the fracture strain. In terms of metal cutting, U_T may be thought of as the energy necessary to separate the layer to be removed from the rest of the workpiece.

Analyzing Equation (4.5), we may conclude that, out of two parameters that define the energy necessary to achieve fracture, only the fracture strain depends on the parameters of deformation, as the ultimate strength of a material may be considered as its mechanical constant (at least at temperatures less than 0.6 T_m, where T_m is the melting point of a material).[2] From the viewpoint of metal cutting, the fracture strain may be considered as the most important characteristic of a ductile material, as it defines the energy involved in the cutting process and thus all other process' parameters. Unfortunately, as previously discussed, in metal cutting, it is believed that the shear flow stress is the primary parameter defining all other process' parameters, and the main focus of past studies has been on the determination of this stress theoretically and/or experimentally. However, the above discussion shows that the shear flow stress cannot be used to define the energy involved in the cutting process, but the existing theories do not leave any room for other mechanical properties or parameters of workpiece material to calculate the parameters of the cutting process — for example, the cutting force.

4.4 Experimental verification

To prove that the proposed approach is the case in the real metal cutting process, the deformation process in metal cutting should be compared with those in other deforming processes at the point of fracture. It can be accomplished by comparing the energy spent in the cutting process with that spent in any other deformation process under similar strains at fracture. In the present study, the cutting process is compared with incremental compression. Incremental compression was selected for comparison with cutting for two reasons. First, in incremental compression, friction losses are negligibly small and all energy is spent on plastic deformation. Second, incremental compression is the only standard material test where strains as high as in metal cutting °!y be achieved under controllable conditions (ASTM Standard Test Method F 1624-95).

Four steels selected for the comparison were numbered as follows: 1, plain carbon steel AISI 1040; 2, low-alloy steel AISI 3310H; 3, low-alloy steel AISI 4130; 4, austenitic stainless steel AISI 30400. This numbering will be kept throughout the further discussion.

4.4.1 Incremental compression

A Computer Controlled Material Testing System 647 was used for the experiments. Following the methodology proposed by Rozenberg and Rozenberg,[31]

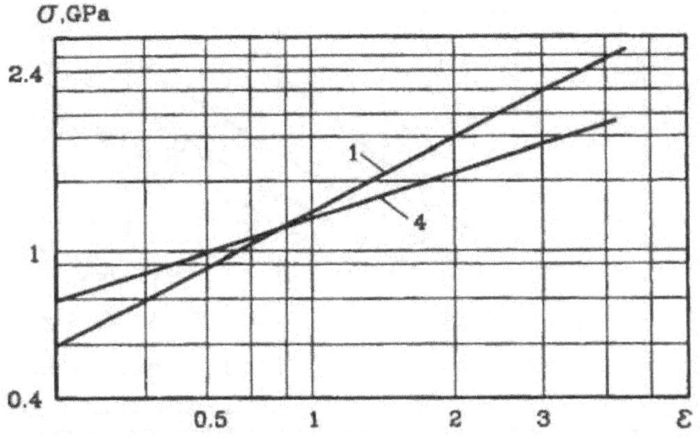

Figure 4.2 Axial stress in compression vs. true strain for materials 1 and 4.

for each chosen material the experimental relationship "axial stress σ-strain ε" under high strains were obtained. Figure 4.2 shows the results for steels 1 and 4. As such, the following high fracture strains for the steels under study were achieved in the incremental compression tests: $\varepsilon_1 = 7.2$; $\varepsilon_2 = 5.1$; $\varepsilon_3 = 4.4$; $\varepsilon_4 = 4.9$. These were calculated using the known formula proposed by Zorev:[7]

$$\varepsilon = 1.5 \ln \frac{l_0}{l_1} \tag{4.6}$$

in which l_0 is the initial length of a specimen before any load is applied; l_1 is the deformed length of the specimen.

Under the achieved high deformations, for all of the materials used in the study, the flow curves were obtained in the form of a power expression,[2,7] $\sigma = K_\sigma \varepsilon^m$, where K_σ and m are constants for a given material.

As mentioned above, in incremental compression, the friction losses are negligibly small so that all energy is spent on the plastic deformation. Hence, the expression for the incremental work done in compression would be

$$dW_{com} = F_i \, dl_i \tag{4.7}$$

Here, F_i is the instantaneous compression force, which is defined as:

$$F_i = \sigma A_i \tag{4.8}$$

Here A_i is the instantaneous cross-sectional area of the specimen, $A_i = V/l_i$; l_i is the instantaneous length of the specimen, and V is the volume of the specimen. Substituting Equations (4.6) and (4.8) into Equation (4.7), one may obtain:

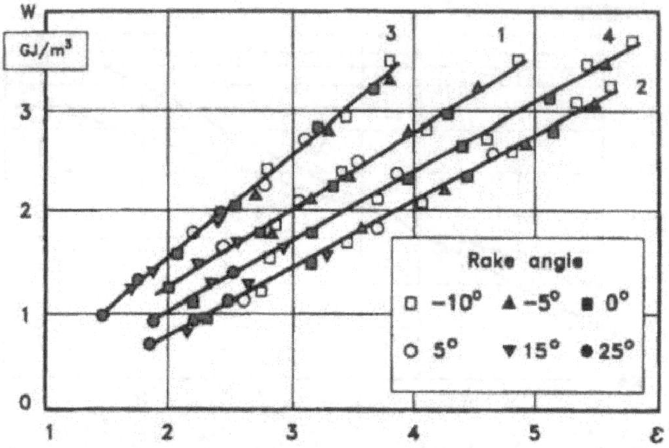

Figure 4.3 Comparison of the specific work done in compression with that in cutting. The results obtained in incremental compression are shown by solid lines and those in metal cutting are shown by symbols.

$$dW_{com} = K_{\sigma} V 1.5^{m} \left(\ln \frac{l_0}{l} \right)^{m} \frac{dl}{l} \tag{4.9}$$

The total work per unit volume done in compression is obtained by integrating Equation (4.9) in the limits from initial length l_0 to final length l_1 to produce:

$$W_{com} = \frac{K_{\sigma} \varepsilon^{m+1}}{1.5(m+1)} \tag{4.10}$$

Using this equation and the experimentally obtained values of K_{σ} and m, the total work per unit volume done in compression W_{com} as a function of the true strain ε_T calculated using Equation (4.6) is shown in Figure 4.3, with solid lines for each material (1 to 4) used in the experiment.

4.4.2 *Metal cutting*

The total work per unit volume (specific work) done in cutting W_{cut} may be defined as follows:[5,7]

$$W_{cut} = \tau_f \, \varepsilon_f \tag{4.11}$$

Here, τ_f is the shear stress at fracture, and ε_f is the final true shear strain.

Two important conclusions follow from Equation (4.11). First, the specific work done in cutting depends only on the fracture true shear strain, because the shear stress at fracture of the strain-hardened workpiece material depends also on this strain. Second, if the high temperature and strain rate that occurred in cutting will affect τ_f then they must affect W_{cut}.

Cutting experiments were performed to obtain the total work per unit volume done in cutting. They included the determination of the cutting force, chip compression ratio, and shear strain. The experimental setup and methodology are discussed in Chapter 6. For comparison, the results of the cutting experiment are plotted by their respective symbols in Figure 4.3, where the results of the incremental compression test are shown as solid lines.

A comparison of the incremental compression test results (solid lines in Figure 4.3) with those obtained in the cutting tests (symbols in Figure 4.3) shows that, regardless of the cutting regime (including the feed, cutting speed, tool material, and tool geometry, as well as the type of cutting fluid used), the specific work done in cutting is the same as that done in incremental compression. This supports the proposed approach. Another interesting fundamental result has been obtained — namely, neither temperature nor strain rate affects the resistance of workpiece material in cutting because the incremental compression was conducted at the room temperature with a very small strain rate, whereas the strain rates in cutting were much higher and the temperature in the deformation zone reached 300° to 500°C.

4.4.3 Influence of high temperatures

Now let us consider the effect of high temperatures in the machining zone on the resistance of workpiece material to cutting. It is known[2] that the stress-strain curve and the flow and fracture properties derived from the tension test strongly depend on the temperature at which the test was conducted. In general, the strength decreases and ductility increases as the test temperature is increased. This property is widely used in hot-rolling, drawing, and other bulk deformation processes to reduce the energy consumption and to increase the workability of workpiece materials.[2] The same principle applies to machining with workpiece preheating where the workpiece is preheated by an external heat source, for instance, by a plasma arc.[2,31] Therefore, it seems to be quite logical to assume that high temperatures occurring in cutting may reduce the shear stress at fracture of workpiece material.

To study the influence of temperature on the properties of workpiece material in metal cutting, the cutting process is assumed to be adiabatic.[5,7] This assumption has a solid foundation as far as orthogonal cutting is concerned, because, at the high speeds used in cutting, the transfer of heat in the direction of motion occurs primarily by transportation, and the conduction term can be neglected. This means that heat generated in the deformation zone and the average temperature θ_s in this zone are proportional to the specific work of metal removal W_{cut} being done by shearing. As such, the average temperature can be calculated as:[7]

$$\theta_s = \frac{W_{cut}}{J\rho_w c_p} + \theta_0 \tag{4.12}$$

where J is the mechanical equivalent of heat; ρ_w is the density of workpiece material; c_p is the average specific heat; and θ_0 is the temperature prior to deformation.

It has been shown that the specific work in metal cutting (W_{cut}) may reach 3 GJ/m^3 or even more.[7] Thus, the temperature θ_s in the deformation zone may reach 300° to 500°C. It is known that the stress-strain curve for a given material assumes lower and lower values as the temperature increases, thus at such high temperatures the shear flow stress of the workpiece material should decrease, as well.[2] This principle combined with Merchant's model for chip formation constitutes a logical background for the numerous thermomechanical models for workpiece material previously discussed. However, the results shown in Figure 4.3 do not conform to this assumption. Therefore, it is of prime importance to understand why the influence of temperatures in metal cutting seems to be at odds with the results of material testing studies.

It is true that the yield and ultimate strength of workpiece material decrease with temperature. However, that can occur *if and only if* that material is kept at that temperature for a certain period of time.[32] Therefore, to discuss the influence of temperature, Rosenberg and Rosenberg[31] proposed estimating the period of time necessary for a microvolume of workpiece material to pass through the deformation zone.

It follows from previous discussion (see Chapter 3, Figure 3.9) that a microvolume of the layer being cut, passing through the deformation zone, changes its velocity from the cutting speed v to the chip velocity $v_1 = v/\zeta$, where ζ is the chip compression ratio. Thus, the average velocity of the microvolume is 0.5 $v(1 - \zeta)$. Therefore, the time necessary to pass the deformation zone having the width of h would be

$$T = \frac{h}{0.5v\left(1+\frac{1}{\zeta}\right)} \tag{4.13}$$

As the width of the deformation zone is $h = 0.5a$,[25] where a is the uncut chip thickness, it is possible to estimate the time necessary for a microvolume to pass the deformation zone for a typical cutting regime. When the workpiece is made of a plain carbon steel, a typical cutting regime is as follows: v = 120 m/min = 2 m/s; ζ = 2.5; a = 0.2 mm. Thus, the estimated time is T = 0.000071 s. When the workpiece is made of a high-strength, low-alloy steel, the typical cutting regime may be v = 120 m/min = 2 m/s; ζ = 1.3; a = 0.05 mm. As such, T = 0.000014 s. As seen, the time necessary for a microvolume to pass the deformation zone is extremely short. As a result, there can be no temperature influence on the mechanical properties of the microvolume of workpiece material.

Besides a very short heating time, there are two other strong reasons why the temperature in orthogonal cutting plays no role in changing the resistance of workpiece material to cutting:

- It is well known that the structural transformation temperatures in metals increase dramatically with the heating rate.[2] In metal cutting, the heating rate reaches hundreds of thousands or even millions of degrees per second and significantly increases the structural transformation temperatures of workpiece material.
- Heat generation in the deformation zone follows the plastic deformation in this zone. Due to the practically used cutting speeds being much higher than that of heat conduction, the microvolume entering the deformation zone is not yet heated; thus, no temperature elevation occurs. Heat is generated later as a result of the microvolume deformation; therefore, the microvolume is heated even over a shorter time than calculated in the examples.

Now the concept of mutual compensation in metal cutting may be revised. As discussed above, Spaans[25] suggested that, in metal cutting, an increase in the shear flow stress due to strain hardening is compensated by the corresponding decrease of this stress due to thermal softening so that the shear flow stress remains the same as in the standard material test. This suggestion becomes very convenient to explain a significant difference between the results obtained from the known theories of metal cutting and those obtained in practice, even though no experimental proof to this suggestion has been reported. The experimental data presented in Figure 4.3 contradict this suggestion. Moreover, to achieve an even greater difference in conditions between cutting and compression, a technical copper was chosen as the workpiece material, as suggested by Rozenberg and Rozenberg.[31] Figure 4.4 shows the results of the comparison between the incremental compression and cutting. In case of incremental compression (solid line in Figure 4.4) the strain rate was 10^{-3} s^{-1} (no heating). In cutting at very low cutting speeds (■ in Figure 4.4) this rate was 100- to 100,000-fold higher (negligible heating). In cutting at high cutting speeds (O in Figure 4.4), the strain rate (according to Oxley[19]) was in the range of 10^3 to 10^5 s^{-1} (high temperature). It can be seen that the "strain-specific work" relations are identical even though the range of the experimental conditions is broad. Therefore, there are no grounds to believe that mutual compensation (or counterbalancing) of strain hardening and thermal softening is the case in metal cutting.

To conclude the discussion, the results of microhardness tests will be presented. It is known that the microhardness (*HV*) of the deformed material is uniquely related with the preceding deformation[33] and with the shear

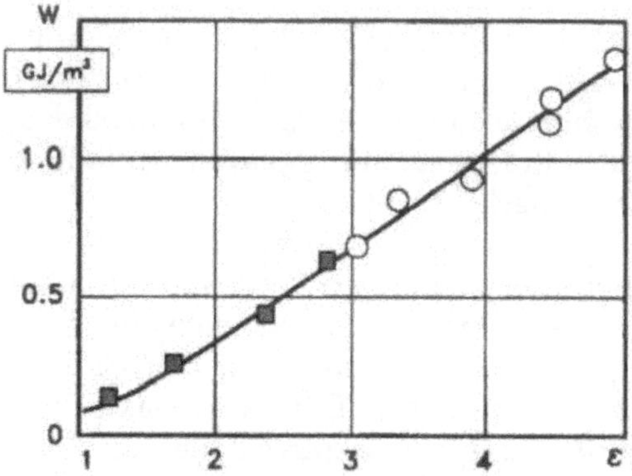

Figure 4.4 Specific work done in compression (solid line) and in cutting (■, small cutting speeds; O, high cutting speeds) of a technical copper vs. the true shear strain.

stress[31] gained by the specimen under consideration at the last stage of deformation. This makes it possible to obtain information about the extent and distribution of deformation and shear stress on the basis of microhardness measurements (discussed in Chapter 5). Performing such measurements, the comparison of stresses and deformations in incremental compression with those in cutting can be made.

An incremental compression test was carried out using specially prepared specimens made of steel AISI 1010. The preparations included shaping, residual stress relaxation by tempering at 180 to 200°C, examination of the microstructure of each specimen, and measurement of the microhardness of each specimen over the entire cross-section.

For this test, the shear flow stress is related with microhardness as:[31]

$$\tau = 0.185\,HV;\ 1 \le \varepsilon \le 5 \tag{4.14}$$

The test result is shown in Figure 4.5 by a solid line.

For the metal cutting test, the specimen made of steel AISI 1010 was used. The specimen preparation, structure, and hardness were the same as for the compression tests. Samples of the deformation zone with the partially formed chip were obtained using a specially designed, computer-triggered, quick-stop device. Using sequential hardness traverses, the strain and stress distributions in the deformation zone were obtained from very early to advanced stages of workpiece material deformation to even further advanced stages, as the chip was separated from the workpiece. The shear stress at fracture was defined as the maximum measured shear stress. The results for different tool geometry and tool feed are shown in Figure 4.5 by their respective symbols. Nearly perfect agreement between the results of the incremental compression test and the cutting experiment supports the proposed approach, and neither

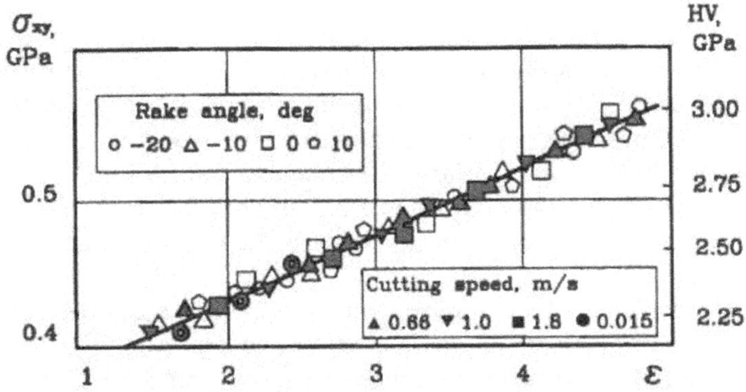

Figure 4.5 Comparison of machining and incremental compression test results. The solid line represent results obtained in the compression test and the symbols represent the results of the cutting test.

high temperatures nor high strain rates affect the resistance of workpiece material to cutting under orthogonal conditions.

4.5 *The cause of significant discrepancies in reported results*

Since Merchant introduced his model,[4] it is a popular belief that orthogonal cutting is a convenient model for studying the mechanics of metal cutting. To conduct a study with this model, one should accept a number of assumptions well analyzed by Shaw.[5] The problem is that the results obtained with an orthogonal model are used to make some practical conclusions and to use them in process and/or tool design featuring non-orthogonal conditions. Here, one practical example is given explaining a significant scatter in the experimental results obtained in the studies of thermomechanical behavior of workpiece materials.

It is known from the second law of thermodynamics that heat always flows spontaneously from a hot to a cold region in a body. Because the machining zone is a heat source, heat leaving the machining zone creates a certain dynamic (time-dependent) temperature field around this zone. It is understood that for the energy transfer by thermoconductivity, there is no priority direction. As might be expected, this heat expansion affects the workpiece material if and only if the velocity of the heat expansion is equal to or greater than the cutting speed with respect to the workpiece.

Normally, cutting speeds encountered in practices employing modern tool materials are much higher than the velocity of heat expansion. Thus, for true orthogonal cutting, there should not be any influence of the heat expansion on the workpiece material. However, there are two basic reasons explaining why such an influence has been noticed and reported in many studies:

- The cutting speeds used in metal cutting studies with orthogonal models are usually much lower than those used in industry, especially when a quick-stop device is used;[25] therefore, the velocity of heat expansion may be less, equal to, or may even exceed the cutting speed depending on a particular combination of the chosen cutting regime and respective material properties. It is clear that in the last two cases, the heat generated in cutting will affect the properties of workpiece material.
- End-tube turning is commonly used to model orthogonal cutting.[5,8,21,29] In the author's opinion, such modeling creates a problem. Originally, in orthogonal cutting the workpiece was assumed to be infinitely long, and the cutting tool moved over it at the cutting speed. It is understood that if the cutting speed is greater than that of heat expansion, heat plays no role in orthogonal cutting. The velocity of heat expansion when a heat source moves is characterized by the Peclet number:

$$Pe = \frac{v a_1}{w_w} \tag{4.15}$$

Here, v is the velocity of a heat source (the cutting speed, m/s); a_1 is the uncut chip thickness (m); and w_w is the thermal diffusivity of workpiece material (m^2/s).

For $Pe > 10$, the source moves faster than heat can expand, and this is common for practical cutting conditions. It is important to emphasize here that in true orthogonal cutting, the tool never passes the same or even a neighboring point of the workpiece more than once. However, in tube end turning, this is not the case. The cutting tool comes to the same point after each revolution. The residual heat from the previous revolution affects the shear flow stress of the current revolution and so on. This influence depends not only on the physical properties of workpiece material and the cutter, but also on the cutting regime used, dimensions of the workpiece, etc. Unfortunately, these factors have never been accounted for in the previous studies. To demonstrate a significant difference between orthogonal cutting and tube end turning, the Peclet number is calculated for each of these processes under identical modeling cutting conditions, which were selected to be as follows: workpiece material, steel AISI 1040; thermal diffusivity of the workpiece material $w_w = 6.67 \times 10^{-6}$ m^2/s; cutting speed $v = 1.5$ m/s; uncut chip thickness $a_1 = 0.5 \times 10^{-3}$ m. Additional conditions for end tube turning include a cutting feed $f = a_1 = 0.5$ mm/rev; mean diameter of the tube of $d_{mt} = 0.1$ m. For orthogonal cutting, using Equation (4.15), $Pe = 112.4$ (as $Pe >> 10$, there is no influence of the residual heat); for tube end turning, $Pe = 0.17$ (as $Pe << 10$, there is a great influence of the residual heat). In addition to the residual heat, the residual stress unavoidably gained by the machined surface on the preceding pass may affect the shear flow stress on the current pass.

The foregoing analysis offers reasonable explanations for the existence of many different models for the thermomechanical behavior of metals in cutting.

4.6 Nature of plastic deformation of polycrystalline materials

Because it has been proven that the strain at fracture has to be defined to predict metal cutting performance and, in order to achieve better operations, the cutting conditions that minimize this strain should be controlled, the following consideration attempts to reveal the influence of process parameters on the strain at fracture. As discussed before, the main shortcoming of the mathematical theory of plasticity is that it does not consider the real structure of polycrystalline materials. As a result, it cannot be used to define the fracture strain of a material.

It is assumed that fracture mechanics provides another way to deal with the strain at fracture. It is intended as a multidisciplinary engineering topic that has a foundation in both mechanics and material science;[34] however, it originates from the Griffith theory, which assumes that a deformed body initially has a crack. Because of the prominence of the Griffith theory, it has been natural for metallurgists to use their microscopes to search for Griffith cracks in metals. However, based on observations up to magnifications available with the electron microscope, there is no reliable evidence that Griffith cracks exist in metals in the unstressed conditions.[2] Although there is a considerable amount of experimental evidence to show that microcracks can be produced by plastic deformation, no reliable model for microcrack nucleation is available.

The foregoing consideration of the possible ways to deal with deformation in metal cutting shows that another fresh approach should be used to derive, at least to the first approximation, the theoretical expression for the strain at fracture in order to predict metal cutting performance.

4.6.1 Fracture and its correlation with deformation

The fracture strain depends on the characteristics of a material itself and the parameters of the deformation process used (Figure 4.6). It is instructive to discuss the correlation between fracture strain and other limiting characteristics of metals, as it is important that the process engineer understand the nature of fracture in various materials subjected to different conditions. The same material at two different temperatures may give rise to different fracture processes. Furthermore, the type of fracture depends on whether the material is subjected to a fluctuating or static load.

Work dealing with the fracture of engineering materials is directed toward defining the conditions under which materials should not fracture or

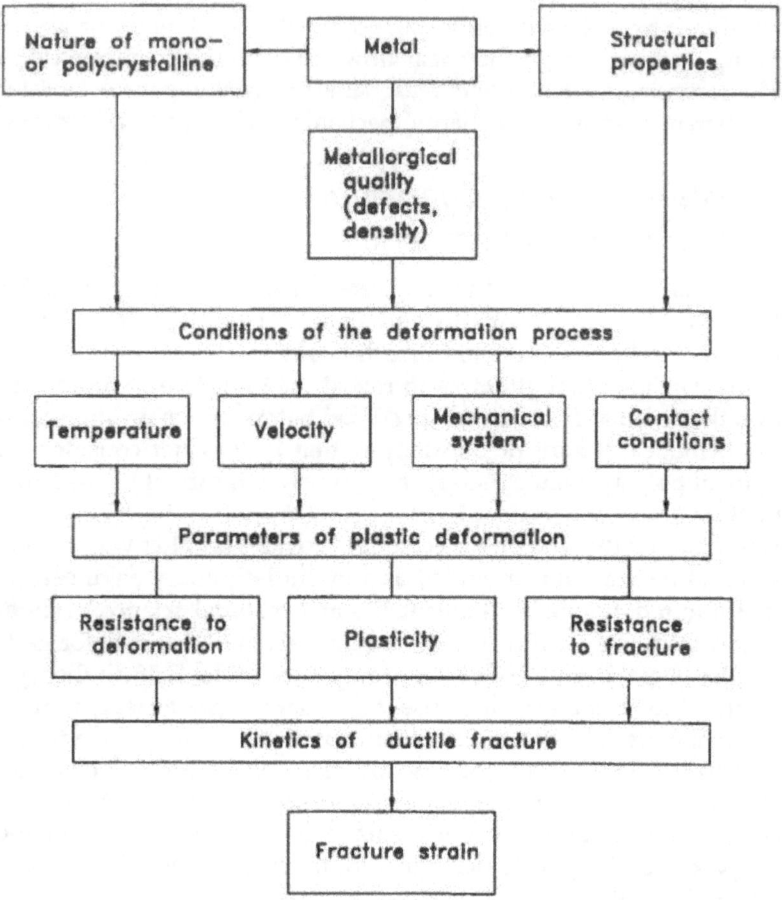

Figure 4.6 Major interactions between the fracture strain and other structural deformation parameters.

fail in service. In contrast, in metal cutting, the opposite effect is desired — how to fracture a given material with minimum efforts. Therefore, in metal cutting applications, studies of fracture processes in materials should be conducted in this direction. Such studies might give valuable insight into why materials behave as they do, why some materials exhibit more resistance to cutting than others, and what we really mean by stiffness, strength, brittleness, and toughness when these characteristics are considered in terms of metal cutting. It is mainly by means of such studies that better metal removing processes can be developed.

Earlier studies characterized the failure stress of a material by a critical stress, such as the "ultimate tensile strength" which was assumed to be a material property just as, for instance, its density or the modulus of elasticity is.[35]

Several failure theories, which have been developed during the past 200 years, are based on the concept that the stress conditions causing failure in a standard tensile test of a bar specimen are also responsible for failure in a component subjected to combined loads.[36] These works utilize the failure criterion which may be referred to as the stress criterion. The maximum normal stress theory is perhaps the simplest failure theory which describes failure as occurring when the maximum tensile or compressive stress exceeds the uniaxial tensile or compressive strength of the material. This theory is generally suitable for brittle materials.

Another group of failure theories utilizes the concept that yielding of a material occurs when the distortion energy in the tensile test associated with the shape change is equal to the distortional energy in the component that experienced multiaxial loading. These works utilize the maximum distortion energy, or von Mises' yield criterion, which is in a better correlation with actual test data.[36] This criterion is widely used today and defines only the very beginning of plastic deformation, which is good enough in the calculations used in the design. However, this criterion cannot be used to define the fracture strain ε_f, that is, the total amount of plastic deformation at failure. Therefore, it is not applicable in the analysis of the metal cutting process.

4.6.2 Plasticity and stress relaxation

Plasticity of a material can be defined as its ability to undergo irreversible plastic deformation when a sufficient external load is applied. Here, the word "ability" assumes that certain plastic mechanisms of stress relaxation exist in the material.

Plastic deformation, including its final stage (fracture), takes place under complicated non-homogeneous stress and structural changes and a transient temperature field and is accompanied by at least five different energy fluxes: release of elastic energy, crack formation, heat flow, mass transfer (pure diffusion), and movement and multiplication of dislocations. The probability of each energy flux to predominate under given conditions may be estimated by Einstein's formula:[37]

$$p \sim \exp\left(\frac{S_{cur} - S_{eq}}{k}\right) \sim \exp\left(-\frac{J_d t_p}{k}\right) \tag{4.16}$$

where S_{cur} and S_{eq} are the system energy in the current and equilibrium state respectively; k is the Boltzmann constant; J_d is the density of dissipated energy flux equal to the derivative of the entropy of the system with respect to the temperature, that is dS/dT; and t_p is the time necessary for relaxation which characterizes the duration of the restoration of the system's equilibrium state and depends on the nature of the restoration process.

According to Equation (4.16), the probability of predominance of one of the five listed irreversible processes with respect to the others is defined by both the density of dissipated energy flux J_d and the time necessary for relaxation t_p. When two or more processes have equal densities of dissipated energy, the process possessing the lower relaxation time t_p has a higher probability.

Maxwell was probably the first who interpreted the relaxation time as a measure of plasticity.[38] The relaxation of stress in plastic deformation may be represented as:

$$\sigma_t = \sigma_0 \, exp\left(-\frac{t}{t_p}\right) \approx \sigma_0 \, exp\left(-v_p t\right) \tag{4.17}$$

where σ_0 and σ_t are the applied initial and current stresses, respectively; t is the current time; $t_p = 1/v_p$ is the period of time during which the stress is reduced in e ($\approx$ 2.71) times; and v_p is the velocity of relaxation per unit time.

Kurnakov[37] used a specially designed mechanical test to measure the relaxation time and showed that this characteristic can be used as a measure of plasticity (brittleness) of solids. According to him, the longer the relaxation time (smaller v_p), the slower the elastic deformation transforms into plastic deformation, the higher brittleness of the material, and vice versa. When the strain rate, as a measure of the velocity of deformation, is lower than the velocity of relaxation per unit time, the material exhibits plastic behavior, and when the strain rate exceeds v_p the same material exhibits brittle behavior.

Taking the ratio of Equation (4.17) written for two independent instances in time t_1 and t_2 and taking natural logarithms of both sides, we obtain:

$$v_p \approx \frac{\ln(\sigma_1/\sigma_2)}{t_1 - t_2} \sim c_\varepsilon \, m_\varepsilon = \frac{\ln(\sigma_1/\sigma_2)}{\varepsilon_1 - \varepsilon_2} \tag{4.18}$$

where m_ε is known as the strain rate sensitivity,[2,39] and c_ε is a constant.

It follows from Equation (4.18) that the strain rate sensitivity may be considered as one of the plasticity criteria. This conclusion is also supported by the results of mechanical tests at high pressure[39] which revealed the direct correlation between the strain rate sensitivity m_ε and the elongation e_f. However an important warning follows from the results of studies conducted on pressing, upsetting, extrusion, cutting, etc. — m_ε does not depend on a particular mechanical system, whereas the fracture strain does significantly.[40] It has been also proven that an increase (decrease) in m_ε corresponds to an increase (decrease) in ε_f even though particular values of ε_f, depending on the state of stress, homogeneity of deformation, and many other mechanical factors, may vary over a broad range. Experimental relationships $\varepsilon_f = f(m_\varepsilon)$ obtained over the test temperature range of 0.1 to 0.8 T_m (T_m being the melting temperature) in the tensile tests ($\dot{\varepsilon} = 10^{-1}\ s^{-1}$) for different materials are shown in Figure 4.7. One important conclusion may be drawn from this figure. Within each particular type of crystal structures (face-centered cubic, body-centered cubic, and hexagonal close-packed), metals of higher density exhibit higher fracture strains than those of lower even though both have approximately the same technical purity and the strain rate sensitivity.

4.6.3 *Defect population in metals*

It is known that materials possess low fracture strength relative to their theoretical capacity because most materials deform at much lower stress levels due to the defects that are microstructural in origin and induced during the manufacturing process.[36] The defects and impurities in metals are classified in Table 4.1.

The defects affect the density of metals. For example, the theoretical density of copper ($\rho_{Cu(T)} = 8988\ kg/m^3$) is higher than that of monocrystal by 0.4% and that of polycrystal by 0.52%. The density of a ferrite structure of 0.0001% impurity is $\rho_f = 7874\ kg/m^3$ and is higher by 0.33% than that of a pearlite structure ($\rho_p = 7848\ kg/m^3$) and that of a cementite structure by 2.7% ($\rho_c = 7662\ kg/m^3$). As the discussed difference in density occurs mainly due to the presence of defect population in a real material having a certain real structure, this difference might have direct correlations with fracture kinetics and strain.[37]

The presence of macrodefects is always undesirable from the design point of view, as it lowers the levels of mechanical properties including strength and ductility; however, these increase machinability as the energy expended in cutting lowers. This is widely used in well-known free-machining steels which are produced with the addition of 0.1 to 0.3% sulfur or 0.1 to 0.35% lead or combinations of both to reduce cutting forces, cutting temperatures, and consequently tool wear rates.[41]

The role of micro- and submicrodefects is dual. On the one hand, the strength of most metals increases with an increase in their concentration to a certain limit. Moreover, ductility and thus fracture strain increase with increasing defect mobility. On the other hand, when the concentration of micro- and submicrodefects exceeds the critical limit, the strength of most metals decreases, as these defects promote the development and propagation of micro- and macrocracks.

Although the latter may reduce fracture strain, significantly increasing machinability, there are very few studies on the behavior of metals having supercritical microdefect concentrations, as it is of no importance in the design practice where the opposite effect is desired. In the author's opinion, use can be made of such defects, taking into consideration the healing effect and the actual state of stress.

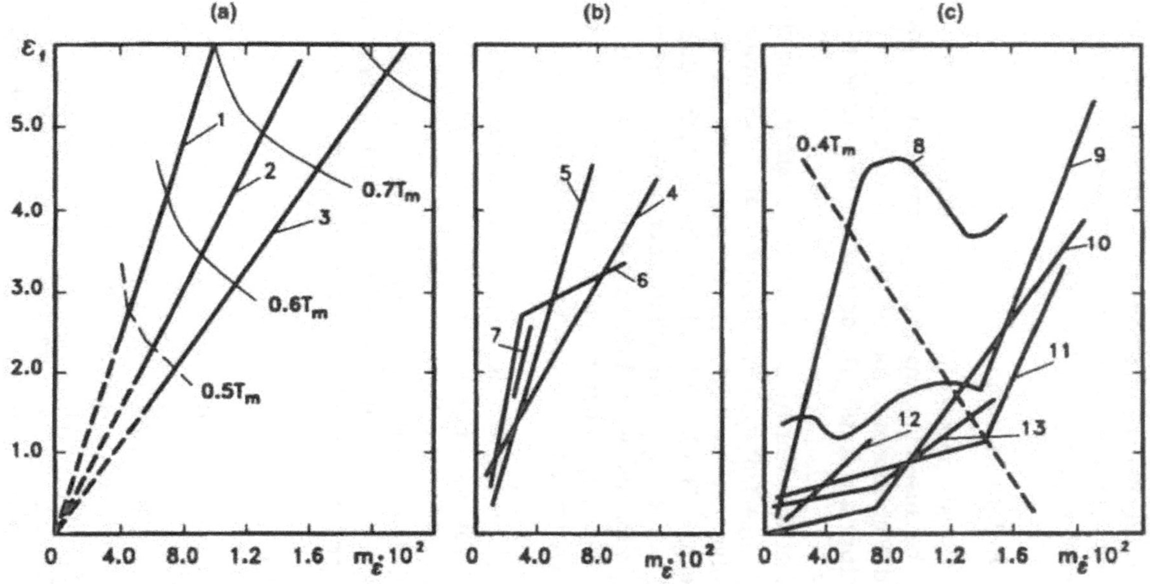

Figure 4.7 Effect of the strain rate sensitivity on the fracture strain in the tensile test over the test temperature range 0.1 to 0.8 T_m. **(a)** Face-centered cubic metals: 1, silver; 2, copper; 3, aluminum. **(b)** Body-centered cubic metals: 4, iron; 5, vanadium; 6, molybdenum; 7, chromium. **(c)** Hexagonal close-packed metals: 8, cadmium; 9, titanium; 10, zinc; 11, cobalt; 12, magnesium; 13, lanthanum.

Table 4.1 Defects and impurities in metals

Type	Minimum size	Point and line defects		Surface defects
		+ defect	− defect	
Submicrodefects	(1 … 5) $\cdot 10^{-10}$ m	Interstitial impurity atoms, substitutional impurity atoms	Vacancies of different types	
	(5 … 10) $\cdot 10^{-10}$ m	Clusters, Guinier-Priston zones, gas subnuclei, dislocation	Concentrations of vacancies, dislocation loops	Surface of the material, disclinations, imperfections in the crystal structure, grains boundaries, interphase boundaries, submicrocracks
	(50 … 2000) $\cdot 10^{-10}$ m	Microblisters, fine precipitates tates throughout the alloy (for example, carbides in tempered steels)	Submicropores	
	(0.2 … 1000) $\cdot 10^{-7}$ m	Fine inclusions dispersed throughout steels (for example, non-metallic-metallic inclusions and carbides in tempered steels)	Micropores	Residual microstresses, microcracks, blasters, non-fusion zones, variable grain size, overburned and overheated zones
Macrodefects	> 10^{-3} m	Non-metallic inclusions, shrinkage cavities, porosity, blasters	Pores	Residual macrostresses, cracks (including quenching), machining marks, (gouges, burns, tears, scratches, etc.)

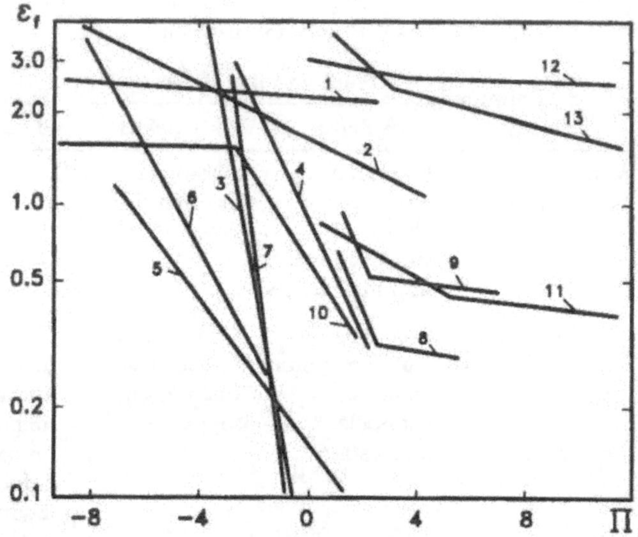

Figure 4.8 Effect of Π-factor on the fracture strain: 1, niobium; 2, iron; 3, tungsten; 4, molybdenum; 5, beryllium; 6, magnesium; 7, zinc; 8, tin alloy; 9, brass; 10, brass alloy; 11, tin bronze; 12, deformed lead; 13, cast lead.

4.6.4 Influence of the state of stress

The state of stress in the body which undergoes plastic deformation affects the fracture strain. For example, it is known that hydrostatic compression increases and tension decreases the fracture strain;[37] however, this general knowledge is not sufficient to control the deformation process. Therefore, a certain generalized parameter characterizing the state of stress should be selected to study the correlation between the state of stress and the fracture strain.

Figure 4.8 shows the relationships between the fracture strain and of the state of stress represented by the Π-factor:

$$\Pi = \frac{3I_1(\sigma)}{2\sqrt{I_1^2(\sigma) - 3I_2^2(\sigma)}} \tag{4.19}$$

where $I_1(\sigma)$ and $I_2(\sigma)$ are the stress invariants which may be expressed in terms of principal stress σ_1, σ_2, and σ_3 as:

$$\begin{aligned} I_1 &= \sigma_1 + \sigma_2 + \sigma_3 \\ I_2 &= -\left(\sigma_1\,\sigma_2 + \sigma_2\,\sigma_3 + \sigma_3\,\sigma_1\right) \end{aligned} \tag{4.20}$$

It was mentioned in Chapter 3 that Lode investigated the validity of the yield criteria[42] using some thin-walled tubes made of steel, copper, and nickel subjected to various combinations of uniaxial tension and internal hydrostatic pressure. In doing this, he devised a sensitive method to determine the effect of the intermediate principal stress on yielding. This may be explained as follows.

Tresca[43] suggested that yielding occurs when the maximum value of the extreme shear stress in the material equal to half the difference between algebraic maximum (σ_1) and minimum (σ_3) principal stresses,

$$\tau_{max} = \tau_2 = \pm 1/2(\sigma_1 - \sigma_3) \tag{4.21}$$

attains a critical value. As seen, the Tresca yield criterion requires the maximum and minimum principal stresses to be known in advance. When applied for yielding in uniaxial tension when $\sigma_1 = \sigma_y$, which is the uniaxial yield stress of the material, $\sigma_2 = \sigma_3 = 0$, the criterion gives:

$$\sigma_1 - \sigma_3 = \sigma_y \tag{4.22}$$

from which, assuming that $\sigma_1 \geq \sigma_2 \geq \sigma_3$,

$$(\sigma_1 - \sigma_3)/\sigma_y = 1 \tag{4.23}$$

The intermediate principal stress, σ_2, can thus vary from a maximum value $\sigma_2 = \sigma_1$ to a minimum value $\sigma_2 = \sigma_3$ without apparently affecting the yield criterion expressed by Equation (4.23). To characterize the influence of the intermediate principal stress, σ_2, Lode introduced the parameter:

$$\mu_L = \frac{2\sigma_2 - \sigma_3 - \sigma_1}{\sigma_1 - \sigma_3} = \frac{\sigma_2 - \dfrac{\sigma_1 + \sigma_3}{2}}{\dfrac{\sigma_1 - \sigma_3}{2}} \tag{4.24}$$

which is known as the Lode stress parameter.[44]

Equation (4.24) can be rearranged so that:

$$\sigma_2 = \frac{\sigma + \sigma_3}{2} + \mu_L \frac{\sigma - \sigma_3}{2} \tag{4.25}$$

The von Mises yield criterion[44] in terms of the principal stresses is

$$(\sigma_1 - \sigma_2)^2 + (\sigma_2 - \sigma_3)^2 (\sigma_3 - \sigma_1)^2 = 2\sigma_y^2 \tag{4.26}$$

and, if σ_2 from Equation (4.25) is substituted, then after rearranging and simplifying the von Mises yield criterion becomes:

$$\frac{\sigma_1 - \sigma_3}{\sigma_y} = \frac{2}{\left(3 + \mu_L^2\right)^{1/2}} \tag{4.27}$$

When $\sigma_2 = \sigma_3$, Equation (4.24) shows that $\mu_L = -1$, and when $\sigma_2 = \sigma_1$, $\mu_L = +1$. Because $\sigma_1 \geq \sigma_2 \geq \sigma_3$, it follows that $-1 \leq \mu_L \leq +1$. When $\mu_L = -1$, the principal stress are σ_1, $\sigma_2 = \sigma_3$ which is uniaxial tension $(\sigma_1 - \sigma_3)$ with hydrostatic stress σ_4. When $\mu_L = +1$, the principal stresses are $\sigma_1 = \sigma_2$, σ_3 which is uniaxial compression.

In addition to the stress parameter, μ_L, defined by Equation (4.24), Lode also introduced the plastic strain parameter, υ_L, defined by:

$$\upsilon_L = \frac{2d\varepsilon_2^p - d\varepsilon_3^p - d\varepsilon_1^p}{d\varepsilon_3^p - d\varepsilon_1^p} = \frac{d\varepsilon_2^p - 1/2\left(d\varepsilon_3^p + d\varepsilon_1^p\right)}{1/2\left(d\varepsilon_3^p - d\varepsilon_1^p\right)} \tag{4.28}$$

where $d\varepsilon_1^p$, $d\varepsilon_2^p$, and $d\varepsilon_3^p$ are plastic strain increments in the principal directions.

Analyzing stress-strain relations proposed by Prandtl for plane strain deformation, Reuss[45] assumed that the plastic strain increment is, at any instant of loading, proportional to the instantaneous stress deviation and the shear stress such that

$$\frac{d\varepsilon_x^p}{\sigma_x'} = \frac{d\varepsilon_y^p}{\sigma_y'} = \frac{d\varepsilon_z^p}{\sigma_z'} = \frac{d\gamma_{xy}^p}{\tau_{xy}} = \frac{d\gamma_{yz}^p}{\tau_{yz}} = \frac{d\gamma_{zx}^p}{\tau_{zx}} = d\lambda \tag{4.29}$$

or, more compactly, in tensor notation:

$$d\varepsilon_{ij}^p = \sigma_{ij}' \, d\lambda \tag{4.30}$$

where σ_{ij}' is the deviator stress tensor and $d\lambda$ is a scalar non-negative constant of proportionality which is not a material constant and may vary throughout the stress history. As before, the superscript "*p*" denotes the plastic strain increment.

Considering the principal directions, Equation (4.30) can be stated as:

$$\frac{d\varepsilon_1^p}{\sigma_1'} = \frac{d\varepsilon_2^p}{\sigma_2'} = \frac{d\varepsilon_3^p}{\sigma_3'} = d\lambda \tag{4.31}$$

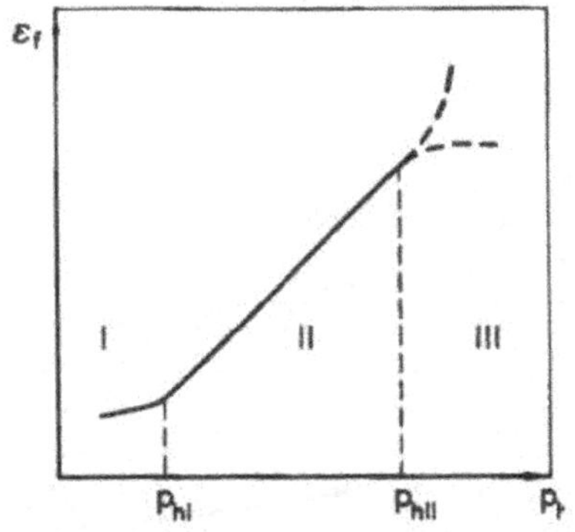

Figure 4.9 Generalized dependance of the fracture strain ε_f on the hydrostatic stress σ_m.

If Equation (4.31) is correct, then μ_L should be numerically equal to υ_L. From Equations (4.31) and (4.25):

$$\upsilon_L = \frac{2d\varepsilon_2^p - d\varepsilon_3^p - d\varepsilon_1^p}{d\varepsilon_3^p - d\varepsilon_1^p} = \frac{2\sigma_2' - \sigma_3' - \sigma_1'}{\sigma_3' - \sigma_1'} = \frac{2\sigma_2 - \sigma_3 - \sigma_1}{\sigma_3 - \sigma_1} \quad (4.32)$$

Thus, $\upsilon_L = \mu_L$. To prevent confusion, it should be appreciated that the Lode strain parameter υ_L is not related to Poisson's ratio.[44]

A generalization of the available experimental material on deformation of metals results in the following conclusions:

- Fracture strain depends significantly on the characteristic of the state of stress, as shown in Figure 4.8. Therefore, by changing this state, the fracture strain and thus the energy consumption per unit volume of the layer to be removed in cutting can be minimized. In the author's opinion, the easiest way to do this is to change the geometry of the cutting tool used to achieve the necessary μ_L. This approach appears to be new grounds for selecting both the cutting geometry and the cutting regime.
- The dependence of the fracture strain ε_f on the hydrostatic stress p_h may be characterized by three distinct zones (Figure 4.9):
  - (I) Zone of insignificant dependence, wherein ε_f is independent or depends insignificantly on p_h until a certain limit p_{hI}. This behavior is observed in tensile tests of brittle materials (cast iron, chromium).
  - (II) Zone of significant, approximately linear, dependence of ε_f on p_h. This behavior has been observed in testing of all metals.

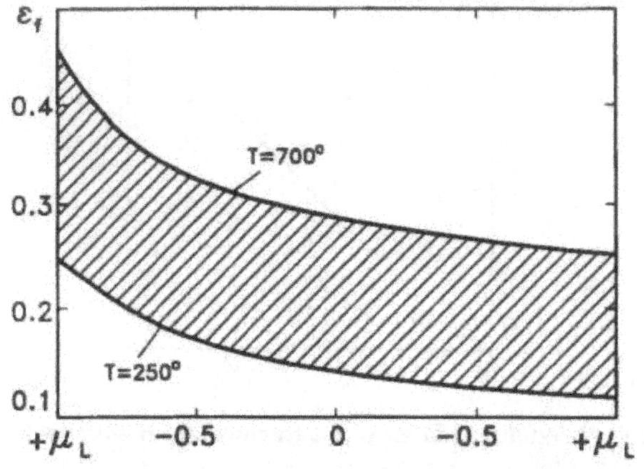

Figure 4.10 Changes in the effect of the Lode stress parameter on the fracture strain.

(III) Zone of a parabolic increase in plasticity. The boundary between zones II and III (the limit p_{hII}) is not one line as shown in Figure 4.9; rather, it is a small zone. The exact behavior of metals in zone III is not yet known.

- The temperature under which the test is carried out has a significant influence on the fracture strain; however, over the range of temperatures involved in cutting, the temperature may not significantly affect the dependence of the fracture strain on the Lode stress parameter, as shown in Figure 4.10.

4.6.5 Generalizations

A detailed analysis of a great deal of work on plastic deformation of metals, concerning, in particular, the nature of the fracture strain, results in the following generalizations:

1. The physical sense of the fracture strain is much broader than any available characteristic of plasticity accepted in the mathematical theory of plasticity or in metal forming, as the fracture strain correlates the ability of a material to undergo plastic deformation, kinetics of internal defect formation, and developments such as crack formation, accumulation of porosity, etc. Moreover, the fracture strain, considered as a probability characteristic of plastic deformation, includes the probability of fracture depending on the structure, internal energy, and state of stress in a material. Unfortunately, there are few studies relating the fracture strain and the direct characteristics of ductile fracture.
2. A great variety of different characteristics and criteria are used to characterize the fracture strain, and the choice of each depends upon

the required accuracy and specific problem under consideration. In any case, the fracture strain is a complex function which in turn depends upon a number of functions and state factors.

3. Because the mathematical theory of plasticity considers only the macroscopic behavior of a plastically deforming solid in a uniform state of complex stress, it cannot help in studies of the nature of the fracture strain.
4. Multiple factors and functions affecting the fracture strain may be grouped in four categories: (a) defect population, (b) stress relaxation, (c) structural-energy strength level, and (d) state of stress and deformation. However, none of these is a part (at least, directly) of the known characteristics of plastic deformation.
5. Multiple common particular relationships between fracture strain and temperature, rate of strain, porosity, purity, etc. obtained experimentally for common metals cannot be derived using any known theoretical expression for the fracture strain. Although there are a number of available generalized parameters which may characterize experimental conditions, such as state of stress and deformations (for example, Lode stress and strain parameters), strain rate, homologous temperature, etc., the known experimental relationship for the fracture strain does not account for these parameters. Moreover, these relationships were obtained within rather narrow ranges of experimental parameters and expressed through special formats (natural, average, true, relative, etc. strain) using different scales (natural, semi- or log-log) that makes it very difficult or even next to impossible to compare these parameters among themselves and with the known theoretical results.
6. The strain rate sensitivity of a metal, considered as its ability to relax stresses, appears to be the most objective characteristic of ductility of polycrystalline materials. All other characteristics of ductility used, such as elongation, reduction of area at fracture, fracture strain, etc.[2] may be thought of as the consequences of the strain rate sensitivity. However, the strain rate sensitivity does not account for a number of the important parameters of the deformation process such as defect population in metals, the state of stress involved, the grain structure of the deformed material, etc.; thus, when considered alone, it appears to be irrelevant for computing the fracture strain in a complex practical deformation process.
7. To a first approximation, the strain at fracture may be considered as the ratio of the two functions:

$$\varepsilon_f = \frac{f_1(C_{def})}{f_2(C_f)} \tag{4.33}$$

where $f_1(C_{def})$ is a function accounting for the material's ability for stress relaxation and may be thought of as $f_1(m_G)$; $f_2(C_f)$ is a function accounting for the probability of fracture and thus depends on the defect population in the material, its microscopic structure, the state of stress, etc.

Consequently, to derive a theoretical expression for computing the fracture strain, the functions f_1 and f_2 should be known.

4.7 Fracture of ductile polycrystallines

Fracture is the separation or fragmentation of a solid body into two or more parts due to stress.[2] Fracture can be considered to be made up of two components— crack initiation and crack propagation — and can be classified into two general categories: ductile fracture and brittle fracture. The kinetics of brittle fracture are well known[2,36] and are characterized by a rapid rate of crack propagation with no gross plastic deformation. The tendency for brittle fracture increases with decreasing temperature and increasing strain rate. In the design of engineering structures, brittle fracture is avoided at all costs, because it occurs without warning and usually produces disastrous consequences. In metal cutting, it is the best type of fracture, as it minimizes the consumed energy and tool wear and has the highest efficiency and reliability. For example, it is known that the machining of cast irons or brittle brasses would never present any significant problems as compared with those associated with the machining of austenitic stainless steels. Before proceeding further, it is very important to distinguish what kind of fracture, brittle or ductile, takes place in cutting of ductile materials.

4.8 Mechanism of fracture in metal cutting

The fact that fracture is a real phenomenon of metal cutting even in machining of ductile materials has been duly recognized by Starkov[46] and Shaw.[47,48] Considering dislocations as the linear lattice defects that are responsible for nearly all aspects of plastic deformation of metals, Starkov assumed that application of the penetration force to the workpiece from the cutting tool increases the number of dislocations in the workpiece material and increases and facilitates dislocation mobility. As a result, ductile fracture takes place in metal cutting.[46] However, this consideration is rather qualitative, as the origin of dislocations and the mechanisms of their multiplication are not yet quantitatively related to the structure of a real polycrystalline with random orientation of individual crystals.

Shaw,[47,48] studying an elastic-plastic finite element stress field based on an assumed continuum and experimentally observed chip geometry and cutting forces, has found it to be inconsistent with physical conditions that must pertain along the shear plane (constant stress on the shear plane equal to the shear flow stress of the heavy pre-strain hardened work material).

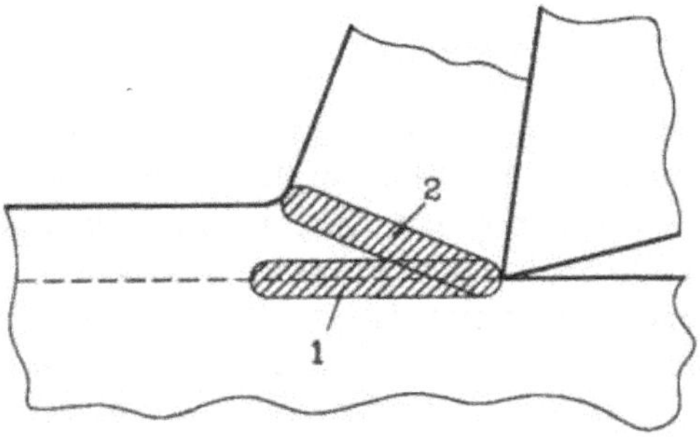

Figure 4.11 Zones where the limiting stress can occur.

Shaw has concluded that the material does not behave as a continuum and that microcracks along the shear plane play a significant role just as they do on the tool face. Although this very important finding explains many known contradictory results, it has not been noticed by other researchers.

To deal with fracture in metal cutting, two important question should be answered. The first addresses the possible place of fracture and the second requires the determination of fracture (brittle or ductile) that occurs in metal cutting.

4.8.1 Possible regions of fracture

An analysis of the dynamics of the deformation zone for different cutting tool geometries and mechanical properties of workpiece material (presented in Chapters 2 and 3) shows that the limiting (for a given work material) stress in the workpiece material may occur in two regions (Figure 4.11): Region 1, along the surface separating the workpiece and the layer being removed, and Region 2, along the surface of the maximum combined stress (Chapter 2). Region 2 is commonly considered in the metal cutting studies while little attention has been payed to Region 1.

Region 1 is the region of high stresses, as (1) the cutting edge causes concentration of the shear stress in this region; (2) the chip-cantilever "tears off" a layer of metal being removed from the workpiece, creating significant tensile stress (as shown in Chapter 5); and (3) the deformation level here is high due to high intensity of the work material's flow even in the presence of a crack (i.e., without mechanical connection between the parts of work materials located above and below the separation line).

Evidence from multiple experiments shows that fracture in metal cutting always starts from Region 1, and further development depends on a given combination of the mechanical properties of workpiece material, tool geometry, and cutting regime used.

4.8.2 Experimental evidence

Another way to prove that fracture exists in the metal cutting process is through metallographic observations of the deformation zone and partially formed chip. It was found, however, that the multiple reported observations are not sufficient to do the job, and the reasons for that have to be discussed.

While the modern researcher has many investigative techniques at his disposal, the optical reflection microscope remains the most effective means for examining the structures of workpiece material, deformation zone, and chip. In particular, the chip is the only testimony which reflects the history of the loading, strain, and strain rate of the cutting process, and because any proposed cutting model considers the structure of the deformed and partially deformed chip, special attention has been dedicated to the chip morphology to verify the existence of the proposed model. With skill and experience, there is always much more to be gained. The recognition of significant features combined with an appreciation of physical metallurgy provides a powerful basis for rationalization and diagnosis.[49]

While metallography remains a versatile tool, modern metallurgy makes use of a wide variety of physical and mechanical properties and interactive radiation measurements in the definition of states of a metal as a polycrystalline. In sober appraisal, there is little room for controversies as to the relative merits of various tools. The fact stands out that no tool is self-sufficient, for each describes different aspects of the nature of a material. Considered separately, these aspects may not be adequate but in joint appraisal may permit unambiguous interpretation. It remains, therefore, to identify properly the limitations as well as the capabilities of metallurgical studies.

Metallography of the chip was first presented by Ernst.[50] Based on the results of his metallographic study, Ernst has identified three basic types of chips found in metal cutting. Although no magnification, sample preparation technique, cutting regime, and workpiece material have been reported, Merchant[4] considered these results as classical examples. Merchant did not provide any explanations as to why these examples did not match his own model shown in the very same paper.[4] Since then, these examples have been used in many of the works on metal cutting and in textbooks.[6]

In his book, Trent[51] presented a great number of microstructures of the formed and partially formed chips to support his qualitative description of the metal cutting process. Unfortunately, they are not related to any particular model of chip formation. As a result, there is no justification for the magnifications selected, chemical treatment of different phases used, and deformation mechanisms.

In the author's opinion, metallography can be considered as a kind of visual art unless one or more of the following conditions is justified:

- The original structure of workpiece material and the chip structure are compared.

- The chip structure is considered at the macro- and micro-levels, depending on the particular phenomenon to be investigated, as each phenomenon can be observed at a certain magnification and requires a different preparation technique for each specimen.
- The structures of the chip and deformation zone are compared with those predicted theoretically to verify the theory used in the analysis.
- A microhardness survey is performed on the structure to reconstruct the distributions of stress and strain gained by different regions of the chip and deformation zone at the last stage of deformation. The obtained results are compared then with those obtained theoretically.

Unfortunately, the known metallographical studies in metal cutting only partially meet these conditions. The reported structures have never been compared with the initial structures of workpiece materials so that it is next to impossible to distinguish the changes produced by the cutting process itself. Previous studies on macrohardness of the chip do not account for the grain location in the chip that may result in inadequate results since the microhardness of the grain boundaries (pearlite) is higher then that of grains (ferrite). Therefore, these boundaries should be avoided in the measurements. To do this, the measurements should be taken with the aid of a special microscope.

In examining the cutting process, it is not satisfactory simply to stop the machine by hand; therefore, a method of quickly arresting this operation is needed. This can be performed with a quick-stop device. Quick-stop devices have become an essential tool for any fundamental research of the cutting process. The quick-stop technique is one in which the velocity of the tool relative to the work piece is rapidly brought to zero, which can be achieved by arresting a machining operation by disengaging the tool and workpiece so quickly that the resulting static situation accurately represents the actual dynamic situation. Detachment of the tool and the workpiece can be achieved by either accelerating the workpiece from the tool or by accelerating the tool from the workpiece.

The preparation steps for microscopy are of great importance, as the true microstructure may be fully obscured by poor technique or execution. Improperly prepared samples can lead to misleading interpretations. Therefore, special attention is paid to preparation of the samples. The preparation of metallographic specimens should be in accordance with the ASTM E 3-86, ASTM E 340-89, and ASTM E 407-87 standards.

The first step in metallographic analysis is to select a sample that represents the material to be studied. In this type of study, three types of specimens are investigated:

1. Samples to study the initial structure of workpiece material. Three samples from each material are selected from the bulk material at certain distances from the surface of the workpieces to avoid the

influence of the primary process (hot rolling) on the samples' microstructure.

2. Samples of partially formed chips obtained using the quick-stop device.
3. Samples of fully formed chips.

Sectioning of a metallographic sample must be performed carefully to avoid altering or destroying the structure of interest. An abrasive cutoff saw has been used to section the sample that is of interest. To minimize burning and deformation, a lubricant of high flow rate has been used in the cutoff operations.

Mounting facilitates the handling of the specimens, especially when it is very small. Mounting involves placing the specimen in a mold and surrounding it with the appropriate powders. Two types of mountings have been used in this study — hot mounting for the initial structure material and cold mounting for the chip specimens. In hot mounting, the mold (a plastic ring) and its content (specimen and Bakelite powder) are heated under pressure to the thermosetting temperature of the content. Once the powder is cured, the thermosetting mounts are removed and cooled to the ambient temperature. Cold mounting is used for the chip, because the pressure and temperature associated with hot mounting may alter the structure of the tested specimens. In cold mounting, the plastic used to bind the chip is set at room temperature. Epoxy resins have been used.

Grinding is an important phase of the sample preparation sequence because the damage introduced by sectioning must be removed in this phase. If sectioning produces extensive damage, then it cannot be removed by grinding. It is usually better to resection the material in an unaffected area with a more gentle cutting method. Special care is taken to minimize mechanical surface damage, especially when it is important to preserve the surface without any alteration.

Grinding is generally performed by the abrasion of the surface of the specimen against water-lubricated abrasive wheels. The specimens have been subjected to mechanical grinding beginning with 120-grit paper and proceeding to 240-, 320-, 400-, 600-, and 800- grit SiC papers. Scratches and damage of the specimen surface from each grit must be removed by the next finer grinding step. Distilled water is used to clean the specimens before being subjected to the next grit.

The surface damage remaining on the specimen after grinding must be removed by polishing. Polishing of the metallographic specimen generally involves rough polishing and fine polishing. The rough polishing has been performed using a kitten ear cloth with 6-μm and 1-μm diamond paste with lubricating oil, while the fine polishing has been performed with 0.3-μm and then 0.05-μm alumina (α-Al_2O_3) with distilled water. Between any two stages, the specimens are tested on the optical microscope to check if the scratches were removed.

Table 4.2 Etchants recommended for workpiece materials

Material	Etchant
AISI 1045	10 mL nital, 90 mL alcohol
AISI 303	40 mL hydrofluoric (HF), 20 mL nitric acid (HNO_3), 40 mL glecren
AISI 4340	10 mL nital, 90 mL alcohol

Etching includes the process used to reveal the microstructure of a metal. Because many microstructural details are not observable on an as-polished specimen, the surface of the specimen must be treated to reveal such structural features as grains, grain boundaries, slip lines, phase boundaries, and microcracks. To obtain sharply delineated contrast conditions, special care is taken during the polishing stage so that the surfaces are free from all artifacts. Next, the samples are throughly cleaned and subjected to a carefully controlled etching process using the appropriate etchants (Table 4.2). After this stage the samples are ready for metallographic tests. Samples prepared in this way have been studied using a Clemex 1024 Vision system having a dedicated image processor. Clemex Vision 2.2 software was used.

Figure 4.12 shows an example of experimentally obtained results. In this example, the cracks formed in Zone 2 appear as black strips which can be clearly distinguished with magnification. As can be seen, long cracks occur along the surface of the maximum combined stress (see Chapter 2) at an early stage of deformation. Then, when the chip moves closer to the tool, the fractured surface becomes less visible due to severe plastic deformation which "heals" the formed cracks. It is clearly seen that the velocity field is very close to that discussed in Chapter 3 (Figure 3.17) and obeys the proposed velocity distribution relationship expressed in Equations (3.30) and (3.31). The foregoing analysis shows that the microstructure of the fully formed chip could only be analyzed when it is very difficult to distinguish the presence of fracture. Because the resistance to tool penetration decreases on fracture, the further plastic deformation of the fractured fragment reflects the build-up of the penetration force (the combined stress in the deformation zone) necessary for the next act of fracture. Between two successive acts of fracture, the chip formation process takes place as described in Chapter 3.

Figure 4.13 shows the microstructure of the partially formed chip when fracture takes place in Zone 1. Here, the black strips representing fracture can be seen to be parallel to the line that separates the workpiece from the layer to be removed. This was observed in cutting of materials having relatively low ductility (bronze) and in cutting of highly ductile materials when seizure occurs at the tool/chip interface (discussed in Chapter 2).

Figure 4.14 shows fracture in Zone 2. Appreciable plastic deformation occurring prior to crack formation shows that the workpiece material is not brittle.

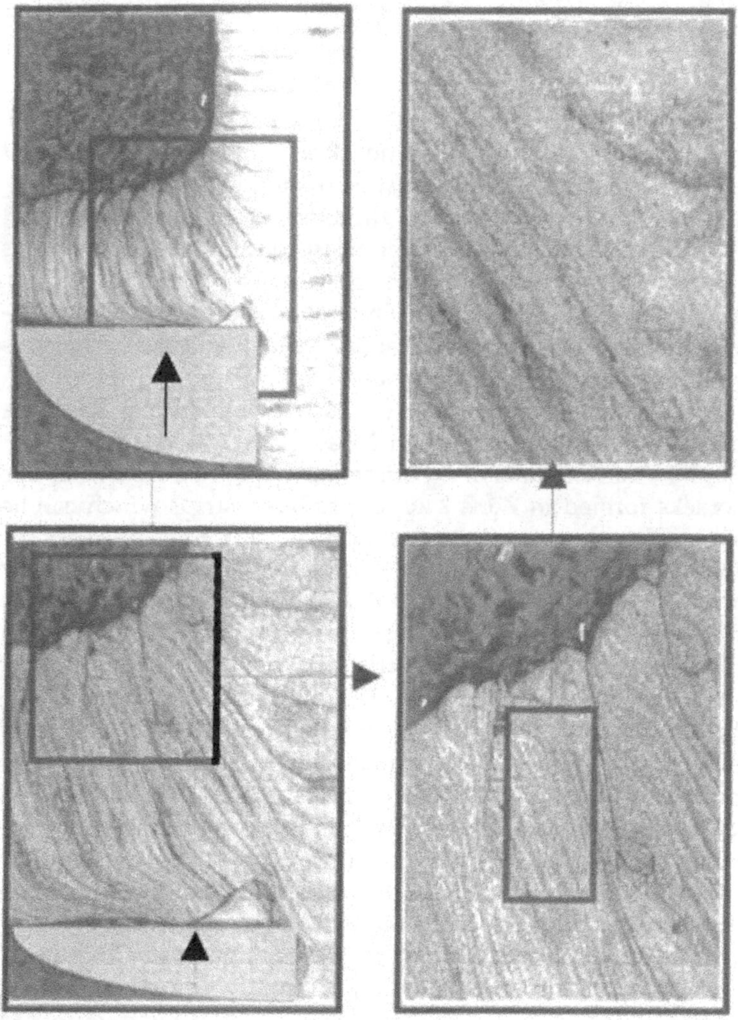

Figure 4.12 Quick-stop micrograph of a partially deformed chip. Workpiece material was steel 4340. Cutting conditions: cutting speed, 75 m/min; feed, 0.1 mm/rev; oil-based cutting fluid, Castrol Almacut 534; cutting tool material, P10; tool geometry, rake angle 0° and flank angle 8°.

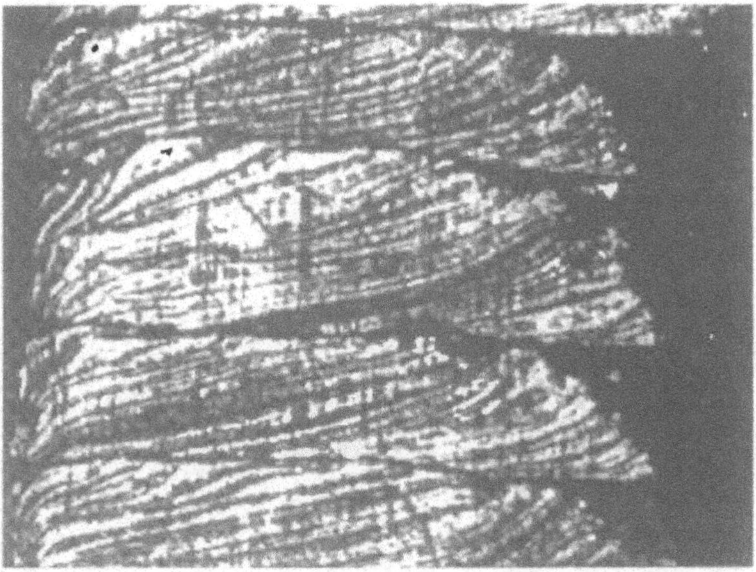

Figure 4.13 Quick-stop micrograph of partially deformed chip. Workpiece material was low carbon high alloy. Cutting conditions: cutting speed, 60 m/min; feed, 0.15 mm/rev; oil-based cutting fluid, Shell Garia H; cutting tool material, M10; tool geometry, rake angle 0° and flank angle 8°.

4.8.3 Type of fracture

As has been pointed out in the above discussion, brittle and ductile fracture may occur in the deformation of ductile materials, such as a mild steel.[2] Therefore, it is important to understand what kind of fracture takes place in metal cutting, because this can then be used. If the strain rate in metal cutting is as high as reported (discussed in Chapter 3), then ductile fracture may not occur as there is no time for microcracks to cut through millions of grains.

It is well known that brittle fractures occur in a transgranular manner, while ductile fracture takes place in an intergranular manner.[2] In the brittle fracture of mild steel, the large microcracks observed in the ferrite grains are invariably associated with fractured carbide particles located somewhere in the grain or in the surrounding grain boundary.[2] Fracture of the carbide particle by the stress field of a pile-up is an essential intermediate event between the formation of a dislocation pile-up and cleavage of the ferrite. The formation of a crack in the carbide phase can initiate cleavage fracture in the adjacent ferrite phase if the local stress is sufficiently high and the time necessary for the stress relaxation is insufficient.

To distinguish what kind of fracture takes place in metal cutting of ductile materials, a special test has been carried out using the difractometer

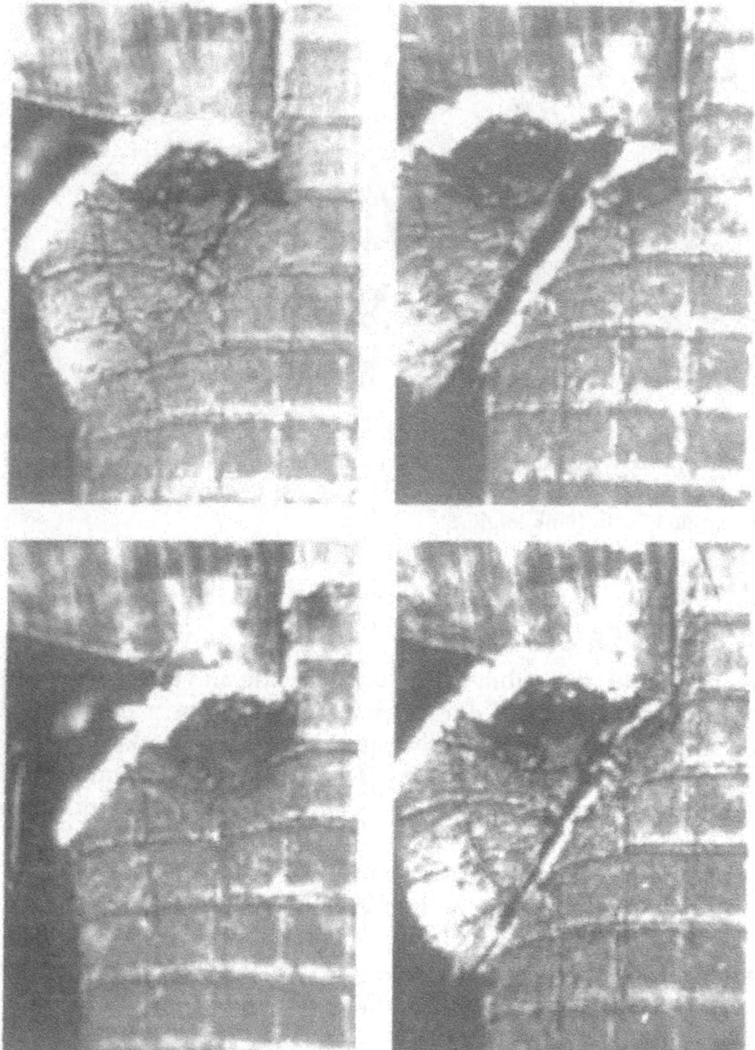

Figure 4.14 Quick-stop micrographs of partially formed chip. Workpiece material was lead bronze, with orthogonal cutting on a shaper. Cutting conditions: cutting speed, 35 m/min; uncut chip thickness, 2.2 mm; dry cutting; cutting tool material, P10; tool geometry, rake angle, 15° and flank angle 8°.

(X-ray) technique.[52,53] Among the principal uses of the X-ray method, the following are of importance in metal cutting studies:

- Determination of the degree of preferred orientation and crystalline structure
- Measurement of certain physical characteristics, such as small crystalline sizes, strain, perfection, lattice disorder, and damage

In the procedure described here, the degree of preferred orientation is used to characterize the type of fracture. This parameter has been determined with a pole-figure device attached to the difractometer to measure intensities for various specimen orientations. The intensities are plotted on a stereographic projection and countered to show the normals to selected lattice planes in various crystallographic directions. Such plots are used to follow structural changes in the chip as compared to the workpiece structure.

The following should be considered prior to the measurements.[54,55] In cutting ductile materials, the plastic deformation in the chip takes place by shearing under the action of the combined stress. The shearing occurs along the surface of maximum combined stress that may be approximated by a plane. The sliding surface does not exist throughout the entire cycle of chip formation; rather, it forms at its end as the result of stress redistribution in this cycle so that the formed chip has a certain structure characterized by the angle of chip structure, as discussed in Chapter 3. Therefore, to determine the type of fracture, it is sufficient to compare the size and orientation of crystals by their slip plane in the original material and in the chip.

The experiments were carried out for different groups of steels (plain low carbon, plane medium carbon, plane high carbon, low alloys, stainless, etc.), nickel- and chromium-based high alloys, molybdenum, aluminum, and high titanium alloys representing all the different crystallographic structures. In all the considered cases, similar results have been obtained, and the results obtained for a chromium-nickel-based high alloy are presented.

Figure 4.15 shows difractometer traces of the alloy. Figure 4.15a shows difractometer traces for the initial material where crystals are randomly orientated, and Figure 4.15b shows difractometer traces for a chip which is a kind of textured material. The chip was obtained in a cutting test conducted using the following regime: cutting speed, 0.1 m/s; cutting feed, 0.1 mm/rev; depth of cut, 5 mm; no coolant. The results of Figures 4.15a,b may be interpreted as follows. The pick intensity corresponding to the slip plane (101) for the chip is significantly smaller for the chip than for the matrix material. As the picks in Figure 4.15 are recommended to be considered as the random intensity,[53] they can be approximated by means of the normal density function:

$$f(x)=\frac{1}{\sqrt{2\pi}\,\sigma_x}\exp\left[-\frac{(a-x)^2}{2\sigma_x^2}\right] \tag{4.34}$$

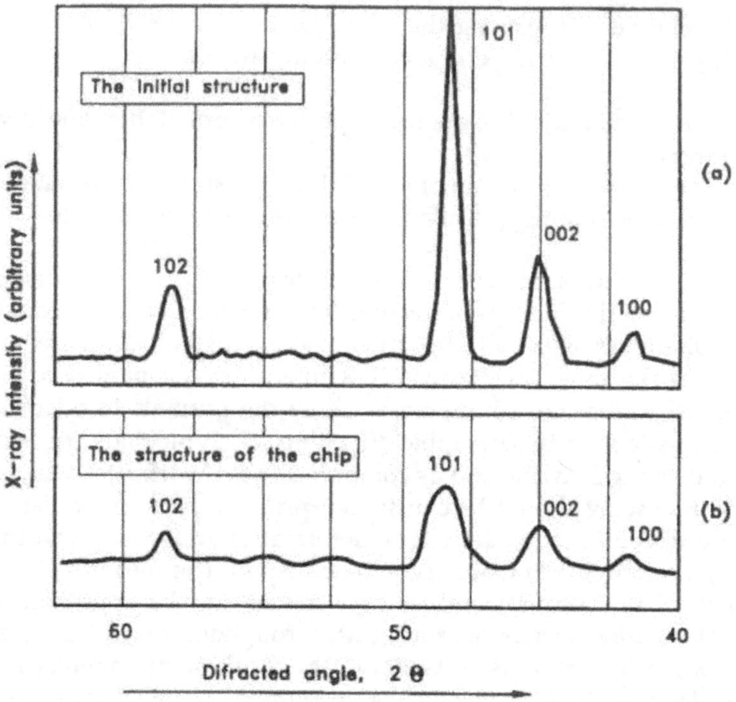

Figure 4.15 Difractometer traces of a high chromium-nickel-based alloy: **(a)** the matrix material, and **(b)** the materials of the chip.

Therefore, the ratio of the height of the picks for the matrix (initial) material and the chip defines the texturing level of the crystals in the chip. The ratio of their average widths defines the grain refining in chip formation. The results of the calculations show that the texturing of crystals in the chip is five times higher than that in the initial structure, and the average size of crystals in the initial structure is 25 times greater than that in the chip.

Because the obtained result is of prime importance, it has been verified using scanning electron microscopy. Figure 4.16 shows fractography of dimpled rapture of a high chromium-nickel-based alloy. The ductile character of fracture has been confirmed by dimensions of the cup-like depressions of 3 to 100 μm. The visible centers at the bottom of the cups indicate the formation of microvoids at the stress concentrators. The wavy cup walls indicate that void growth is accomplished by a process of void coalescence — that is, by elongation of the voids and elongation of the bridges of material between the voids.

The experimental results prove that ductile fracture takes place in metal cutting. Therefore, the following conclusions may be drawn from the results of the workpiece material considerations:

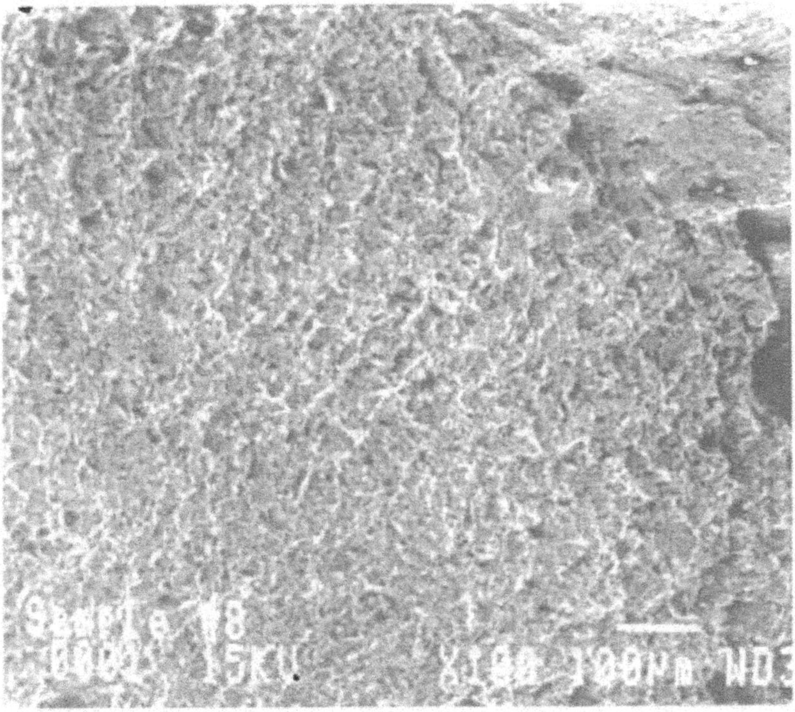

Figure 4.16 Scanning electron microscope view illustrates the definite appearance of a ductile fracture surface. (Courtesy Ms. Priti Wanjara, Department of Metallurgical Engineering, McGill University, Montreal, Canada.)

1. Ductile fracture mechanism and its correlation with the workpiece properties and cutting parameters should be understood in order to predict metal cutting performance.
2. The most important conclusion from this finding regards the state of stress. As ductile fracture occurs exclusively under a combined state of stress,[2] the metal cutting process and thus the chip formation process should be considered as taking place under a combined state of stress, as discussed in Chapters 2 and 3.
3. The results obtained support the conclusion made in Chapter 2 that the strain rate in metal cutting is much smaller than reported in previous studies.

4.9 Fracture strain determination

The analysis above shows that complex relationships exist between the fracture strain, on the one side, and the mechanical and metallurgical properties of the material under deformation, the mechanical system involved, the state

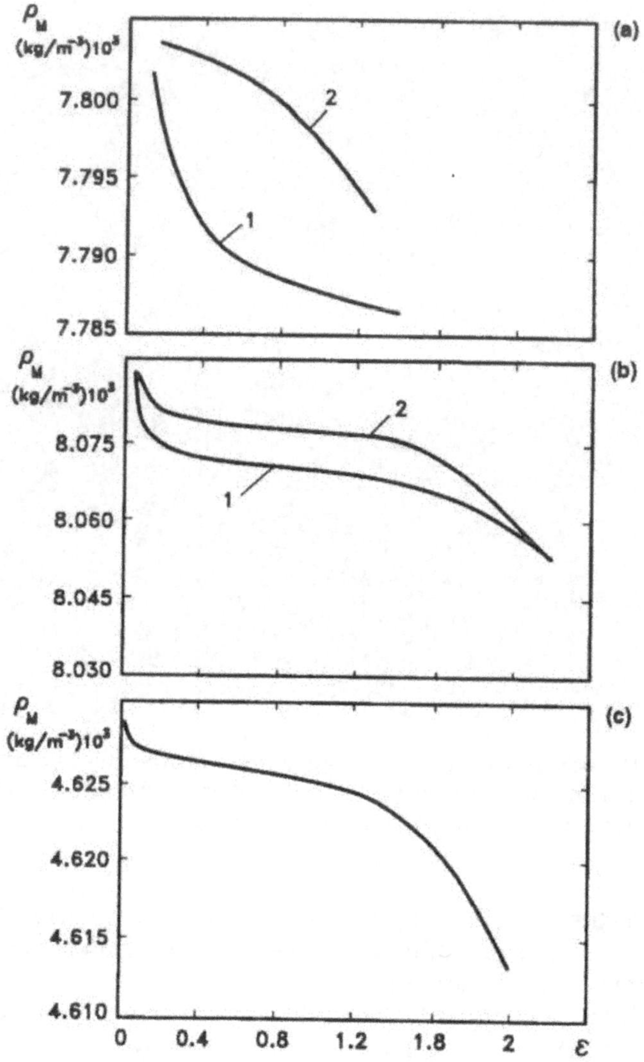

Figure 4.17 Density of material as a function of plastic strain gained in deep drawing (1) and flat-rolling (2) for: **(a)** low alloy steel (0.16% C, 1% Cr, 1% Ni, 1% Si); **(b)** high alloy steel (0.05% C, 18% Cr, 9% Co, 5% Mo, 1% Ti); **(c)** titanium high alloy.

of stress, the strain rate, and the process temperature, from the other side. As has been pointed out above, the mathematical theory of plasticity is phenomenological in nature and attempts to formalize experimental observations of the macroscopic behavior of a plastically deforming solid in a uniform state of complex stress, but its methods cannot be applied to reveal such relationships. The physical explanation of the elastic and plastic properties of metals, from a microscopic viewpoint, in relation to its crystal structure, is the subject of material science. The latter, however, pays little attention to the influence of the state of stress in a plastically deforming solid as well as to other macroparameters of deforming processes. Therefore, it is only logical to assume that further progress in the predictability of mechanical properties of materials in cutting can be achieved by merging the two approaches into a unified theory of plasticity. An attempt has been made in this section toward that direction.

4.9.1 Existent criteria of ductile fracture

Mathematical relationships between the stress and strain at fracture have been developed from several points of view. The various approaches are based on phenomenological concepts, micromechanisms of ductile fracture, and upper-bound methods. Several criteria of fracture are based on the well-known effect of hydrostatic pressure in suppression of ductile fracture and of tensile stress in promoting fracture.[56] Hoffmanner[57] proposed an exponential relationship between the effective strain at fracture ε_{ef} and the reduced average tensile stress transverse to the inclusion alignment of the material:

$$\varepsilon_{ef} = a_{fr} \exp\left(b_{fr} \frac{\sigma_t}{\vartheta}\right) \tag{4.35}$$

where σ_t is the average stress perpendicular to the fiber; ϑ is the effective stress; a_{fr} and b_{fr} are constants determined experimentally for the material of interest. The stress, σ_t, is determined through analysis of the process in question.

A second example is the criterion of Cockcroft and Latham,[58] which suggests that fracture occurs when the tensile strain energy reaches a critical value:

$$\int_0^{\varepsilon_{ef}} \sigma^* d\varepsilon_e = C_M \tag{4.36}$$

where σ^* is the maximum tensile stress, and C_M is the constant for the material at a given temperature and strain rate, determined from a simple tension test.

It is generally agreed that the most important structural features influencing ductile fracture are inhomogeneities such as inclusions and precipitates which are sites of void formation so that the voids grow and join along bands of local shear-strain concentration between particles. The overall effect of particles on the ductility of copper has been demonstrated clearly by Edelson and Baldwin.[59] An important start toward developing an analytical treatment of ductile fracture has been made by McLintock,[2] who used a model consisting of cylindrical holes initially of radius b_0 and average spacing l_0 to obtain the strain at fracture in the following form:

$$\varepsilon_f = \frac{(1-m)\ln(l_0/2b_0)}{\sinh\left[(1-m)(\sigma_a-\sigma_b)/\left(2\sigma_{Ty}/\sqrt{3}\right)\right]} \tag{4.37}$$

for a material with a stress-strain curve given by $\sigma = K_\sigma\varepsilon^m$. In this equation, σ_a and σ_b are the stresses parallel and perpendicular to the axis of the cylindrical holes, respectively, and σ_{Ty} is the true stress flow. While Equation (4.37) is not in close agreement with the limited data available, it certainly predicts the proper variation of fracture strain with the important variables. Equation (4.37) indicates that the fracture strain decreases as the void fraction increases, as the strain hardening exponent m decreases, and as the stress state changes from uniaxial tension to triaxial tension.

In sheet-forming processes, failure usually occurs due to localized thinning and fracture in locations where the stress and strain conditions are critical. An analysis of this localized thinning process, starting at the site of an inhomogeneity in the material, was presented by Marciniak and Kuczynski.[60] From their results, it was possible to conduct a theoretical map of the limiting strains in various sheet-forming processes.

Based on the observation that surface cracking in bulk deformation processes (e.g., forging, rolling, extrusion) is similarly preceded by localized thinning, the analysis by Marciniak and Kuczynski was extended by Lee and Kuhn[61] to include surface fracture. As before, through such analysis it is possible to construct theoretical maps of limiting strains at fracture for a variety of deformation processes. Such maps may be considered as a representation of function $f_2(C_f)$ (Equation (4.33)) and can be constructed to reflect the influence of material properties such as work-hardening rate, strain rate sensitivity, and inclusion content.

A different approach to the formulation of a fracture criterion is based on the upper-bound concept.[62] Flow-velocity fields for the process of interest are formulated along with modifications of the flow field to simulate fracture. The prevailing flow field is that for which the energy associated with the fracture flow field is less than that for sound flow. This approach may be considered as a representation of function $f_1(C_{def})$ (Equation (4.33)).

An attempt to apply fracture mechanics to estimate the fracture of ductile materials has been made by Ravichandran and Vasudevan.[34] It is admitted that the basic fracture relationships are strictly applicable to brittle materials in which the energy dissipation due to plastic deformation is almost negligible. Many structural materials show evidence of plastic deformation and have fracture toughness levels higher than those that can be estimated from the surface energy alone. The authors proposed a modified form of the

Griffith equation for the surface energy G_c to account for this additional contribution to fracture resistance due to elastic deformation as:

$$G_c = 2\gamma_{pf} + \Gamma\sigma_y\delta_c \tag{4.38}$$

or, in terms of fracture toughness:

$$K_c = \sqrt{2E\gamma_{pf} + \Gamma\sigma_y\delta_c} \tag{4.39}$$

where E is Young's modulus, γ_{pf} is the plastic work required to extend the crack wall, Γ is a constant, σ_y is the material yield strength, and δ_c is the critical opening displacement at the crack tip at the onset of fracture. The first term is the energy consumed in the creation of two fracture surfaces and is considered to be independent of microstructure. The second term, $\sigma_y\delta_c$, approximately represents the energy consumed in plastic deformation accompanying fracture, a strong function of microstructure. Because the latter is several times higher than the former, the surface energy term is often ignored in the case of metallic structural materials.

Although the provided explanation is clear, fracture toughness values for a given material should be obtained by testing pre-cracked compact tension or three-point bend specimens following the ASME E399 test procedure or by impact fracture testing. Therefore, no prediction can be made on the basis of Equation (4.39). Moreover, it is not clear why one should believe that the results obtained from a toughness test will be applicable in real working conditions where the states of stress, shape of the part, strain rate, working temperature, etc. may differ from those used in the test. No experimental evidence to support such a claim has been provided.

As a short conclusion for this section, it is admitted that there is no theoretical relationships which can be used to predict the strain at fracture accounting for the real process conditions as per Equation (4.33).

4.9.2 *Basic starting concepts for the analysis of ductile fracture*

An analysis of a great body of works on plastic fracture[2,36–40,56–67] results in a list of the following starting points that can be used as fundamentals for further analysis.

Ductile fracture is preceded by plastic deformation which may be considered as the preparatory stage for fracture occurrence. Therefore, ductile fracture in metal cutting cannot occur without an appreciable amount of plastic deformation. Plastic deformation, considered in such context, includes initial, main, and final stages which can be easily distinguished in the micrographs of the deformed metals.

The initial stage (up to 5%), which can be referred to as the easy glade region, begins with a heterogeneous distribution of low-density dislocations which originate from limited sources and then, with increasing applied load, move along their slip planes with little interference from other dislocations. Most of the formed slip planes terminate within a single crystal (a grain in micrographs), and, as can be observed in micrographs, the traces of these planes have gradually vanishing ends that show the plastic stress relaxation. As such, there is no high stress concentration (no dislocation pile-ups) within the deformed metal and, as a result, there are not many microcracks formed in this stage. The difference in the shape of the slip lines (straight or wavy) indicates different mechanisms of stress relaxation.

In the main stage (>5%), the deformation increases the density of dislocations, creates jogged segments, and forms twins and other regions stress-concentration places. The dislocation motion becomes inhomogeneous, accompanied by extensive formations of dislocation pile-ups as strong obstacles as Cottrel-Lomer locks. Dislocations frequently pile up on slip planes at barriers such as grain boundaries, second phases, or sessile dislocations. The leading dislocation in the pile-up is acted on not only by the applied shear stress but also by interaction forces with other dislocations in the pileup. This leads to a high concentration of stress on the leading dislocation in the pileup. When many dislocations are contained in the pile-up, the stress in the dislocation at the head of a pile-up can approach the theoretical stress of the crystal. This high stress can either initiate yielding on the other side of the barrier or, in other instances, nucleate a crack at the barrier. Dislocations piled up against a barrier produce a back stress acting to oppose the motion of additional dislocations along the slip line in the slip direction. The number of dislocations which can be supported by an obstacle will depend on the type of barrier, the orientation relationship between the slip plane and the structural features of the barrier, the material, and the temperature. The breakdown of a barrier can occur by slip on a new plane, climb of dislocations around the barrier, or generation of high enough tensile stress to produce a crack. Crack formation can be considered as the very beginning of fracture.

Local zones of high stress slow down plastic deformation. As such, in zones of dimensions of 1 to 10 μm, the local stress may be reduced due to secondary plastic relaxation, dislocation climb, activation of neighboring Frank-Read sources, etc. All this results in a reduction of the number of dislocations in the considered zone (up to 4 to 5), delay of new crack formation, and deceleration of existent crack development. Cracks, twins, and slip lines in the relaxed state can be considered to be in a "frozen" state. However, if a local zone of high stress is small (less than 1 μm), the plastic relaxation there may not occur due to an insufficient number of neighboring Frank-Read sources which results in a microcrack explosion. The formed microcracks have dimensions of ~1 μm and porous appearance with a length/width ratio of 2. This has been confirmed by the theoretical and experimental results for pure iron, steel and steel alloys, aluminum, and copper. The discussed microcrack formation may be considered as stress relaxation necessary to continue plastic deformation.

The slip-line density increases with strain. Compared to the initial density, the slip-line density for copper deformed to 21% increases 1.5 times (from 390 to 580 mm/mm^2); titanium deformed, 1.9 times (from 320 to 600 mm/mm^2); and iron at a temperature of 300°C in the interval of deformation from 6 to 13.3%, 3.2 times (from 220 to 700 mm/mm^2).

Temperature of the test T has significant effect on plastic deformation when $T > (0.5 \text{ to } 0.7)T_m$. At such temperatures, most of the metals undergo the transition from transgranular fracture to intergranular fracture. When transgranular fracture occurs, the slip planes are weaker than the grain boundaries, while for intergranular fracture the grain boundary is the weaker component. The temperature at which the grains and grain boundaries have equal strength is defined as the equicoheresive temperature (known as ECT in metallurgy). It is of prime importance to note here that the ECT is not fixed but varies significantly with the state of stress and strain rate. Decreasing the strain rate lowers the ECT and therefore increases the tendency for intergranular fracture. The effect of the strain rate on the strength-temperature relationship is believed to be much higher for the grain-boundary strength than for the strength of the grains. Because the amount of grain-boundary area decreases with increasing grain size, a material with a large grain size will have higher strength above the ECT than a fine-grain material. Below the ECT the reverse is true.

The mechanism of plastic deformation appears to be also the mechanism of plastic stress relaxation in the local zones of high stress concentration (for example, deformation bands, crack tips) — that is, in the zones of high likelihood of crack nucleation which may be thought of as the origin of fracture. Stress relaxation and strain hardening take place simultaneously, both caused by the applied stress.

It is generally recognized that the initiation of voids in the fracture of ductile materials begins in a pile-up of edge dislocations, and the most essential fact is that it may start even when two dislocation meet each other. Therefore, at this stage, plastic deformation and fracture complement each other so that they may be considered as one thermoactivation process having a common activation energy. Careful metallographic studies of ductile fracture show that deformation is governed mainly by the shear stress, while fracture is caused by the normal stress. It is understood that these two are related by the actual state of stress.

Pre-existent defects and impurities of different size (Table 4.1) can be considered as the origins of fracture. Among these, submicro- and microcracks have the greatest contribution to ductile fracture.

The ductile fracture process can be separated into three regimes: (1) initiation of microcracks, (2) microcrack concentration up to a critical level, and (3) formation and growth of macrocracks or mainstream cracks. It is important to note that the first and second stages are reversible — microcracks formed at these stages can be "healed" by dispersion of vacancies even at temperatures less then half of the melting point if the stress is released and the specimen is left unstressed for some time. To facilitate the relaxation, hydrostatic compression or tempering can be used.

The extent of these stages depends on the applied stress conditions, specimen geometry, flaw size, and the mechanical properties of the material. The problem in metal cutting is to reduce the extent of the second stage (which has the chief effect on the fracture strain); thus, this stage should be considered in detail.

At the second stage, neither microcrack "healing" nor growth is taking place. Instead, multiplication of the microcrack's population is the case, until their ever-growing concentration leads to a reduction in the density of the material on $\Delta\rho/\rho_{M0} = 10^{-2}$ to $10^{-3} \approx 0.1$ to 1.0% ($\Delta\rho$ is the reduction in the density of the material; ρ_{M0} is the initial density of the material). This reduction is suggested to be considered as the main physical criterion of fracture of polycrystalline ductile materials. Figure 4.17 shows the effect of strain on the density of specimens made of a low-alloy steel, high-alloy steel, and titanium high alloy. As seen, the reduction in the density of steels lies in the range of 0.35 to 0.41%, while for the titanium alloy it is 0.11 to 0.15%.

The third stage of ductile fracture begins with a high concentration of microcrack multiplication at certain local areas or cross-sections of the specimen where the density of the material reduced by 0.1 to 1.0% (a fair agreement with the measured density of the deformed materials). As such, a number of the microcracks (approximately 10) coalesce to form a mainstream macrocrack; therefore, a strong correlation exists between the reduction in the material density and its fracture.

Plastic deformation has different affects on the various stages of ductile fracture. During the first and second stages, plastic deformation facilitates fracture by promoting the process of void nucleation, growth, and coalescence. Strain-hardening energy increases the energy expedient of fracture by converting "ductile" cracks into "elastic" or "almost elastic" cracks which may propagate under relatively low stresses. In contrast, the third stage of plastic deformation slows down the development of microcracks by rounding their tips and lowering high tip local stresses σ_L so that $\sigma_L < \sigma_c$, σ_c being the cohesive strength of material. As a result, the mainstream crack forms almost abruptly, in its tip $\sigma_L \to \sigma_c$, and the rate of its growth reaches the velocity of sound.

4.9.3 *Relationship between strain at fracture of a material and parameters of current plastic deformation*

As has been pointed out in the previous analysis, plastic deformation and plastic stress relaxation take place simultaneously. Therefore, we may assume that during a certain time interval, the total change in the number of microcracks (or, in general, the defected areas) ΔN (or the change in the density $\Delta\rho$) in a material undergoing plastic deformation may be thought of as consisting of the number of microcracks which may form during the considered interval due to the plastic deformation ΔN_1 (the reduction in the density $-\Delta\rho_1$) and the number of microcracks ΔN_2 which may be "closed" due to stress relaxation (the increase in the density $+\Delta\rho_2$); that is,

$$\Delta N = \Delta N_1 - \Delta N_2 \quad and \quad \Delta\rho = -\Delta\rho_1 + \Delta\rho_2 \tag{4.40}$$

When $\Delta N = 0$ ($\Delta\rho = 0$), there is no increase in the number of microcracks. Two basically different cases are possible here. The first one takes place when the initial number of microcracks is equal to their critical number or, in other words, when the initial density is equal to the critical density ($N_0 = N_{cr}$ or $\rho_0 = \rho_{cr}$). As such, the local normal stresses σ_{loc} are high so that $\sigma_{loc} = \sigma_{ch}$ (σ_{ch} is the theoretical cohesive strength[2]) that results in brittle fracture. The second case takes place when $N_0 = const << N_{cr}$ and $\rho_{0M} >> \rho_{cr}$ and $\sigma_{loc} < \sigma_{ch}$ which results in ideal plastic flow or superplasticity.

Generally, for real metals $\Delta N \neq 0$ and $\Delta\rho \neq 0$, or more accurately $\Delta N > 0$ and $\Delta\rho <> 0$, as, with the beginning of plastic shearing, that is when the strain-hardening coefficient becomes $A > 0$, and the high local stress due to the plastic deformation promotes the formation of microcracks. It is important to note here that $\sigma_{loc} \rightarrow (0.8 \text{ to } 1.0)\sigma_{ch}$ from the very beginning of the plastic deformation. For instance, microcracks have been observed in nickel under 2% of its plastic deformation and not much later for other metals.[37]

According to the results of Scudnov,[37] the terms of Equation (4.42) may be represented through the deformation process and material parameters as:

$$\Delta N_1 = \varepsilon_r\, N\, \Delta\varepsilon_i = a_1\, N\, \Delta\varepsilon_i\, \varphi(M) \tag{4.41}$$

$$\Delta N_2 = a_2\, N\, \Delta t \tag{4.42}$$

In these equations, N is the current number of places of microcrack nucleation; $\Delta\varepsilon_i$ is the current plastic strain; $\varphi(M)$ is a function accounting for the influence of the mechanical system (M) of deformation (state of stress and deformation) on microcrack nucleation.[37] For example, $\varphi(M) = 1$ for simple shearing when the hydrostatic stress, Lode stress, strain parameters — Equations (4.24) and (4.14) — and Π-factor (Equation (4.19)) are equal to zero; Δt is the period of time of the deformation. $\Delta t = \Delta\dot{\varepsilon}$ where $\dot{\varepsilon}$ is the rate of strain defined experimentally; a_1 and a_2 are constants; $\varepsilon_r = a_1\, \varphi(M)$ is the reduction in the density of the material under the mechanical system (M) of deformation. When $\varphi(M) = 1$, $\varepsilon_r = a_1\, \Delta N/(N\, \Delta\varepsilon_i)$ or, equivalently, $\varepsilon_r = \Delta\rho/(\rho_M\, \Delta\varepsilon_i)$, as the increase in ΔN_1 leads to the corresponding reduction in the density $-\Delta\rho_1$, and a decrease in ΔN_2 leads to the corresponding increase in the density $+\Delta\rho_2$.

When a real deformation process is considered, ε_r, besides the discussed influence of the mechanical system of deformation, depends also on a number of properties of a real workpiece material such as its initial structure, the effective number of the zones of microcrack nucleation, and the probability of microcrack formation in these zones. This may be represented as:[37]

$$\varepsilon_r = a_1\,\varphi(M) = (K_1 + K_2\,\varepsilon_i)\varphi(M) = (K_1 + K_2\,\varepsilon_i)P_0(ms,\, cc)\exp[\varphi\,(\mu_L,\, v_L)\Pi] \tag{4.43}$$

where K_1 is a coefficient accounting for the initial number, shape, and distribution of the zones of possible microcrack nucleation in the workpiece, $0 \le K_1 \le 1$; K_2 is the coefficient accounting for the influence of the current deformation on the initial number, shape, and distribution of the zones of possible microcrack nucleation in the workpiece; $0 \le K_2 \le 1$; $P_0(mf, cc)$ is the effective number of the zones of possible microcrack nucleation in the workpiece as dependent on the morphology of the initial structure *mf* and chemical composition of the material *cc*. $P_0(mf, cc)$ can be approximated by the known strain hardening exponent for a material.[37]

Equations (4.41) to (4.43) have been solved simultaneously[37] to yield the expression for the density of the material when strain changes from ε_0 to ε_i with a rate of $\dot{\varepsilon}$:

$$\rho_{wi} = \rho_{mo}\exp\left(\left[\frac{a^{\prime\prime}}{\dot{\varepsilon}} - a_1\,K_1\,\varphi(M)\right](\varepsilon_i - \varepsilon_0) - K_2\,\frac{a_1}{2}\,\varphi(M)\left(\varepsilon_i^2 - \varepsilon_0^2\right)\right) \tag{4.44}$$

where:

$$a^{\prime\prime} \approx \varepsilon\exp(1/m_m) \tag{4.45}$$

is a term showing the velocity of "healing" of microdefects in the material during deformation; m_m is the strain-hardening exponent for the material.

At fracture, the density of workpiece material reaches its critical value $\rho_{M0} + \Delta\rho = \rho_{cr}$ and strain reaches the fracture strain at $\varepsilon_i = \varepsilon_f$. Therefore, Equation (4.45) can be solved relative to the strain at fracture to obtain the relationships for this strain for every particular loading conditions defined by the initial conditions of workpiece material, tool geometry, depth of cut, and cutting regime.

When $K_1 = 0$, $K_2 = 0$,

$$\varepsilon_f = \varepsilon_0 + \frac{\ln(\rho_{OM})/\rho_{cr}}{P_0\,(ms,\, cc)e^{\varphi(\mu_L,\, v_L)\Pi} - a^{\prime\prime}/\dot{\varepsilon}} \tag{4.46}$$

When $K_1 = 0$, $K_2 = 1{,}0$,

$$\varepsilon_f = \frac{\frac{a''}{\varepsilon} + \left[\left(\frac{a''}{\varepsilon}\right)^2 - P_0(ms,\ cc)\, e^{\varphi(\mu_L,\, \nu_L)\Pi}\left(\ln\frac{\rho_{OM}}{\rho_{cr}} + C'\right)\right]^{1/2}}{0.5 P_0(ms,\ cc)\, e^{\varphi(\mu_L,\, \nu_L)\Pi}} \tag{4.47}$$

where:

$$C' = \left[\frac{3}{2} P_0(ms,\ cc) e^{\varphi(\mu_L,\, \nu_L)\Pi} - a'' / \dot{\varepsilon}\right]\varepsilon_0 \tag{4.48}$$

and:

$$P_0(ms,\ cc) \approx \frac{HB}{\sigma_{ult}} - 2 \tag{4.49}$$

When $K_1 = 1{,}0$, $K_2 = 1{,}0$:

$$\varepsilon_f = \frac{1}{\frac{a''}{\dot{\varepsilon}} P_0(ms,\ cc) e^{\varphi(\mu_L,\, \nu_L)\Pi}} - 1 \pm \Delta\rho \tag{4.50}$$

An analysis of Equations (4.46) to (4.50) reveals the following:

1. If the initial density of workpiece material is as low as $\rho_{OM} = \rho_{cr}$ (i.e., when $N_0 = N_{cr}$), the numerator of Equation (4.46) is equal to zero, thus the material is brittle and fractures within the elastic zone. The strain at fracture is $\varepsilon_f = \sigma_{ult}/E$ and cannot be changed by the parameters of the cutting system.
2. If the initial density of workpiece material is higher than critical, the numerator of Equation (4.46) becomes greater than zero; therefore, the workpiece material is ductile when $\varepsilon_f >> \varepsilon_0$. In this case, the strain at fracture ε_f varies over a broad range depending on the parameters of the cutting system. This possibility offers a way to optimize the cutting process by minimizing ε_f, thus minimizing power consumption, cutting forces, tool wear, etc.
3. The easiest way to control ε_f is by the Π-factor (Equation (4.19)) which depends entirely on the geometry of the cutting tool and the cutting regime used. As such, the variation of the state of stress from $-\Pi_{cr}$ to $+\Pi_{cr}$ (Π_{cr} is its critical value at fracture) leads to a decrease of Π_f under any temperature of deformation and strain rate.

4. With changing the Lode stress and strain parameters — μ_L in Equation (4.24) and υ_L Equation (4.32) — over the range from –1 to +1, the strain at fracture reduces. In addition, it was found that the values $\pm\, \Pi_{\sigma}$ reduce significantly.
5. With increasing hydrostatic pressure, the probability of fracture decreases and the "healing" of microdefects takes place, especially noticeable in materials of complicated structures having a number of possible regions of local stress concentration in their structures. Therefore, the strain at fracture increases significantly and explains the known difficulties observed in cutting of high-strength alloys.
6. With increasing initial strain ε_0, the strain at fracture reduces significantly. This conclusion is supported by multiple known methods of machining of high alloys using pre-cold working of the workpiece to be machined.
7. With increasing strain rate, the strain at fracture reduces hyperbolically, and this reduction may be enhanced by high temperature. In the author's opinion, this explains the efficiency of cutting with preheating the workpiece.

References

1. Johnson, W. and Mellor, P.B., *Engineering Plasticity*, John Wiley & Sons, New York,1983.
2. Dieter, G.E., *Mechanical Metallurgy*, 3rd ed., McGraw-Hill, New York, 1986.
3. Honeycombe, R.W.K., *The Plastic Deformation of Metals*, 2nd. ed., Edward Arnold, London, 1984.
4. Merchant, M.E., Mechanics of the metal cutting process, *J. Appl. Phys.*, 16, 267, 1945.
5. Shaw, M.C., *Metal Cutting Principles*, Clarendon Press, Oxford, 1984, p. 200.
6. Boothroyd, G. and Knight, W.A., *Fundamentals of Machining and Machine Tools*, 2nd ed., Marcel Dekker, New York, 1989.
7. Zorev, N.N., *Metal Cutting Mechanics*, Pergamon Press, Oxford, 1966.
8. Stephenson, D.A. and Agapiou, J.S., *Metal Cutting Theory and Practice*, Marcel Dekker, New York, 1977.
9. King, R.I., Ed., *Handbook of High-Speed Machining Technology*, Chapman & Hall, New York, 1985.
10. Hill, R., *The Mathematical Theory of Plasticity*, Oxford University Press, London, 1950.
11. Rubenstein, S., A note concerning the inadmissibility of applying the minimum work principle to metal cutting, *ASME J. Eng. Industry*, 105, 294, 1983.
12. Dewhurst, W., On the non-uniqueness of the machining process, *Proc. Roy. Soc. Lond. A*, 360, 587, 1978.
13. Usui, E., Progress of "predictive" theories in metal cutting, *JSME Int. J. Series III*, 31(2), 363, 1988.
14. Kobayashi, S. and Thomsen, E.G., Metal-cutting analysis-I: re-evaluation and new method of presentation of theories, *ASME J. Eng. Industry*, 83, 63, 1962.

15. Thé, J.H.L. and Scrutton, R.F., The stress-state in the shear zone during steady state machining, *ASME J. Eng. Industry*, 101, 211,1979.
16. Zhang, B. and Bagchi, A, Finite element simulation of chip formation and comparison with machining experiment, *ASME J. Eng. Industry*, 116, 289, 1994.
17. von Turkovich, B.F., Shear stress in metal cutting, *ASME J. Eng. Industry*, 92, 151, 1970.
18. Kececioglu, D., Shear-zone size, compressive stress, and shear strain in metal-cutting and their effects on mean shear-flow stress, *ASME J. Eng. Industry*, 81, 79, 1960.
19. Oxley, P.L.B., *Mechanics of Machining: An Analytical Approach to Assessing Machinability*, John Wiley & Sons, New York, 1989.
20. Ocusima, K. and Hitomi, K., An analysis of the mechanism of orthogonal cutting and its application to discontinuous chip formation, *ASME J. Eng. Industry*, 82, 545, 1961.
21. Stevenson, R. and Stephenson, D.A., The mechanical behavior of zinc during machining, *ASME J. Eng. Mater. Technol.*, 117, 173, 1995.
22. Kececioglu, D., Shear strain rate in metal cutting and its effects on shear-flow stress, *ASME J. Eng. Industry*, 77, 158, 1956.
23. Nakayama, K., Studies on the mechanism in metal cutting, *Bull. Fac. Eng. Yokohama Nat. Univ.*, 5, 1, 1959.
24. Palmer, W.B. and Oxley, P.L.B., Mechanics of metal cutting, *Proc. Inst. Mech. Eng.*, 173, 623, 1959.
25. Spaans, C., A treatise of the streamlines and the stress, strain, and strain rate distributions, and on stability in the primary shear zone in metal cutting, *ASME J. Eng. Industry*, 94, 690, 1972.
26. Klamensky, B.E. and Kim, S., On the plane stress to plane strain transition across the shear zone in metal cutting, *ASME J. Eng. Industry*, 110, 322, 1988.
27. Shih, A.J., Finite element simulation of orthogonal metal cutting, *ASME J. Eng. Industry*, 117, 84, 1995.
28. Song, X., Strain-hardening and thermal-softening effects on shear angle prediction: new model development and validation, *ASME J. Eng. Industry*, 117, 28, 1995.
29. Lei, S., Shin, Y.S., and Incopera, F.P., Material constitutive modeling under high strain rates and temperatures through orthogonal machining tests, in *Manufacturing Science and Engineering*, Vol. 6(2), Proc. 1997 American Society of Mechanical Engineers International Mechanical Engineering Congress and Exposition, 1997, p. 91.
30. DeGarmo, E.P., Black, J.T., and Kohser, R.A., *Materials and Processes in Manufacturing*, 7th ed., Macmillan, New York, 1988.
31. Rosenberg, A.M. and Rosenberg, O.A., The state of stress and deformation in metal cutting (in Russian), *Sverhtverdye Materaily*, No. 5, 41, 1988.
32. Kattus, J.R., Effect of holding time and strain rate on the tensile properties of structural materials, *Proc. Symp. Short-Time High-Temperature Testing*, Los Angeles, March 25–29, American Society for Metals, 1959, p. 67.
33. Weinmann, K.J., The use of hardness in the study of metal deformation processes with emphasis on metal cutting, *Proc. Symp. Material Issues in Machining and The Physics of Machining Processes*, American Society of Mechanical Engineers, Warrendale, PA, 1992, p. 1.

34. Ravichandran, K.S. and Vasudevan, A.K., Fracture resistance of structural alloys, in *ASM Handbook*, Vol. 19, *Fatigue and Fracture*, American Society for Metals International, Metals Park, OH, 1996, p. 381.
35. Jayatilaka, A.S., *Fracture of Engineering Brittle Materials*, Applied Science Publishers, London, 1979.
36. Hertzberg, R.W., *Deformation and Fracture Mechanics of Engineering Materials*, 3rd ed., John Wiley & Sons, New York, 1989.
37. Scudnov, V.A., *Limiting Plastic Deformation of Metals*, (in Russian), Metallurgia, Moscow, 1989.
38. Barzduka, A.M., and Hertsov, L.V., *Stress Relaxation in Metals and Alloys* (in Russian), Metallurgia, Moscow, 1972.
39. Pagh, H.L.D., Ed., *The Mechanical Behavior of Materials Under Pressure*, Elsevier, New York, 1970.
40. Sockholov, L.D. and Scudnov, V.A., *Plasticity of Metals: Analytical Survey* (in Russian), VILS, Moscow, 1980.
41. Mills, B. and Redford, A.H., *Machinability of Engineering Materials*, Applied Science Publishers, London, 1983.
42. Lode, W., Versuch über den einfluss der mitteren hauptspannung auf das fliessen der metalle eisen, kupfer and nickel, *Z. Phys.*, 36, 913, 1926.
43. Tresca, H., Sur l'ecoulement des corps solides soumis à de fortes pression, *C.R. Acad. Sci. Paris*, 59, 754, 1864.
44. Slater, R.A.S., *Engineering Plasticity: Theory and Application to Metal Forming Processes*, Macmillan, London, 1977.
45. Reuss, A., Beruecksichtigung der elastishchen formaenderungen in der plastizitaetstheorie, *Z. Angew. Mach. Mech.*, 10, 266, 1930.
46. Starkov, V.K., *Dislocation Theory in Metal Cutting* (in Russian), Mashinostroenie, Moscow, 1979.
47. Sampath, W.S. and Shaw, M.S., Fracture on the shear plane in continuous cutting, in *Proc. 11th NAMRI Conf.*, Dearborn, MI, 1983, p. 281.
48. Shaw, M.S., personal communications, 1997–1998.
49. Rostoker, W. and Dvorak, J.R, *Interpretation of Metallographic Structures*, 2nd ed., Academic Press, New York, 1977.
50. Ernst, H., Physics of metal cutting, in *Machining of Metals*, American Society for Metals, Metals Park, OH, 24, 1938.
51. Trent, E.M., *Metal Cutting*, Butterworth-Heinemann, Oxford, 1991.
52. Underwood, E.E., Quantitative metallography, in *Metals Handbook*, Vol. 9, *Metallography and Microstructures*, 9th ed., American Society for Metals, Metals Park, OH, 1985, p. 123.
53. Parrish, W., X-ray powder diffraction, in Berner, H.B., Ed., *Encyclopaedia of Material Science and Engineering*, Vol. 7, Pergamon Press, Oxford, 1986, p. 5496.
54. Astakhov, V.P., Shvets, S.V., and Osman, M.O.M., Chip structure classification based on mechanism of its formation, *J. Mater. Proc. Technol.*, 71, 247, 1997.
55. Astakhov, V.P. and Osman, M.O.M., Correlations amongst the process parameters in metal cutting and their use for establishing the optimum cutting speed, *J. Mater. Proc. Technol.*, 62, 175, 1996.
56. Bridgman, P.W., *Studies in Large Plastic Flow and Fracture with Special Emphasis on the Effect of Hydrostatic Pressure*, McGraw-Hill, New York, 1952.

57. Hoffmanner, A.L., The use of workability test results to predict processing limits, in Hoffmanner, A.L., Ed., *Metal Forming: Interrelation Between Theory and Practice*, Plenum Press, New York, 1971, 349.
58. Cockcroft, M.G. and Latham, D.J., Ductility and workability of metals, *J. Inst. Metal.*, 96, 121, 1968.
59. Edelson, S.I. and Baldwin, M.M., Jr., The effect of second phases on the mechanical properties of alloys, *ASM Trans. Q.*, 55, 230, 1962.
60. Marciniak, Z. and Kuczynski, K., Limit strains in the processes of stretch-forming sheet metal, *Int. J. Mech. Sci.*, 9, 609, 1967.
61. Lee, P.W. and Kuhn, H.A., Fracture in cold upset forging — a criterion model, *Metal. Trans.*, 4, 969, 1973.
62. Avitzur, B., Analysis of central bursting defects in extrusion and wire drawing, *ASME J. Eng. Industry*, 90, 79, 1968.
63. Knott, J.F., *Fundamentals of Fracture Mechanics*, Butterworths, London, 1973.
64. Leibowitz, H., Ed., *Fracture: An Advanced Treatise*, Vol. I, *Microscopic and Macroscopic Fundamentals;* Vol. VI, *Fracture of Metals*, Academic Press, New York, 1969.
65. Tetleman, A.S. and McEvily, E., *Fracture of Structural Materials*, John Wiley & Sons, New York, 1967.
66. Colangelo, V.J. and Heiser, F.A., *Analysis of Metallurgical Failures*, 2nd ed., John Wiley & Sons, New York, 1987.
67. Collins, J.A., *Failure of Materials in Mechanical Design*, John Wiley & Sons, New York, 1981.

chapter five

Finite element simulation

5.1 General

The finite element method (FEM) is firmly accepted as a powerful general technique for the numerical solution of a variety of problems encountered in engineering. For linear systems, at least, the technique is widely employed as a design tool. However, similar acceptance for nonlinear situations is not so obvious. Before FEM can be used in design, its accuracy must be proven. The development of improved element characteristics and more efficient nonlinear solution algorithms along with experience gained in their applications allow the technique to be performed with some confidence.

The objective of this chapter is to describe in detail the application of FEM to simulation of the metal cutting process which is a materially nonlinear engineering analysis problem. The physical meaning of each step in analysis and the problems that arise in simulations are the main focus.

5.1.1 FEM in metal cutting

As shown in Chapter 2, FEM seems to be most suitable for analyzing the state of stress and deformation in the cutting system. This method has been applied to simulate machining since 1973 with some success.[1-12] The development of this approach, its history and assumptions, and the techniques involved have been well summarized by Zhang and Bagchi.[9]

Although earlier studies of metal cutting using FEM provided useful insight into the cutting process, they were all based on the oversimplified model of the cutting process known as the shear-plane model. As such, the contact condition at the tool/chip interface was modeled as simple friction, or, at best, the law governing the normal and shear stresses at the tool/chip interface was assumed to be known. Regarding the latter, the distributions of the normal and shear stresses at the tool/chip interface are assumed to be corresponding to generic stress distribution on the rake face of the cutting tool, proposed by Zorev.[13] The physics of chip separation are considered from the point of computational convenience rather than a real process so the separation criteria chosen are quite artificial. All these assumptions and simplifications of the real cutting process, made in order to employ FEM, originate from the lack of a realistic model for this process.

At this point, it is worthwhile to discuss an issue that is sometimes troublesome to many specialists in the field — namely, that FEM simulation is not a substitute for the theory of metal cutting or for the cutting experiment. The comparison of the results of FEM simulation with those obtained theoretically or experimentally makes sense if and only if it is made using the same theoretical fundamentals. It should be clearly understood that the results of FEM simulation cannot be considered as having separate meaning. FEM is a formal computational method and, consequently, its results depend entirely on the input. A good comparison of different conditions used in FEM has been offered in Zhang and Bagchi,[9] who showed, for example, that there are as many different assumptions regarding the tool/chip contact interactions as the number of published papers. It is clear that the results of simulation differ from one paper to the next, and comparison of the modeled results is meaningless as they are obtained under different sets of assumptions.

5.1.2 Aims and layout

The results of Chapter 4 suggest that metal cutting is commonly accomplished by ductile fracture, thus a triaxial state of stress in the deformation zone must be considered. The results of Chapter 2 proved that this state of stress is a combination of the bending stress and shear stress due to compression in the deformation zone. Moreover, it is proven that the chip formation process is cyclical; therefore, a heavy nonlinear problem accounting for these conditions should be considered. Thus, the primary objective of this chapter is to develop a comprehensive finite element analysis of orthogonal metal cutting based on the proposed model of chip formation.

Unlike other texts on linear and nonlinear FEM simulation in metal cutting which have dealt predominantly with computational aspects, this chapter is intended to be more practical; therefore, attention is focused on the physical side of simulation.

Nonlinearity arises in FEM simulation of the metal cutting process from a nonlinear material response which, in turn, can be the result of elastoplastic material behavior or hyperplastic effects of some form. Additional nonlinear characteristics can be associated with temporal effects such as viscoplastic behavior or dynamic transient phenomena. Each of these nonlinearities may occur in each component of the cutting system.

5.2 Computational details of simulation

In FEM simulations, mathematical relationships between the stress and strain for the materials involved have to be known, thus FEM has been efficiently used where elastic deformation is of prime importance. When plastic deformation is considered, FEM has been rarely used because of the lack of a universal mathematical model to correlate the stress and strain in the region above the elastic limit.

Because chip formation is a process of purposeful fracture of work material, it is necessary to pass successive stages of elastic and then plastic deformations to reach this fracture. The existing FEM algorithms assume a linear stress-strain relation. However, they can be used to study plastic deformation using the assumption of a linear strain hardening material which may prove to be inadequate for most of the real materials. To improve the accuracy of modeling, it is possible to represent the stress-strain relationship in a piecewise linear fashion.[14,15]

5.2.1 *Program structure*

For simulations, the FEM program suggested in Owen and Hilton[14] has been used. The program is characterized by a modular approach so that separate subroutines are employed to perform the various operations required in nonlinear FEM simulations. The program consists of nine modules, each with a distinct operational function. Each module in turn is composed of one or more subroutines which are common to several modules. Control of the modules is held by the main or master segment.

The modules, shown schematically in Figure 5.1, are described in relation to their general functions as follows:

1. *Initialization or zeroing module.* This is the first module entered, and its function is to initialize to zero various vectors and matrices at the beginning of the solution process.
2. *Data input and checking module.* This is the second module entered. It handles input data defining the geometry, boundary conditions, and material properties. These data are checked using diagnostic routines, and if errors occur they are flagged and the remainder of the input data are printed out before the program is terminated. For isoparametric elements, Gaussian integration constants and mid-side nodal coordinates for straight-sided elements are also evaluated in this section. Once used, this module is not needed again.
3. *Loading module.* This module organizes the calculation of nodal forces due to the various forms of loading for two-dimensional applications. These include pressure, gravity, and concentrated loading.
4. *Load increment module.* Any materially nonlinear finite element solution must proceed on an incremental basis. Therefore, the function of this section is to control the incrementing of the applied loads evaluated by the loading module. It also ensures that any specified displacement values are also incrementally applied.
5. *Stiffness module.* This module organizes the evaluation of the stiffness matrix for each element. The stiffness matrices are stored in memory

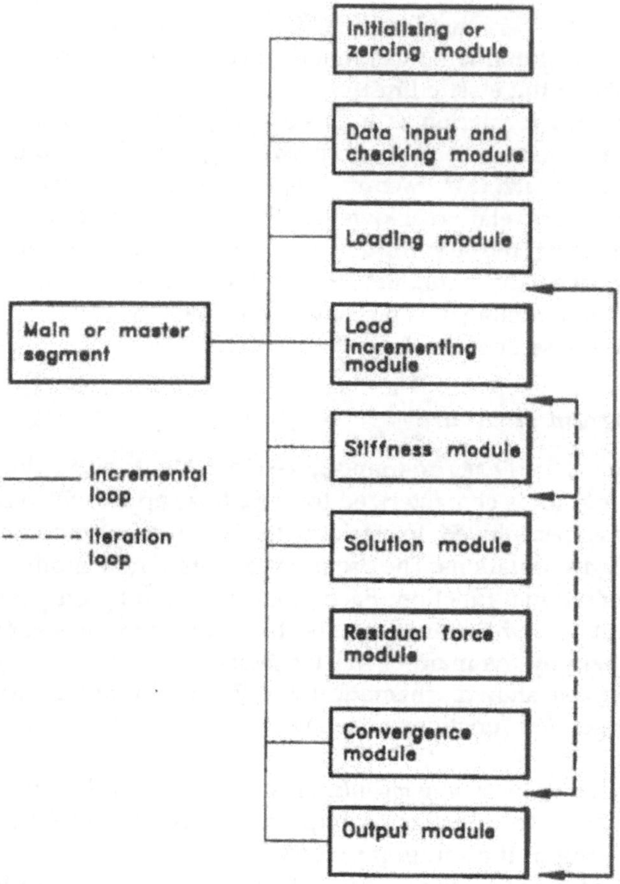

Figure 5.1 Program modules for nonlinear solution codes.

and ordered in the sequence required for equation assembly and reduction.

6. *Solution module*. The general purpose of the routine is to assemble, reduce, and solve the governing set of simultaneous equations to give the nodal displacements and force reactions at restrained nodal points.
7. *Residual force module*. The general purpose of this module is to calculate the residual or "out-of-balance" nodal forces at each stage of the analysis.
8. *Convergence module*. In this module, the convergence of the nonlinear solution is checked against the defined criteria.
9. *Output module*. This module organizes the output of the requested quantities.

The main purpose of the main or master segment is to call the above modules and to control the load increments and iteration procedure according

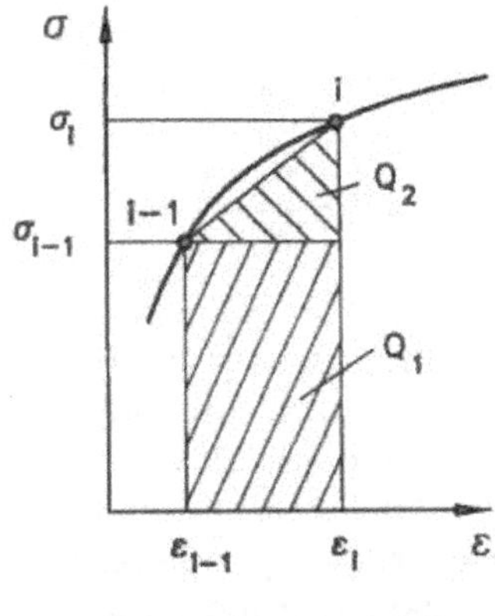

Figure 5.2 A piece-wise linear approximation of a non-linear stress-strain relation.[13]

to the solution algorithm being employed and the convergence rate of the solution process.

5.2.2 *Workpiece and tool materials modeling*

Because the the stress-strain relationships are represented in a piecewise linear fashion, the whole plastic region of the stress-strain relation is divided into small subregions, and within each subregion the stress-strain relation is assumed to be linear.[14,15] It is also known[16] that any given stress-strain state may be reached by incremental loading. Within each increment *I*, Hooke's law can be used when the mechanical constants E_i and μ_i are defined. The program automatically sets the increments of the external load using known values of E_i and μ_i (taken from the input data) and calculates the results of each step.

The FEM program discussed here and others available commercially enable users to to use elements with various mechanical properties. Because the mechanical constants (E_i and μ_i) for a continuous medium under non-uniform plastic deformation are different for the different medium's regions, this medium can be modeled by a set of elements having different E_i and μ_i.

The particular values of E_i and μ_i are defined as follows. The stress-strain diagram of the work material can be represented as a continuous line which consists of linear segments having different inclinations to the horizontal axis (Figure 5.2). The area under the stress-strain diagram represents the specific work of deformation (Chapter 4). Within the elastic limit, the specific work of deformation (the resiliency of the work material) is equal to the area of the triangle (i.e., $A_1 = 0.5 \cdot \sigma_p \cdot \varepsilon_p$). Here, σ_p is the maximum stress for elastic conditions, and ε_p is the elastic strain. As known, $E_1 = \tan \alpha_1 = \sigma_p/\varepsilon_p$. According to the assumption made, when the stress exceeds the proportional limit, the relationship between $d\varepsilon$ and $d\sigma$ corresponds to Hooke's law, but this region of the graph is characterized by a new value of the elastic modulus which is different from E_1. Introduced in this way, the elastic modulus E_i for each segment of number "*i*" can be calculated as:

$$E_i = \frac{\sigma_i - \sigma_{i-1}}{\varepsilon_i - \varepsilon_{i-1}} \tag{5.1}$$

The stress-strain and strain hardening diagrams for different workpiece materials are very well known;[17] unfortunately, such data are not available for cemented carbides used as tool materials. To the first approximation, the necessary data have been obtained using the known mechanical properties of these materials (ultimate stress σ_{UTS}, elasticity modulus E, and Poisson's ratio μ) taking into account that the proportional limit σ_p for carbides is lower than the ultimate stress by 20 to 30%, and the plastic deformation at the point of fracture is about 1.0 to 1.3%.[18]

Consider the procedure used in the computations. According to von Mises' yield criterion, the limiting condition (failure) in a two-dimensional finite element occurs when the total energy per unit volume reaches a predetermined limiting value.[19]

The specific work of deformation at point "*i*" of the stress-strain diagram can be defined as follows:

$$A_i = 0.5\left[\varepsilon_y \sigma_y + \sum_{i=2}^{n}(\sigma_i - \sigma_{i-1})(\varepsilon_i - \varepsilon_{i-1})\right] \tag{5.2}$$

where n is the number of the diagram's points where different values of σ and ε are introduced.

The work done at each stage of loading is defined by the equivalent stress and equivalent strain. In uniaxial tensile, an equivalent stress σ_e is equal to tensile stress σ. The increment of the equivalent stress with an *i*-th increment of the external load is

$$\Delta\sigma_e = 0.707\sqrt{(\Delta\sigma_x - \Delta\sigma_y)^2 + \Delta\sigma_y^2 + \Delta\sigma_x^2 + 6\,\Delta\sigma_{xy}^2} \tag{5.3}$$

Here, $\Delta\sigma_x$, $\Delta\sigma_y$, $\Delta\sigma_{xy}$ are the corresponding changes in the normal and shear stress, respectively.

The increment of strain within each element is

$$\Delta\varepsilon_e = \frac{\Delta\sigma_e}{E_i} \tag{5.4}$$

Therefore, both the incremental stress and strain can be defined when the system experiences a new load increment. Within the considered element, the total specific work done after a load increment is

Table 5.1 Selected mechanical properties of materials at the points of the approximated stress-strain diagram

Plain carbon steel (0.45% C)			Plain cast iron			Carbide P10 (79% WC, 15% TiC, 6% Co)			
σ_c (GPa)	ε	μ	σ_c (GPa)	ε	μ	σ_c (GPa)	σ_b (GPa)	ε	μ
0.36	0.0017	0.30	0.21	0.0017	0.23	2.25	0.86	0.0017	0.20
0.50	0.0350	0.32	0.45	0.0038	0.23	2.65	1.03	0.0040	0.23
0.58	0.0900	0.34	0.63	0.0055	0.24	2.86	1.10	0.0070	0.26
0.64	0.1750	0.36	0.80	0.0075	0.25	3.00	1.15	0.0120	0.30
0.68	0.3300	0.38	1.00	0.0100	0.26	—	—	—	—
0.70	0.5500	0.40	1.10	0.0120	0.27	—	—	—	—

$$a = a_{i-1} + a_n + a_m \tag{5.5}$$

where a_{i-1} is the specific work done before an i-th load increment (the area under the graph up to point "$I - 1$" in Figure 5.2); $a_n = \Delta\sigma_{e(i-1)}\Delta\varepsilon_j$ and $a_m = 0.5(\Delta\sigma_i\Delta\varepsilon_i)$ are areas Q_1 and Q_2, respectively (Figure 5.2).

If, after a certain load increment, the total specific work a is found to be greater than the specific work of deformation A_i ($a > A_i$), then a new set of elastic constants (E_{i+1} and μ_{i+1}) for the considered element are introduced in calculations at the next step of the load increment. This calculating process continues until the condition ($a > A_i$) is found to be valid for the last point of the stress-strain diagram. It is understood that this point corresponds to the strain at fracture, calculated using data of Chapter 3 for particular tool geometry (Π-factor), workpiece material (purity, mechanical properties including the strain rate sensitivity), and cutting regime (strain and strain rate). It is assumed that at this point the work material exhausts its resistant ability and, as a result, failure occurs.

Table 5.1 shows the pertinent mechanical properties of a few materials corresponding to the points where the graph, approximated by the "stress-strain" diagram, changes its slopes. Poisson's ratio μ at each point is established by its reference value at point 1 taking into consideration that μ changes linearly and cannot exceed a value of 0.5.[19]

5.3 *Applications of FEM*

Because it has been proposed that we consider the process of metal cutting as taking place in the cutting system, the main system properties (the system time frame and dynamic interactions of the components of the system) should be respected in FEM simulations. That is to say that the components of the system should be modeled simultaneously, accounting for all their

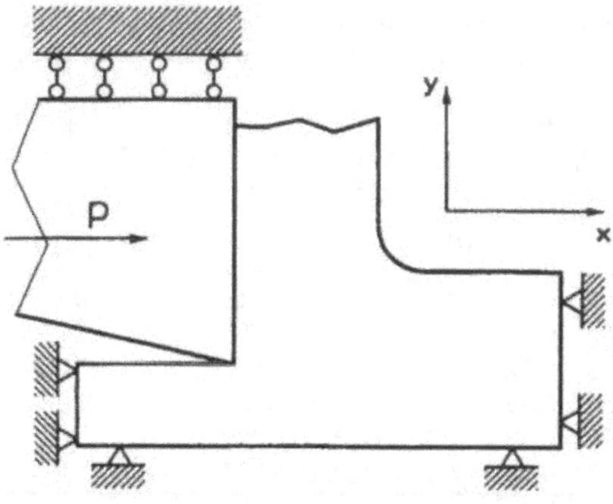

Figure 5.3 FEM model of orthogonal cutting.

dynamic interactions. Such modeling seems to be free from the common idealizations of a real cutting process. Indeed, no preliminary experimentation or assumption is needed to determine the value and distribution of the contact stresses along the tool/chip/workpiece interfaces. Naturally, one might calculate these stresses when the solution is advanced.

Figure 5.3 shows the model with the selected boundary conditions. The cutting system is divided into 880 finite triangle elements. The finite elements with the mechanical properties of a work material constitute the workpiece and chip. The elements with the mechanical properties of a tool material and with a possibility for the displacement in the x-direction constitute the cutting tool.

The external load is applied to the cutting tool at a certain distance from the cutting zone, either as a single or distributed force that corresponds to the real conditions of tool mounting on the tool post of a machine tool. It is understood that the choice of any of these two models does not change the stress distribution in the cutting wedge but affects only the stress distribution in the body of the tool, which is not considered here.

In order to advance the cutting wedge into the workpiece, a certain feed rate should be applied. To achieve this, the displacements of certain nodes located on the free surface are blocked in the y-direction, but not restricted along the x-direction. Within the workpiece, the displacements of nodes located at a distance where the deformations are negligibly small are restricted in both directions, thus rigid clamping of the workpiece is assumed (Figure 5.3).

To complete the mathematical description of the cutting system, the contact dynamic problem at the tool/work interface should be considered and solved. Among the various ways to solve such problems, two are of common use. The essence of the first one is an iteration procedure to balance the normal and shear forces applied to the nodes in the contact zone. The second method includes the consideration of a so-called "third body" placed between the contact surfaces. This third body consists of elements having special properties which differ from those of the workpiece and tool materials. Such a representation constitutes a boundary layer which enables one to simulate the contact friction and sliding. The analysis of the algorithms, based on the these approaches, shows that at the present stage these algorithms still have demonstrative rather than applied usefulness. They suffer from a lack of versatility and serve mainly to estimate the results of known analytical solutions. The relative displacements allowed in these solutions correspond to micrometers or even micrometer fractions. In contrast, the cutting process involves large displacements of work metal passing through all deformation stages up to failure. Bearing in mind these restrictions, we may conclude here that such algorithms and solutions can hardly be applied to metal cutting.

Therefore, at the present time, it is possible to analyze only the beginning of the entire chip formation cycle, showing how it starts. Unfortunately, a lack of specific knowledge concerning the contact problem severely hampers quantitative treatments of the chip formation cycle, thus, at this point, it seems impossible to consider the questions of how it proceeds and how it is completed. However, we hope that progress in the solution of the contact problem will continue with the development of numerical methods that will make answering such questions possible within a few years.

5.4 *Results and discussion*

Simulation of the orthogonal machining of plain carbon steel AISI 1045 was carried out with a 0.12-mm/rev cutting feed and 1-mm undeformed chip thickness. The tool material selected was tungsten carbide P10 (79% WC, 15% TiC, 6% Co). In the simulations, the tool rake angle was varied within the limits from $-18°$ to $18°$. The simulation of the cutting tool advancing into the workpiece was carried out by applying an incremental load. The "size" of the increment was chosen to be $\Delta P_z = 250$ N. The calculations were stopped when the specific work of deformation for a certain element became equal to the specific work at the breaking point in the standard tensile test of the work material. Five steps were required to reach that level under the considered conditions. As was expected, the maximum shear stress $\sigma_{xy} = 0.7$ GPa was found in front of the cutting point. A comparison shows that the same stress is required to cause failure in case of a punch penetration into a semi-infinite solid. Indeed, it is known that the failure stress in the direction of punch penetration is $\sigma = (1 + \pi)\sigma_p$; hence, $\sigma_{xy} = 0.5\sigma = 0.5(1 + \pi)0.36 = 0.74$ Gpa, which coincides with the computational results. Therefore, according to the calculations, the force *P* was found to be 1250 N at the "cut in" instant and corresponds with the experimental results.[18]

5.4.1 *Modeling of the deformation zone*

To demonstrate the significance of the chip as a component of the cutting system, two basically different models have been studied. The first one assumes: (1) the chip formed in cutting does not affect the cutting process, thus having no effect on the stress and deformation in the cutting zone; and (2) there is no sliding of the chip over the tool rake face. The results of calculations made under these assumptions show that the stress distribution in the tool point region corresponds to the classical stress distribution under a punch penetration into a semi-infinite solid. Therefore, the workpiece material undergoes compression and, as a result, the stress at the tool point increases, at least theoretically, to infinity. Nevertheless, chip formation does not occur even if an extremely high load is applied. Consequently, if there is no sliding of the work material over the rake face, as assumed, then simulation of a punch penetration into a semi-infinite solid actually takes place. The considered model is not fully representative of a real punch-penetrating process due to the absence of the balancing forces acting on the external surface of the workpiece. However, the permitted penetration was chosen to be so small that the balance was achieved due to the assumption of the rigid "clamping" at the tool/workpiece interface along the rake face.

At the initial stage of the load application to the cutting system, the maximum stress in the work material occurs in the vicinity of points *A* and *B* because these points are the stress concentrators (Figure 5.4). Then, increasing the applied load, the workpiece material in the vicinity of point *B* undergoes a higher stress than in the vicinity of point *A*.

Analysis of the state of stress in the layer being cut under incremental loading shows that the maximum shear stress moves away from the tool rake face with each successive load increment. The layers, which are deformed to a higher degree at the initial stages of the load increments, are characterized by lower "moduli of elasticity" (according to the selected method introduced above). With tool penetration, the maximum increment of shear stress occurs in the microvolumes located farther and farther from the cutting edge. Even though the deformation zone develops from the cutting edge, the described phenomenon explains the way in which the stress field becomes so uniform that such a uniform state of stress is referred to as the hydrostatic pressure in many studies.[13,20,21]

Summarizing the results of the conducted simulation with the first model, it is necessary to emphasize once again that when there is no sliding of the work material over the tool rake face, the deformation model corresponds to a case of simple compression, which cannot cause chip formation. This is further evidence of the unity and mutual influence of the cutting system's components: cutting tool, workpiece, and chip. The cutting process cannot occur at all without any one of these components.

The second model permits the sliding of the partially formed chip cantilever over the tool rake face. On sliding, the chip interacts with the rake face

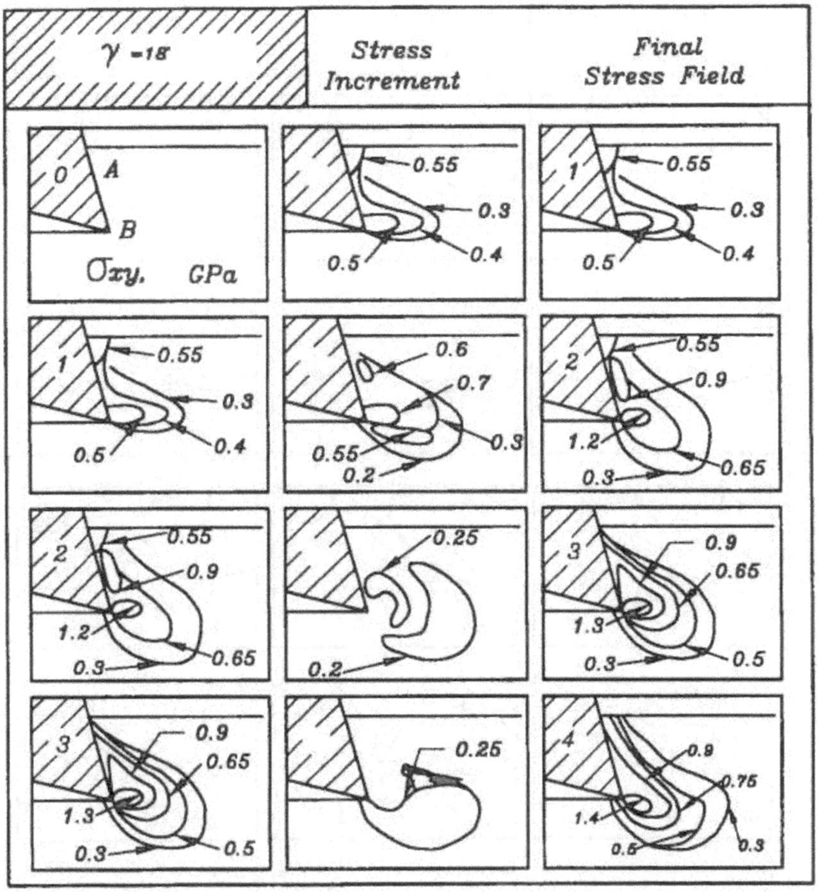

Figure 5.4 Stress contours from finite element simulation using the first model.

along a certain distance commonly referred to as the contact length. As discussed in Chapter 2, this interaction results in chip deformation and, as a result, a certain force is applied to the chip. In turn, the components of this force, transmitted through the chip to the deformation zone, change the state of stress in this zone.

A comprehensive analysis of the second model including all three components of the cutting system shows that the shear stress in the cutting zone develops as a result of the mutual action of the compression and bending (Figure 5.5). At the initial stage of loading, the maximum stress was found in the vicinity of the tool point and in the transition zone where the outside workpiece surface turns into the chip-free surface. The location of deformation centers corresponds exactly with the assumption of the mutual action of compression and bending.

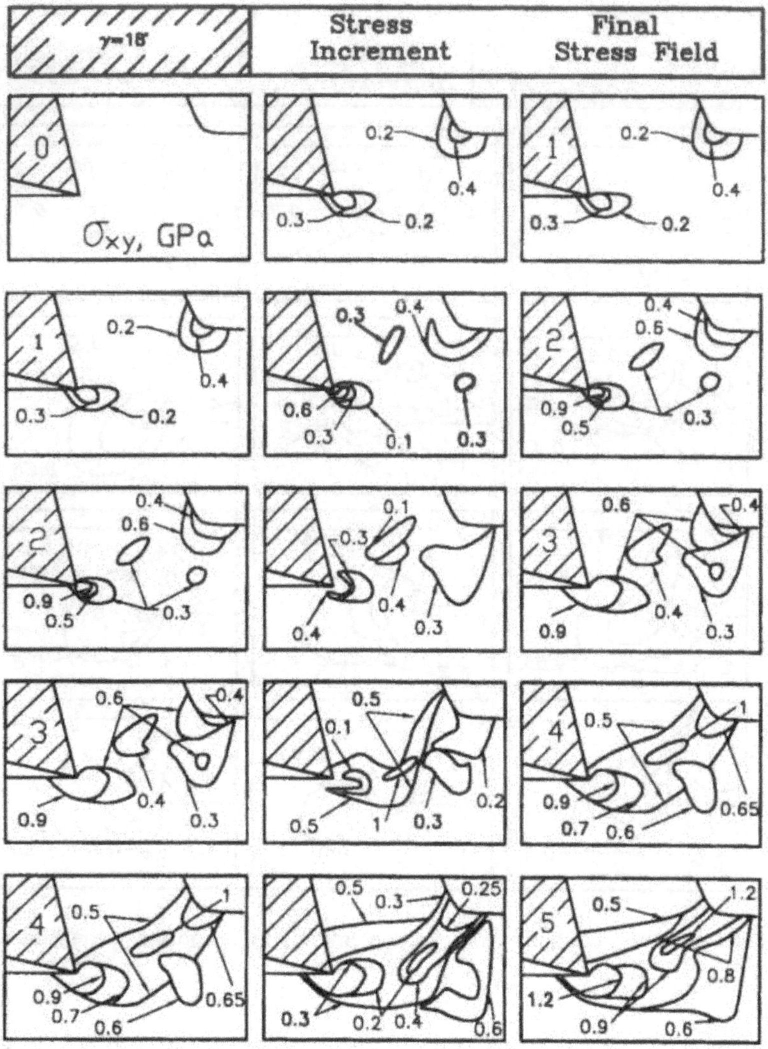

Figure 5.5 Stress contours from finite element simulation using the second model.

The plastic deformation due to compression develops from the cutting edge while that due to bending has its maximum on the free surface of the chip at the place of the chip cantilever support (as discussed in Chapter 2). This is in agreement with the results of calculations. As one might expect, the maximum stress is developed at the cutting edge at each load increment and in the zone where the workpiece outside surface turns into the chip-free surface. In contrast, at final load increments, the stress concentration is not developed in the immediate vicinity of the cutting edge. Similar results were obtained in the preceding analysis of the first model.

The fact that maximum stress at final load increments does not occur in the nearest vicinity of the cutting edge requires explanation. Consider the "yielding" regions of workpiece material where the yield strength is achieved as a result of the previous load increments. Here, the stress increases at a lower rate than in other less deformed neighboring regions even though displacements are almost the same. The accumulation of the stresses and strains, formed independently at each load increment, leads to an increase in the dimensions of the zone of plastic deformation. When dimensions of this zone reach a certain limit, failure of the layer being removed occurs and causes chip formation.

The calculated field of the shear stresses in the deformation zone is similar to the modeled field of slip lines obtained in Chapter 2 by graphical superposition of the fields of slip lines due to compression and bending (Figure 2.13). The considered dynamics of the stresses and deformations in the deformation zone along with the experimental results presented in Chapter 2 enable revision of the basic concepts in metal cutting such as the shear plane and shear angle, as well as the primary and secondary deformation zones.[13,20,21]

To begin, consider the shear plane model widely used in metal cutting studies since late in the last century (Figure 2.1).[13] According to this model, the chip is considered to be formed from a process of simple shearing which is approximately confined to a single plane expanding from the cutting edge to the workpiece surface ahead of the tool. As shown in the present study, the redistribution of stresses in the machining zone takes place during the entire cycle of chip formation until the stress level achieves its limiting value in a certain volume of work material, which results in the formation of a sliding surface (fracture). It is true that, when machining of a ductile material is considered, the sliding surface can be approximately represented by a plane and is referred to as a shear plane, as the plastic deformation takes place by shearing (Chapter 2); however, this plane does not exist throughout the entire chip formation cycle. Rather, it appears only at its final stage, but the surface of the maximum combined stress preludes the sliding plane (Chapter 3). Therefore, one might follow up the modification of the shear stress on this surface (approximated by a plane) by applying an incremental load to the cutting wedge. Results of such a procedure are shown in Figure 5.6, where the zero-point on the horizontal axis corresponds to the cutting edge and the segment 0–1 corresponds to the shear plane length. Curve 5 in this figure represents the final shear stress level, while curves 1 to 4 represent the sequence of its evolution. Therefore, the shear plane should be considered as a result of the completion of a chip formation cycle by a certain method, one of the possible results of interactions of the components of the cutting system. It is also worthwhile to mention here that even in cutting of ductile materials the sliding plane may form in the direction of the separation line between the layer to be removed and the remainder of the workpiece under a certain arrangement of the cutting system's components.[22,23]

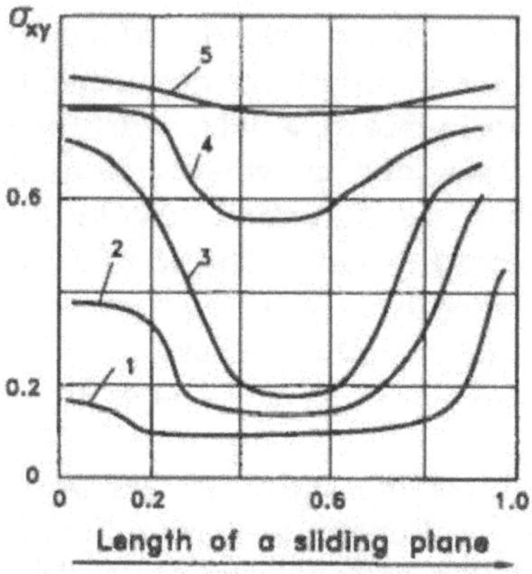

Figure 5.6 The modification of the shear stress, acting along the direction of the future failure, during a cycle of the chip formation. Curves numbered 1 to 5 correspond to the load increments.

The preceding results show that the concepts of the steady-state "primary" and "secondary" deformation zones are not in agreement with a real state of stress and strain in metal cutting. According to these concepts, the work material is deformed twice: first in the primary deformation zone occurring around the shear plane and then in the secondary deformation zone continuously attached to the tool rake face. As discussed above, the stresses as well as strains in the chip formation zone have a dynamic nature and, therefore, cannot be represented by the steady-state primary and secondary deformation zones.

The concepts of "primary" and "secondary" deformation zones have been used in many studies of metal cutting because the cause-effect sequence has been inverted. By this is meant that the displacement and velocity of a microvolume of the workpiece, taken as initial parameters, are assumed to be towards the cutting wedge, then these kinematic parameters are used to calculate the forces and stresses acting on this volume along its course. In reality, neither displacements nor velocities are the causes of the stresses and deformations; rather, the reverse is true. The force applied to the cutting tool results in internal forces (stresses) acting on microvolumes, causing their displacements, velocities, etc. As a result, each microvolume does not move along a smooth trajectory with constant velocity towards or over the cutting edge, but rather it moves in the space by a continuously changing trajectory affected by the instantaneous stress distribution and the displacements of neighboring microvolumes.

The magnitudes and directions of the forces acting within the deformation zone are defined by: (1) the way in which the forces from the cutting tool are applied, (2) the dimensions of the components of the cutting system, and (3) properties of work material in the deformation zone. The latter depends to a large extent on the deformation history within a cycle of chip formation.

Because the force is directed from the cutting tool, the mechanical energy supplied to the machining zone first affects the layers of the work material attached to the tool rake face. The interaction of the cutting system's components and their dimensions affects the way in which the energy propagates into the workpiece; therefore, the formation of the plastic zone is a matter of cutting process development.

The energy supplied to each microvolume of the deformation zone is transmitted through the intermediate layers affected by the deformation work. The deformations of these layers determines the dimensions of the plastic zones, which, in turn, affect the length of tool/chip plastic contact.

5.4.2 Modeling of the tool/chip interface

5.4.2.1 Known results

When metal is cut, the cutting force acts primarily through a small area of the rake face which is in the contact with the chip. It is of interest in cutting force determination, the theory of tool wear, and the mechanics of chip formation, therefore, to determine the contact conditions at the tool/chip interface.

In the known simulations of the contact conditions, the tool/chip interaction is modeled as simple friction. It is implied in the Merchant approach[24] that the contact between the tool and the chip is a sliding contact where the coefficient of friction is constant. As a result, the distributions of the interface stresses are assumed implicitly to be uniform. It is true that, in general, the coefficient of friction for sliding surfaces remains constant within wide ranges of the relative velocity, apparent contact area, and normal load. In contrast, for metal cutting the coefficient of friction varies with respect to the normal load, the relative velocity, and the apparent contact area. The coefficient of friction in metal cutting was found to be so variable that Hahn[25] doubted whether this term served any useful purpose. Moreover, Finnie and Shaw[26] have concluded that a coefficient of friction is inadequate to characterize the sliding between chip and tool and recommended discontinuing use of the concept of coefficient of friction in metal cutting. Discussing the results of this work, Kronenberg showed that the concept of the coefficient of friction in metal cutting studies can be misleading. Trent[27] has been presenting evidence supporting this point for the last 30 years. Unfortunately, further studies in metal cutting continue to use the coefficient of friction in the consideration of the tool/chip interface.

According to Zorev,[13] normal stress distribution at the tool/chip interface may be closely approximated by a power function reaching its maximum on the cutting edge. Alternatively, Palmer and Oxley[28] suggested that, because the chip does not make contact with the tool rake face over a certain region adjacent to the cutting edge, the normal stress simply does not exist there. After examination of many experimental results, Hsu[29] concluded that the distributions of the normal and the shear stresses at the tool/chip interface correspond to those proposed by Zorev.

Takeyama and Usui[30] studied the stress distribution along the tool/chip interface by measuring the cutting forces on tools with various contact lengths. Their results show that both the shear and the normal stress distributions are uniform over the interface. Later on, Usui and Takeyama[31] improved the accuracy of their experiments by using a photo-elastic cutting tool instead of a photo-elastic workpiece. According to their measurements, the distribution of the shear stress corresponds to the model suggested by Zorev, while the normal stress had very unusual distribution with a "plato" region at a certain distance from the cutting edge.

A plato-type distribution for the normal and shear stresses has been reported by Kato et al.[32] In this study, they used the modernized version of the split cutting tool proposed by Kamskov[33] and then used by Loladze[34] and Hsu.[29] A critical analysis of these and other studies related to stress distribution beyond those analyzed above[20,21,35–39] shows that it is quite possible that actual stresses and stress distributions may have not yet come to light due to a significant scatter in the results obtained.

Based upon a system approach in metal cutting and the proposed model for chip formation introduced in Chapter 2, and on the results of finite element simulations discussed in Section 5.4.1, this section aims to understand the nature of the significant scatter in the reported stress distributions at the tool/chip interface by studying the dynamics of stress formation and distribution at this interface.

5.4.2.2 *Simulation of stress distribution at tool/chip interface*

Finite element simulation makes it possible to establish the distributions of the normal and shear stress at the tool/chip interface, as well as the dynamics of these distributions within a chip formation cycle. It is instructive to trace the above-mentioned distributions by applying FEM in the sense in which it has been introduced above — namely, using the incremental loading. The results obtained are now considered.

5.4.2.2.1 *Shear stress distribution.* At the cut-in period before chip formation, the shear stress is distributed almost uniformly along the contact surface (Figure 5.7), in agreement with the previously introduced mechanism. This stress increases with each load increment (the increment "size" was chosen to be equal to 250 N as described in Section 5.4.1. Because calculations were made using rake angles $\gamma = -18°$, $\gamma = 0°$, and $\gamma = 18°$, it can be concluded that the obtained relations between the shear stress and the applied load are valid for a wide range of tool rake angles.

At the beginning of each new chip formation cycle, the shear stress distribution within the tool/chip interface has two distinguished regions (Figure 5.8). The stress distribution in the first region is similar to that at the cut-in period (i.e., the shear stress distributes uniformly and increases with the applied load). Furthermore, both the length of this region and the stress

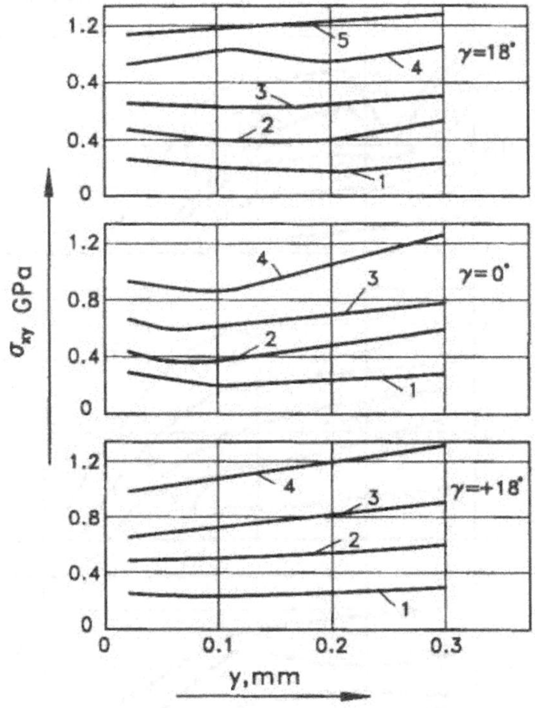

Figure 5.7 Dynamics of the shear stress distribution at the cut-in period. Curves 1 to 4(5) correspond to increasing load with the increment of 250 N, respectively.

magnitude do not change with the rake angle. In contrast, in the second region, the shear stress decreases at a rate that is a function of rake angle.

A dynamic comparison of the shear stress during the cut-in period with that of the stabilized cutting shows that the shear stress on the tool face during the cut-in period should be almost 30% higher than that in stabilized cutting to achieve the same effect (for example, to form the stress level corresponding to the beginning of failure in a certain region of the deformation zone). This result explains a reduction in the tool life in the interrupted cutting.

5.4.2.2.2 *Normal stress distribution.* During the cut-in period when the chip has not yet formed, the normal stress distribution shows strong dependence on the rake angle (Figure 5.9). With rake angles of $\gamma = +18°$ and $\gamma = 0°$, the normal stress increases with each load increment (curves 1 to 4), and the distribution of this stress may be represented by a decreasing, close-to-linear function having its maximum at the cutting edge (Figure 5.9). When the rake angle is negative, the distribution of the normal stress is identical to the above consideration only within the elastic limit. As soon as this limit is achieved,

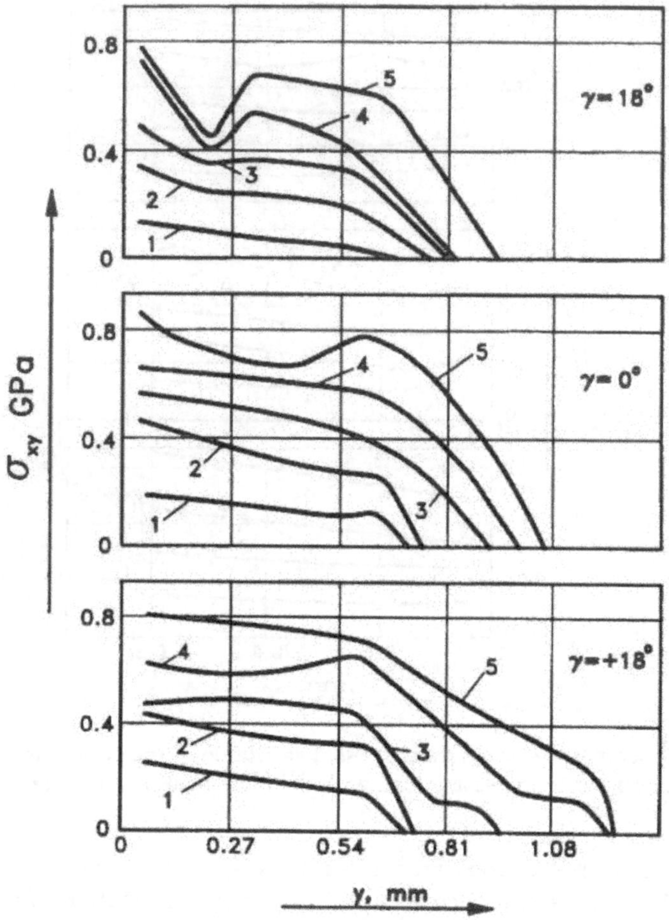

Figure 5.8 Dynamics of the shear stress distribution at the beginning of a new chip formation cycle. Curves 1 to 5 correspond to increasing load with the increment of 250 N, respectively.

the function $\sigma_n = f(y)$ is represented by a curve with a maximum at a certain distance from the cutting edge.

At the beginning of each new chip formation cycle, the presence of the chip affects the normal stress distribution. Here, for positive rake angles and within the proportional limit, the function $\sigma_n = f(y)$ corresponds to the known relation $\sigma_n = \sigma_{max}(y/c)^n$ (Figure 5.10).[13,29,30] In cutting with high negative rake angles, a function $\sigma_n = f(y)$ assumes a plato-type shape at the beginning of work material failure. Therefore, from the earliest stages of cutting (the cut-in period or the beginning of a chip formation cycle), all reported distributions of the shear and normal stresses at the chip/tool interface may occur. As it was with velocity diagrams (Chapter 3), there are no contradictions in

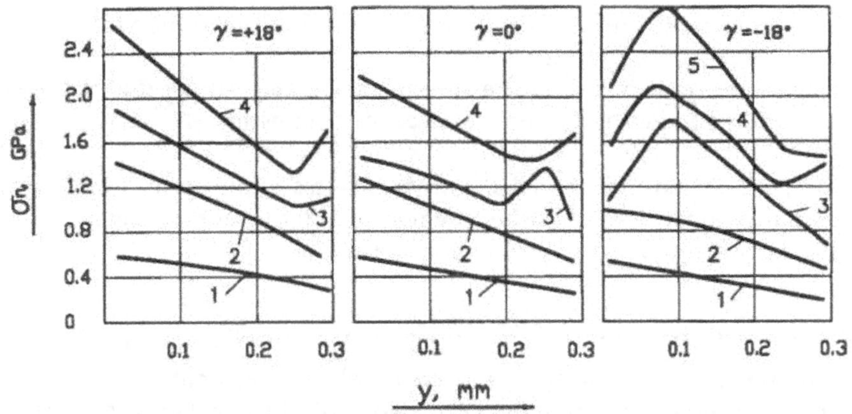

Figure 5.9 Dynamics of the normal stress distribution at the cut-in period. Curves 1 to 4(5) correspond to increasing load with the increment of 250 N, respectively.

the reported data. They are simply considered at different instants over a chip formation cycle and under different cutting conditions.

A comparison of the stress calculation for the cut-in period and for steady-state cutting shows that similarly to the shear stress, the normal stress level at the tool/chip interface in the cut-in period is 25 to 30% higher that the normal stress in the stabilized cutting. This phenomenon may be explained by:

- An increase in the tool/workpiece interface area in steady-state cutting; as the mechanical properties of the workpiece material are identical in

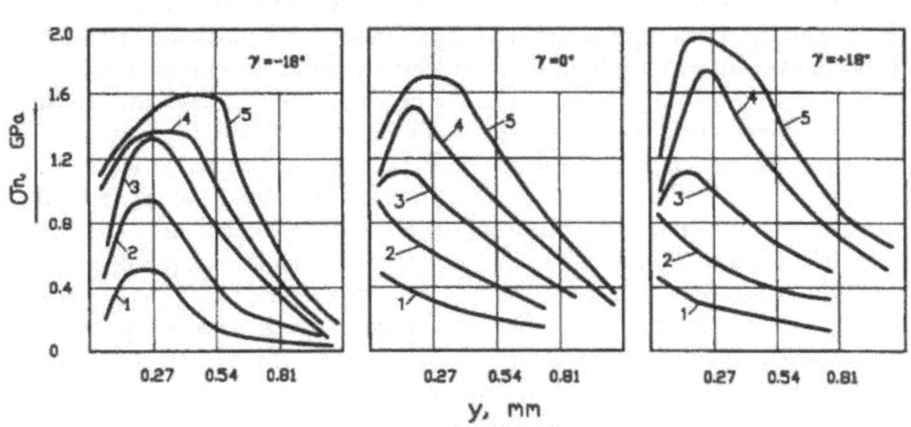

Figure 5.10 Dynamics of normal stress distribution at the beginning of a new chip formation cycle. Curves 1 to 5 correspond to increasing load with the increment of 250 N, respectively.

both cut-in and stabilized cutting, the same penetration force causes a lower contact stress.
- A decrease in the failure stress in stabilized cutting due to the combined action of the shear and bending stresses.

Therefore, the normal as well as the shear stress acting on the tool face have a variable rather than a steady-state nature in terms of their magnitude and distribution within each cycle of chip formation.

5.4.3 *Contact processes at the tool/chip interface*

Evidence shows that seizure is a feature of the metal cutting tribology.[13,27] Seizure or atomic bonding between two contacting metals is possible as a result of their mutual deformation. Although the exact nature of seizure is not yet known, the experimental evidence shows that for seizure to occur, mechanical contact of two surfaces and energy imposition in the form of contact pressure and heat are necessary. When the heat energy is introduced between contacting surfaces, seizure begins to occur at certain contact temperatures. Specifically, the beginning of seizure is related to the contact temperature above one third the melting point of the material having the lower melting point.[18,40] Also, seizure does not occur at contact temperatures above one half the melting point. Therefore, seizure, or more precisely, its resulting built-up edge, is observed within a certain range of the cutting speed. As might be expected, the range limits correspond to the above-mentioned temperature points at the tool/chip interface.

When seizure occurs, the shear strength of the tool/chip bonded interface exceeds the shear strength of the workpiece material. As such, the chip moves over a certain surface of failure situated within the work material. In metal cutting, such a phenomenon is referred to as "internal friction".[13] Because the shear strength of the workpiece material may be assumed constant, then the distance between the tool/chip interface and this sliding surface depends only on the stress distribution in the machining zone.

It is widely recognized and sincerely believed that seizure may reduce tool wear.[13,21,27,34,41,42] In this connection, there is a lot of speculation about the "protective function of the built-up edge" even at the level of undergraduate textbooks. A few explanations and supportive evidences have been provided. As such, the presence of the built-up edge on the rake face is considered to be the most solid argument. Eventually, using a special quick-stop device, one may obtain the real shape and size of the built-up edge for given cutting conditions.[13,41,42] Based on such post-process observations, it is commonly believed that the built-up edge somehow cuts the workpiece material, serving in this sense as the cutting edge extension and in doing so protecting the flank from wear.

However, the described idealistic picture does not hold when the tool life is considered. Difficulties arise when one seeks a correspondence between the maximum tool life and maximum seizure or, more precisely, the stable built-up edge. Experimental evidence provided by many researchers indisputably proves that the cutting speed (and so the contact temperature at the tool/chip interface) corresponding to the maximum tool life is 2 to 5 times higher than the cutting speed corresponding to maximum seizure or the built-up edge.[18] Moreover, no trace of seizure is observed under the maximum tool life;[13,40] therefore, this old contradiction between the tool life and the protective function of the built-up edge must be explained.

It is true that the built-up edge can be observed on the tool rake face in the region adjacent to the cutting edge, and the mechanical properties of the built-up edge are high enough to protect this edge from wear. However, the logical question arises about existence of the built-up edge while cutting. The answer to this question is not straightforward. On one hand, the built-up edge is harder than the work material as it has been severely deformed. It is known that hardened steel can cut identical non-hardened steel. On the other hand, such cutting is not stable, and the cutting regime used is much lower than that with the tool materials. Therefore, the likelihood that the built-up edge cuts the workpiece at the same regime as the tool material is quite low.

Analysis of experimental data, micrographs of the built-up edge, studies of its structure and types of bonding, and transmission electron microscope examinations enable us to suggest that the built-up edge is a post-cutting phenomenon. As discussed in Chapter 2, the deformation zone adjacent to the tool rake face is dynamic and ever-changing. The shape and location of this zone depends on the properties of the work and tool materials, chosen cutting regime, contact conditions at the tool/chip interface, and tool geometry. It was also discussed in Chapter 2 that this zone is almost always "coated" (from the chip side) by the material which would be the chip contact layer through its spreading over the tool/chip interface. Because the size of the deformation zone is continuously changing during the chip formation cycle, one may gain an incorrect impression about the dynamics of this zone by analyzing its micrographs. From the micrographs, it appears that the material flows over the deformation zone considered as the built-up edge. However, a simple analysis of the stress distribution in this zone, slip-line directions, and microhardness scanning analyses show that this is not the case, although it has been discussed in this manner since the last century. The system analysis presented in Chapter 2 shows that the deformation zone pushes the chip ahead which deforms it in the direction of the maximum combined stress. If the cutting process is interrupted using a quick-stop device, part of the deformation zone adjacent to the rake face is observed but has nothing to do with the built-up edge as a real phenomenon in metal cutting. Simply, when the contact conditions at the tool/chip interface exist such that the adhesion forces between the deformation zone are higher that those between the deformation zone (or, more presicely, between the coating of the deformation zone) and the rest of workpiece material, this zone can be observed in a quick-stop test or even after the process is terminated in a common way. If the reverse is true, the so-called built-up edge is not observed.

5.5 Comparison with machining experiments

Comparison of the calculated and experimentally obtained data is an essential part of any FEM modeling. Because the calculated and the experimental data commonly have different formats, a solid base for such a comparison has to be established at this stage. Specifically, the correlation between the in-process and post-process phenomena has to be identified. Unfortunately, this is not the case in known FEM simulations of the cutting process. For example, Komvopoulos and Erpenbeck[5] and Lin and Lin[8] converted the results of FEM simulation of orthogonal metal cutting into the cutting force and then compared them with the experimental cutting force obtained from turning experiments. The type, geometry, and material of cutting inserts; the cutting regime used; and the method of the cutting force determination have not been reported. Strenkowski and Moon[7] compared the results of FEM simulation of temperature distribution in orthogonal cutting with those obtained experimentally using the work-tool thermocouple method. As discussed in Chapter 6, this experimental method allows the measure of the average integral temperature on the tool/chip interface. It is not clear how to compare this temperature with the temperature distribution at the tool/chip interface. Zhang and Bagchi[9] attempted to verify the results of FEM simulations using the experimental results on the cutting force and shear strain distribution in the chip obtained by Wienman and von Turcovich.[43] Among the experimental conditions used in these experiments, only the tool rake angle is reported. It is also not clear how to compare the experimental results of strain distribution with the simulated shear stress distribution. Therefore, to compare the results of FEM simulations with those obtained experimentally, a suitable methodology should be found.

The use of the technique of photo-elasticity was not considered as it involves so many uncertainties and approximations that it may only be used to obtain a qualitative description of the stress distributions in the machining zone. It is understood that such a technique cannot be used to verify analytical results.

Employing a technique involving the use of a split cutting tool,[38] the stress distribution along the tool/chip interface — without significant influence of the so-called primary deformation zone — is obtainable. However, it is not clear how the clearance on the tool rake face would influence the contact conditions at the tool/chip interface.

A promising avenue for the study of stress distribution is the microhardness scanning procedure.[38,43–50] In this procedure, a polished specimen is systematically indented until a sufficiently detailed map of harness numbers is obtained. Although this technique is well defined, difficulties arise when one seeks an explanation for the obtained results. By this it is meant that, because it has been established that the chip formation process is cyclical, it might be expected that the stress and strain distribution in the chip will strongly depend on the instant of time within a chip formation cycle at which the cutting process has been interrupted to obtain a "frozen picture" for the study. Therefore, for the discussed technique to be useful, two major aspects should be identified. First, a correlation between microhardness and the process parameters has to be established. Second, a comparison has to be made between the theoretical and experimental results obtained for the same instant within a chip formation cycle.

As discussed in Chapter 4 the microhardness (*HV*) of the plastically deformed material is uniquely related to the shear stress gained at the last stage of deformation (Figure 4.5) as follows:[48]

$$\sigma_{xy} = 0.185\,HV \tag{5.6}$$

Because this correlation does not depend on the cutting conditions (geometry, regime, temperature, etc.),[18] it is possible to obtain information about the level of shear stress using the results of microhardness measurements.

To obtain samples of a partially formed chip at a defined instant in a chip formation cycle, a computer-triggered, quick-stop device has been used. A command for triggering was given from the force transducer at the beginning of a chip formation cycle that allows it to freeze the stress picture corresponding to that instant. Such an innovation enables one to compare the calculated and the experimental results at the same instant of time.

The orthogonal cutting tests were carried out on a lathe. A two-component force dynamometer was designed on the base of a two–component load washer, Kistler Type 9065. The dynamometer design and its static and dynamic calibrations are similar to those discussed in Hayajneh et al.[51] and Chapter 6.

The output of the dynamometer was connected to a computer-controlled data acquisition system which was triggered off the quick-stop device at the beginning of a chip formation cycle. The instant of this beginning was recognized by the minimum value of the penetration force as registered by the dynamometer.

Three sets of the cutting tools having rake angles of +18°, 0°, and –18° were prepared with a 7° clearance angle. The tool material was P10. The geometry parameters of tools were controlled according to standard ANSI B94.50–1975. Tolerances for all angles were ±0.5°. The roughness, R_a, of the face and flank did not exceed 0.25 μm and was measured according to standard ANSI B46.1–1978. Each cutting edge was examined at a magnification of 15× for visual defects as chip of cracks.

AISI 1045 HR ASTM A576 RD steel was primarily used in experiments to compare the calculated and experimentally obtained results. In addition, workpieces made of AISI 303 ANN CF RD ASTM A582 93 and AISI 4340 HR ANN RD ASTM 322 91 steels were also used to verify tendencies in the results. The composition, element limits, and the deoxidation practice were chosen to comply with the requirements of standard ANSI/ASME B94.55M–1985. To simulate the true orthogonal cutting conditions, the special specimens were prepared as discussed in Chapter 6. After being machined to the configuration, the specimens were tempered at 180 to 200°C to remove the residual stresses. The hardness of each specimen was determined over the entire working part. Cutting tests were conducted only on the bars where the hardness was within the limits of ±10%. Special parameters of the microstructure of the bars such as the grain size, inclusions count, etc. were determined for the initial workpieces structures (shown in Figures 5.11a, 5.12a, and 5.13a) using quantitative metallography.

The test regime was set so that the conditions would be close to those used in calculations. A time constant (a system lag) for the quick-stop device was determined, and the triggering was shifted by programming the data acquisition system.

The samples of the chip, obtained in cutting experiments, are shown in Figures 5.11b, 5.12b, and 5.13b. A microhardness survey of the chips was undertaken to map out different zones through the chip section. The microhardness tests were carried out according to the ASTM E384–84 standard. In this study, a Leco M-400-G2 microhardness tester was used. This tester supports the mounting specimens and permits the indenter and specimen to be brought into contact gradually and smoothly under a predetermined load. The tester has a feature such that no rocking or lateral movement of the indenter or specimen is permitted while the load is being applied or removed. A measuring microscope is mounted on the tester in such a manner that the indentation may be readily located in the field of view. The following steps were performed for the microhardness tests:

1. Verification of the microhardness tester by making a series of indentations on standardized hardness test blocks to confirm that the tester has not become maladjusted in the intervals between periodic routine checks assuring reliable results.
2. Preparation of the specimens as mentioned above.
3. Supporting the specimens so that the test surface was normal to the axis of the indenter.
4. Selecting the correct objective by adjusting the optical equipment.
5. Deciding on the test parameters where the test load was fixed by means of two dials towards the top right of the machine and the loading speed was set in conjunction with the test load to give the largest measurable indentation. A 25-g load was selected so that tests could be performed very close to the edges.
6. Using a highly polished, pointed, and square-based pyramidal diamond with face angles of 136° to indent the specimen.

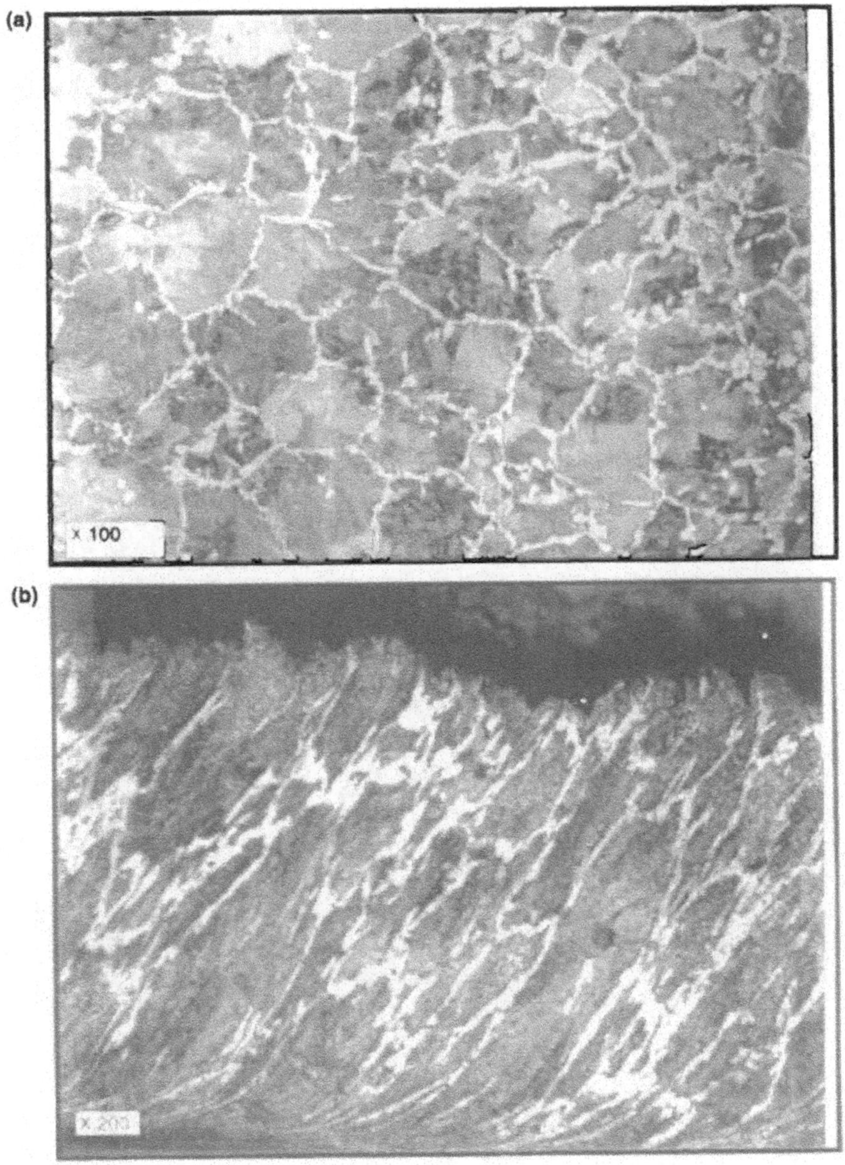

Figure 5.11 Micrographs of the initial structure of **(a)** AISI 1045 steel and **(b)** the chip structure obtained in orthogonal cutting test with cutting speed of 60 m/min, equivalent cutting feed of 0.2 mm/rev. Etched with 10 mL nital, 90 mL alcohol.

7. Measuring the diagonals of the indentation by a micrometer screw.
8. Reading the microhardness value from the hardness display.

(a)

(b)

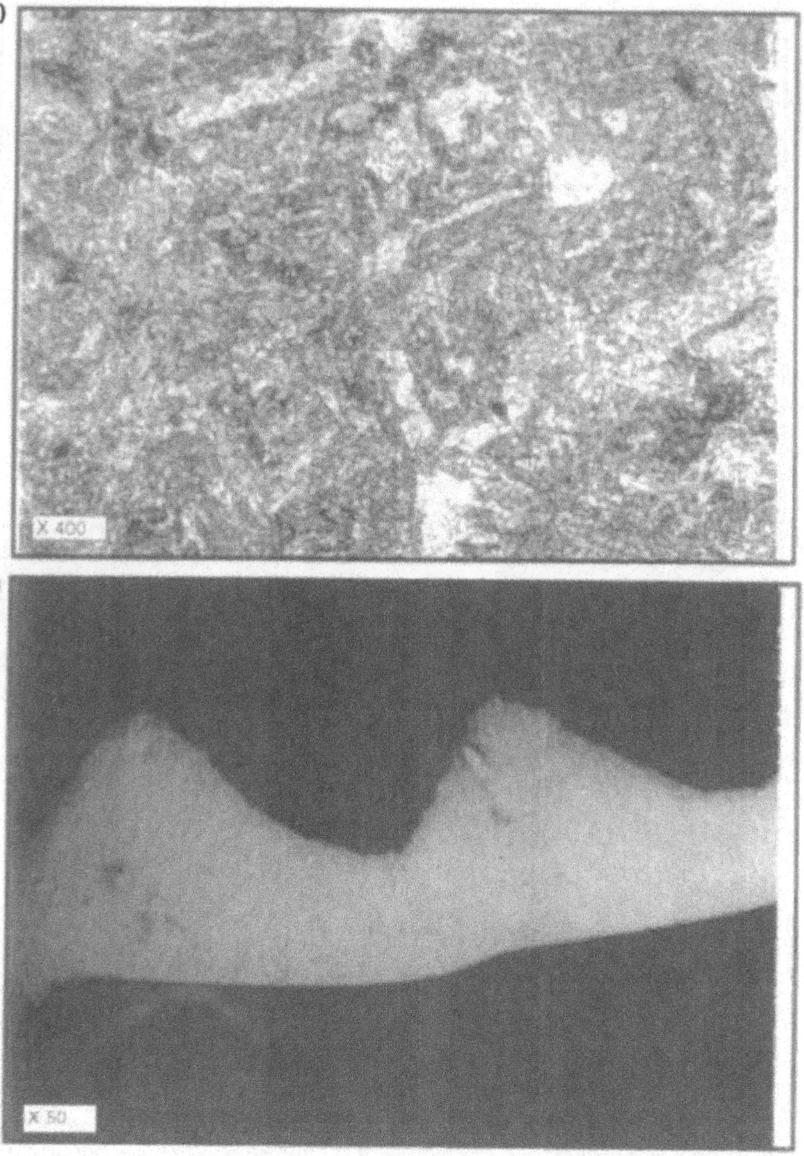

Figure 5.12 Micrographs of the initial structure of **(a)** AISI 4340 steel and **(b)** the chip structure obtained in orthogonal cutting test with cutting speed of 60 m/min, equivalent cutting feed of 0.2 mm/rev. Etched with 10 mL nital, 90 mL alcohol.

A carefully selected number of test coupons were designed for microhardness scanning in order to determine both the hardness level and hardness distribution inside the deformation zone on a plane normal to the cutting edge. Experimentally obtained microhardnesses at each point of

(a)

(b)

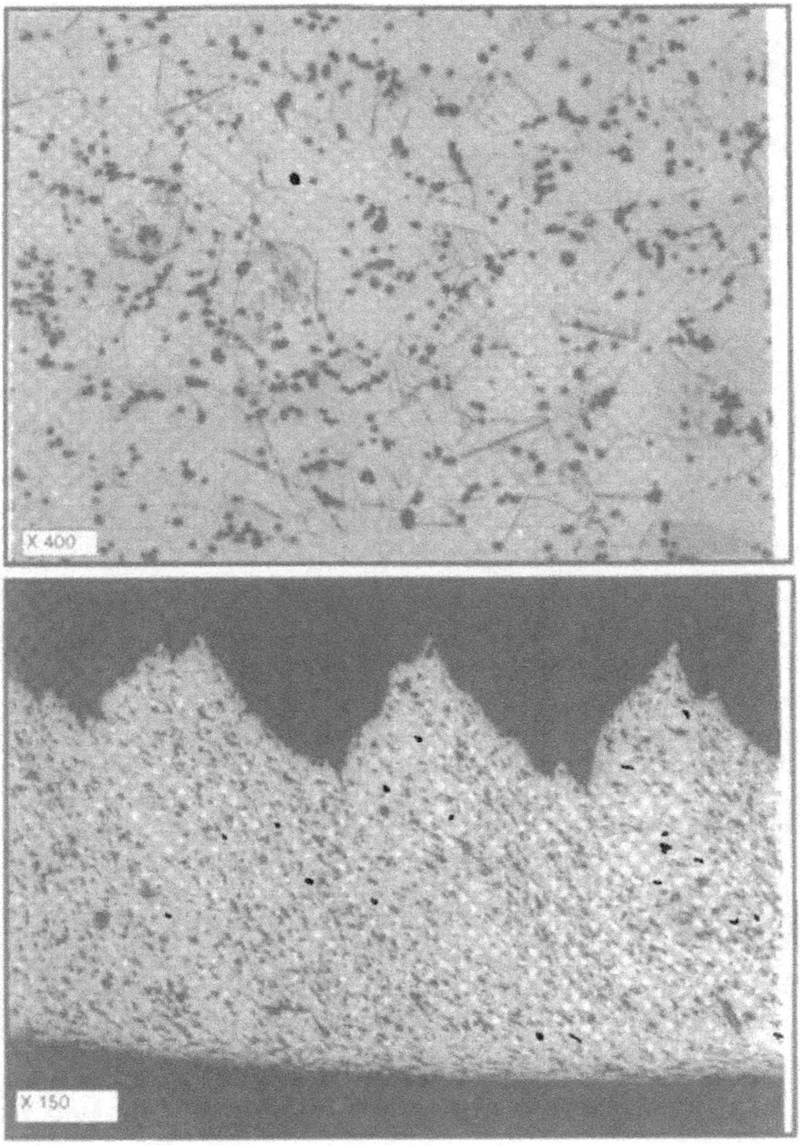

Figure 5.13 Micrographs of the initial structure of **(a)** AISI 303 stainless steel and **(b)** the chip structure obtained in orthogonal cutting test with cutting speed of 60 m / min, equivalent cutting feed of 0.2 mm / rev. Etched with 40 mL hydrofluoric acid (HF), 20 mL nitric acid, 40 mL glecren.

indentation were recalculated into the stress using Equation (5.6). The results of the comparison of measured and calculated stress distribution are shown in Table 5.2. The small differences between the predicted and the experimental

Table 5.2 Comparison between calculated and experimentally obtained shear stresses (GPa)

Workpice material	Chip formation cycle	Place of measurement/calculation					
		Close to contact zone		Middle section		Close to free surface	
		Pred.	Exp.	Pred.	Exp.	Pred.	Exp.
AISI 1045	Beginning	0.30	0.32 ± 0.08	0	0	0.40	0.38 ± 0.05
	Middle	0.90	1.00 ± 0.10	0.40	0.43 ± 0.07	0.60	0.62 ± 0.06
	End	1.20	1.30 ± 0.07	0.90	0.98 ± 0.02	0.90	1.1 ± 0.05
AISI 4340	Beginning	0.32	0.31 ± 0.09	0	0	0.43	0.46 ± 0.06
	Middle	1.00	1.05 ± 0.09	0.50	0.56 ± 0.08	0.72	0.78 ± 0.09
	End	1.40	1.48 ± 0.09	1.10	1.26 ± 0.07	0.95	1.28 ± 0.04
AISI 303	Beginning	0.26	0.24 ± 0.03	0	0	0.25	0.22 ± 0.09
	Middle	0.75	0.70 ± 0.07	0.32	0.29 ± 0.05	0.40	0.37 ± 0.08
	End	0.95	0.87 ± 0.09	0.75	0.71 ± 0.09	0.78	0.76 ± 0.04

Note: Pred. = predicted; Exp. = experiment.

results show no regular trend, and this was expected. It is partially explainable by unavoidable non-homogeneity of the workpiece and tool materials. Even though the standard mechanical tests of all samples used in experiments were carried out using the *t*-test, it would seem reasonable to assume the existence of certain scatter between the behavior of the workpiece material in the standard tensile test and in cutting, where combined rather than simple deformation takes place. However, the prediction accuracy is at the acceptable level for complicated experiments in metal cutting.

References

1. Carroll, J.T. and Strenkowski, J.S., Finite element models of orthogonal cutting with application to single point diamond turning, *Int. J. Mech. Sci.*, 30(12), 899, 1988.
2. Iwata, K., Osakada, K., and Terasaka, Y. Process modeling of orthogonal cutting by the rigid-plastic finite element method, *ASME J. Eng. Mater. Technol.*, 106, 132, 1985.
3. Klamencki, B.E., Incipient Chip Formation in Metal Cutting — A Three-Dimension Finite Element Analysis, Ph.D. thesis, University of Illinois, Urbana-Champaign, IL, 1983
4. Strenkowski, J.S. and Carroll, J.T., A finite element model of orthogonal metal cutting, *ASME J. Eng. Industry*, 107, 349, 1985.
5. Komvopoulos, K. and Erpenbeck, S.A., Finite element modeling of orthogonal metal cutting, *ASME J. Eng. Industry*, 113, 253, 1991.
6. Dokainish, M.A., Elbestawi, M.A., Polat, U., and Tole, B., Analysis of stresses during exit in interrupted cutting with chamfered tools, *Int. J. Mach. Tools Manuf.*, 29(4), 519, 1989.

7. Strenkowski, J.S. and Moon, K.-J., Finite element prediction of chip geometry and tool/workpiece temperatures distributions in orthogonal metal cutting, *ASME J. Eng. Industry*, 112, 313, 1990.
8. Lin, Z.C. and Lin, S.Y., A coupled finite element model of thermo-elastic-plastic large deformation for orthogonal cutting, *ASME J. Eng. Industry*, 114, 218, 1992.
9. Zhang, B. and Bagchi, A., Finite element simulation and comparison with machining experiment, *ASME J. Eng. Industry*, 116, 289, 1994.
10. Shih, A.J., Finite element simulation of orthogonal metal cutting, *ASME J. Eng. Industry*, 117, 84, 1995.
11. Lin, Z.C., Pan, W.C. and Lo, S.P., A study of orthogonal cutting with tool flank wear and sticking behavior on the chip-tool interface, *J. Mater. Proc. Technol.*, 52, 524, 1995.
12. Obikawa, T. and Usui, E., Computational machining of titanium alloy — finite element modeling and a few results, *ASME J. Manuf. Sci. Eng.*, 118, 208, 1996.
13. Zorev, N.N., *Metal Cutting Mechanics*, Pergamon Press, New York, 1966.
14. Owen, D.R.J. and Hilton, E., *Finite Elements in Plasticity*, Pineridge Press, Swansea, U.K, 1980.
15. Bathe, K.-J., *Finite Element Procedures*, Prentice Hall, Englewood Cliffs, NJ, 1996.
16. Hertzberg, R.W., *Deformation and Fracture Mechanics of Engineering Materials*, 3rd ed., Wiley, NewYork, 1989.
17. *Metal Handbook*, Vol. 1, 10th ed., American Society for Metals, Metals Park, OH, 1990.
18. Astakhov, V.P. and Shlafman, N.L., *Mathematical Modeling of Machine Tools and Machine Tool Complexes* (in Russian), O.P. University Press, Odessa, 1992.
19. Dieter, G.E., *Mechanical Metallurgy*, 3rd ed., McGraw-Hill, New York, 1986.
20. Oxley, P.L.B., *Mechanics of Machining: An Analytical Approach to Assessing Machinability*, John Wiley & Sons, New York, 1989.
21. Shaw, M.C., *Metal Cutting Principles*, Clarendon Press, Oxford, 1984.
22. Astakhov, V.P., Shvets, S.V., and Osman, M.O.M., Chip structure classification based on mechanism of its formation, *J. Mater. Proc. Technol.*, 71/2, 247, 1997.
23. Astakhov, V.P. and Osman, M.O.M., An analytical evaluation of the cutting forces in self- piloting drilling using the model of shear zone with parallel boundaries. Part 1. Theory, *Int. J. Machine Tools Manuf.*, 36(11), 1187, 1996.
24. Merchant, M.E., Mechanics of metal cutting process, *J. Appl. Phys.*, 16, 267, 1945.
25. Chao, B.T. and Trigger, K.J., Cutting temperatures and metal cutting phenomena, *ASME J. Eng. Industry*, 73, 777, 1951.
26. Finnie, I. and Shaw, M.C., The friction process in metal cutting, *ASME J. Eng. Industry*, 78, 1649, 1956.
27. Trent, E.M., *Metal Cutting*, Butterworth-Heinemann, London, 1991.
28. Palmer, W.B. and Oxley, P.L.B., Mechanics of orthogonal machining, *Proc. Inst. Mech. Eng.*, 173(24), 623, 1959.
29. Hsu, T.S., A study of the normal and shear stresses on a cutting tool, *ASME J. Eng. Industry*, 88, 51, 1966.
30. Takeyama, H. and Usui, E., The effect of tool-chip contact area in metal cutting, *ASME J. Eng. Industry*, 80, 1089, 1958.
31. Usui, E. and Takeyama, H., A photoelastic analysis of machining stresses, *ASME J. Eng. Industry*, 82, 303, 1960.
32. Kato, S., Yamaguchi, K., and Yamada, M., Stress distribution at the interface between tool and chip in machining, *ASME J. Eng. Industry*,94, 683, 1972.

33. Kamskov, L.F., On the external friction in cutting of plastic metals (in Russian), *Vestnic Machinostroeniya*, No. 5, 68, 1959.
34. Loladze, T.N., *Wear of Cutting Tools* (in Russian), Mashgiz, Moscow, 1958.
35. Chandrashekaran, H. and Kapoor, D.V., A photoelastic analysis of tool-chip interface stress, *ASME J. Eng. Industry*, 87, 495, 1965.
36. Childs, T.H. and Mahdi, M.I., On the stress distribution between the chip and tool during metal turning, *CIRP Ann.*, 38(1), 55, 1989.
37. Kufarev, G.L., *Chip Formation and Surface Finish in Oblique Cutting* (in Russian), Frunze, Mektep, 1970.
38. Poletica, M.F., *Contact Stresses on Working Surfaces of Cutting Tools* (in Russian), Mashinostroenie, Moscow, 1969.
39. Vidosic, J.P., *Metal Machining and Forming Technology*, The Ronald Press Company, New York, 1964.
40. Astakhov, V.P. and Osman, M.O.M., Correlations amongst process parameters in metal cutting and their use for establishing the optimum cutting speed, *J. Mater. Proc. Technol.*, 62, 175, 1996.
41. Takeyama, H. and Ono, T., Basic investigation of built-up edge, *ASME J. Eng. Industry*, 90, 335, 1968.
42. Boothroyd, G. and Knight, W.A., *Fundamentals of Machining and Machine Tools*, 2nd ed., Marcel Dekker, New York, 1989.
43. Wienman, K.J. and von Turcovich, B.F., Mechanics of tool-workpiece engagement and incipient deformation in machining of 70/30 brass, *ASME J. Eng. Industry*, 93, 1079, 1971.
44. Rozenberg, A.M. and Yeremin, A.N., *Elements of Metal Cutting* (in Russian), Mashgiz, Moscow, 1956.
45. Zlatin, N. and Merchant, M.E., The distribution of hardness in chips and machined surface, *ASME J. Eng. Industry*, 94, 683, 1972.
46. Dell, G.D., *Stress Determination in Plastic Regions Using Microhardness Distribution* (in Russian), Machinostroenie, Moscow, 1971.
47. Weinmann, K.J., The use of hardness in the study of metal deformation processes with emphasis on metal cutting, in *Proc. Symp. Material Issues in Machining and The Physics of Machining Processes*, American Society of Mechanical Engineers, Warrensdale, PA, 1992, p. 1.
48. Rozenberg, A.M., and Rozenberg, O.A., State of stresses and deformations under metal cutting (in Russian), *Sverhtverdye Materialy*, No. 5, 41, 1988.
49. Kishawy, H.A. and Elbestawi, M.A., Effect of process parameters on chip morphology when machining hardened steel, in *Manufacturing Science and Engineering*, Vol. 6(2), Proc. 1997 American Society of Mechanical Engineers International Mechanical Engineering Congress and Exposition, November 16–21, Dallas, TX, 1997, p. 21.
50. Jain, V.K., Kumar, S., and Lal, G.K., Effect of machining parameters on the microhardness of chips, *ASME J. Eng. Industry*, 111, 220, 1989.
51. Hayajneh, M.T., Astakhov, V.P., and Osman, M.O.M., An analytical evaluation of the cutting forces in orthogonal cutting using the dynamic model of shear zone with parallel boundaries, *J. Mater. Proc. Technol.*, 82(1–3), 220, 1998.

chapter six

Methodology of experimental studies in metal cutting

Experimentation has been described as "an art that must be learned but cannot be taught."[1] Such a statement is probably true if it means that experimentation is the total act of discovery and acquisition of new knowledge. Some experimenters, with the most meager and most questionable data, leap to great new concepts, while others, after an exhaustive investigation, miss the obvious. Yet, if one cannot learn the art of discovery, one can surely learn the art of *preparing* for discovery. An experimenter can be trained to exclude or account for the random effects of the environment, to plan and space the testing sequence, to evaluate possible errors and their cumulative effects, to check and cross-check the developing data, and to lay out these data in an orderly and revealing manner. Then, if discovery, great or small, is possible, it is most likely to be made. It is this part of the art of experimentation that forms the experimental methodology part. Basic principles and rules of the experimental methodology in metal cutting as well as instrumentation design, calibration, and application are the main focus of this chapter.

6.1 Similarity methods in metal cutting

6.1.1 General

Experimental methods must evolve and become more delicate, refined, and generalized to match the increasing rate at which theories develop and scientific information accumulates. Because, in metal cutting studies, the experiment remains the essential instrument of knowledge, particularly important are the theory of similarity and the theory (and practice) of simulation in their new broad sense, in which they make it possible to condense experimental information, in which they are the basis for experimentation and give guidance on the formulation of tests. Similarity and simulation facilitate a unified description of a great variety of machining operations.

The theory of similarity permits the proper generalization of the results from a series of cutting tests conducted over a rather limited range of test conditions to a wide class of physically similar processes. Simulation in its various forms is essential for the formulation of experiments for a variety of different purposes. Although experimentation is still very useful in studying metal cutting, experimental methodology seems to be the most mistreated issue. The books on metal cutting published in the last 30 years are not of much help, as they offer a very general description of the measurement principles of cutting parameters but do not offer experimental methodology useful for laboratories and shops. Therefore, it would be convenient at this time to give a brief introduction to the relation between theory and experiment, the role of models, the development of simulations, and the types of simulations.

The growth of any science, and particularly metal cutting, begins with experimental investigation. From observation and experiment to the theoretical mode of thinking and then to the specially restricted production process — such is the path of technical and scientific growth.

In the development of science, an important role has been, is, and will be played by scientific hypothesis — definite predictions based first of all on very little experimental data, detailed observations, and guesses. Sometimes hypotheses anticipate experimental investigations (the atomic hypothesis, for example), which in ancient philosophy had primarily a poetic character and was verified only after the passage of many centuries from its conception.

In technical science, it is required that hypotheses which are put forward should be rapidly and completely verified and translated into theory, which can be done only after appropriate generalization and, usually, a specially constructed experiment.

Generalization of experimental data can be obtained rapidly if experiments are designed and results analyzed correctly, which can be guaranteed by the theory of similarity and simulation. In the definition, formulation, and verification of the correctness of hypotheses, analogy as a method of inference is very important. The theory of similarity treats analogy as a type of simulation and indicates a method of using analogy in scientific and practical investigations.

The volume of information obtained during the study of phenomena by methods of simulation must be put in order. This is achieved with the aid of the theory of similarity by means of which, from the given characteristic of one phenomenon, judgments can be formed about large groups of phenomena which are in some sense similar to the first one. The similarity of phenomena assumes that data concerning a process obtained in studying one phenomenon can be extended to all phenomena similar to the given one. The characteristics of any phenomenon in a group of similar phenomena can be obtained by simply "scaling" the known characteristics of the one phenomenon to the characteristics of any other in the group, reducing tremendously in this way the necessary number of experiments. The theory of similarity can be applied to:

- The analytic search for relations, correlations, and solutions of actual problems.
- The analysis of the results of experimental investigations and the testing of various industrial equipment when these results are represented in the form of general criteria.
- The creation of models — equipment reproducing phenomena in other equipment (the original) that is usually larger and more expensive than the model.

6.1.2 Basics of similarity

The study of the properties of similar phenomena and the methods of establishing similarity comprise the content of the theory of similarity.[2] Before proceeding to the problem of similarity, it must be stipulated that it may be efficiently applied only when the phenomenon under study is treated as a system having certain individual components. To each change in the state of a system, generally occurring in space and time, there corresponds one or several processes. During the occupancy of the process, the quantities characterizing the state of the system are altered. These quantities are called the parameters of the process. The system in which the process occurs consists of components which are characterized by their parameters or parameters of the system; therefore, the system approach in metal cutting discussed in the previous chapters is mandatory if the theory of simulation is to be applied to study the cutting process.

In the study of cutting systems, the components' parameters are the tool cutting angles (in the tool-in-hand and setting systems), mechanical and physical properties of workpiece and tool materials, parameters of the cutting regime used, etc. The parameters of the cutting system are cutting temperature, cutting force, chip velocity, tool wear, etc.

The parameters of the components of the system can either be considered as constant during the course of the whole process or changing in time and space. If the system parameters change as one of more of the dependent parameters of the process being studied varies, then they are said to be non-linear, and accordingly the system is called non-linear.

In order to describe processes, it is necessary to introduce a coordinate system in which the mathematical expression for the process can be given, usually in the form of equations linking the parameters of the process with the system parameters. The form of the equation for the process depends on the chosen coordinate system. Thus, a phenomenon is the totality of the processes which can be described by equations linking the parameters of the processes with the system parameters in a chosen coordinate system.

The theory of similarity is of considerable importance in modeling various phenomena. This modeling is used to replace the study of the natural phenomenon of interest by the study of analogous phenomenon in a model (having well-defined parameters of the components) under special laboratory conditions (where it is relatively easier to measure the system parameters with necessary accuracy).

Modeling is normally based on the analysis of physically similar phenomena. Physical similarity may be considered as a generalization of geometrical similarity. Geometrical similarity assumes that two geometric figures are similar if the ratios of all corresponding dimensions are identical. If the similarity ratio and the scale are known, then simple multiplication of the dimensions of one geometric figure by the scale factor yields the dimensions of the other.

By analogy, two phenomena are similar to one another if the characteristics of one can be obtained from the assigned characteristics of the other by a simple conversion, which is analogous to conversion from one coordinate system (or system of units) to another. Therefore, the scaling factor should be known to accomplish the conversion.

The numerical characteristics of two different but similar phenomena can be considered as the numerical characteristics of the same phenomenon expressed in two different system of units. All the dimensionless characteristics (dimensionless combinations of dimensional quantities) of a set of similar phenomena have the same numerical value. Of great importance is that the reverse is also correct —that is, if all dimensionless characteristics of two phenomena are identical, then the phenomena are physically similar even though they can take place in considerably different physical systems. The latter property is significant for metal cutting for which a variety of physical systems such as turning, milling, broaching, drilling, etc. exist. Because the cutting theory is commonly derived using one model of chip formation and the experimental results are obtained using one machine and one process (turning, drilling, etc.), a key question here is how to apply the obtained theoretical and experimental results to each particular system. The use of similarity is the proper way. As such, a set of similar cutting processes constitutes a mode of cutting.

The similarity of two phenomena can sometimes be understood in a broader sense by assuming that the above definitions refer only to certain special physical systems of parameters. These parameters define the phenomena completely and thus are sufficient to find any other characteristics, although the latter could not be obtained by simply scaling when transforming from one to the other similar phenomenon. For example, any two ellipses can be considered to be similar in this sense only when, for both ellipses, the Cartesian coordinates are directed along their major and minor axes. Then the coordinates of any point of the considered ellipses can be obtained in terms of the coordinates of points of some particular ellipse (affine similarity) by simple conversion.

After the system of parameters defining the particular class of phenomena has been established, the similarity conditions can be established. In fact, let a phenomenon be defined by n parameters, some of which can be dimensionless and others dimensional physical constants. Furthermore, it can be assumed that the dimensions of the variable parameters and physical constants are expressed by means of k fundamental units. It is evident that the maximum number of the independent dimensionless characteristics of the phenomenon is $n - k$. Therefore, a certain basic system, which defines all the remaining quantities, can be selected from all the dimensionless quantities formed by the characteristics of the phenomenon.

The foregoing considerations suggest that the necessary and sufficient condition for two phenomena to be similar is that the numerical values of the dimensionless combinations forming the basic system are constant. This condition is referred to as the similarity criteria.[2] When the similarity conditions are fulfilled, the scale factors for the corresponding quantities are needed to calculate the conditions of the full-scale phenomenon from model data.

As an example, consider a long boring bar for deep-hole drilling.[3] One of the major problems with such bars is their low torsional rigidity under cutting forces and their own weight. Because the bars are made over a significant range of lengths, similarity in their design must be established. Because the bar is made of homogeneous material, its elastic properties are determined by two mechanical constants, namely Young's modulus E, Pa (= N/m^2), and the dimensionless Poisson's ratio μ_p. Because the weight of the bar is an essential characteristic, the specific gravity $\gamma_g = \rho_m g$, N/m^3 must appear as a characteristic parameter. The cutting forces applied to the bar and its weight act on the bar. The magnitude of the resultant force is $F_{c\Sigma}$, N. The stress τ_b, N/m^2 develops in the boring bar under the action of its weight and a given cutting force distribution. In order to define all the model dimensions, it is sufficient to assign a certain characteristic dimension B_h, m. It can be assumed that τ_b is the maximum value of some stress components or, in general, a certain stress component acting on the critical cross-section of the bar. Then, the system of characteristic parameters will be

$$E, \mu_p, B_h, \gamma_g = \rho_m g, F_{c\Sigma}, \tau_b \tag{6.1}$$

The combination τ_b/E is dimensionless; consequently:

$$\frac{\tau_b}{E} = f\left(\mu_p, \frac{E}{\rho g B_h}, \frac{F_{c\Sigma}}{E B_h^2}\right) \tag{6.2}$$

Now, if the model of the boring bar and the actual bar are made of the same material, then $E = const$; therefore, the stress in the corresponding points will be identical in mechanically similar states. Hence, failure occurs both in the model and in reality for similar states. If the resultant cutting force acting on the bar is large compared to the weight of the bar, then the parameter $\gamma_g = \rho_m g$ and, therefore, $E/(\rho_m g B_h)$ are not essential. As such, Equation (6.2) becomes:

$$\frac{\tau_b}{E} = f\left(\mu_p, \frac{F_{c\Sigma}}{EB_h^2}\right) \tag{6.3}$$

and the similarity conditions reduce to the two conditions:

$$\mu_p = const \quad or \quad \frac{F_{c\Sigma}}{EB_h^2} = const \tag{6.4}$$

Hence, it means that the resultant cutting force must be proportional to the square of the linear dimensions when modeling with material properties conserved.

All the forms of similarity are subjected to certain general laws which are usually called theorems of similarity. There are three theorems. The first and second theorems are obtained by starting from the assumption that phenomena whose similarity is already known are addressed in the current discussion. Both theorems establish relations between the parameters of similar phenomena, without indicating methods for determining whether the phenomena are similar or how to realize similarity through the construction of models. The answer to this last question is given by the third theorem, known as the Kirpichev-Gukhman theorem, which defines the necessary and sufficient conditions for the similarity of phenomena as the corresponding parameters in the uniqueness conditions being proportional and the similarity criteria of the phenomenon studied being equal.

The second theorem states that every complete equation of a physical process, formulated in a specific system of units, can be represented in the form of an equation between similarity criteria (i.e., non-dimensional ratios constructed from the parameters occurring in the equation). The first theorem states that similar processes are describable by homogeneous equations.

6.1.3 *Applications in metal cutting studies*

6.1.3.1 *Optimum cutting temperature — Makarow's law*

Silin[4] was the first, and unfortunately so far the last, to attempt to apply the basic concepts of the theory of similarity to metal cutting. The objective of his investigations was twofold. First, in applying the theory of similarity to metal cutting it was assumed that the single shear plane model of chip formation would represent the cutting process adequately so that this model could be used to obtain the general theoretical relationships between cutting parameters.

Second, the theory of similarity has been utilized to generalize the results of experimental studies.

Maybe the greatest advantage of the Silin approach is the use of physical optimization of the cutting process based on a breakthrough discovery made by Makarow.[5,6] The basic idea of the discovery can be formulated as the First Metal-Cutting Law (Makarow's law):

> For each given combination of tool and workpiece materials, there is the cutting temperature, referred to as the optimum cutting temperature, at which the combination of minimum tool wear, minimum stabilized cutting force, and highest quality of the machined surface is achieved. This temperature is invariant to the way it has been achieved.

The cutting temperature is determined as the average integral temperature on the tool/chip interface so that it can be measured by the tool-work thermocouple technique.

A cutting regime corresponding to the optimum cutting temperature (OCT) is referred to as the optimum cutting regime and, as the optimum temperature is invariant to the way it has been achieved, there is an infinite number of possible optimum regimes. Therefore, the independence of OCT on the geometry of the cutting tool, the coolant used in machining, cutting speed, cutting feed, etc. makes it possible to optimize the cutting process using a physical criterion of optimization.[6] Figure 6.1 illustrates the physics behind OCT.[6]

OCT can be defined using a tool-life test where the volumetric tool wear per unit length of the tool path is established as a function of the cutting temperature. The temperature corresponding to the minimum of tool wear is considered as the OCT. However, a complete tool-life test is expensive and time consuming, although it has to be carried out only once for a given combination of tool and workpiece materials. It has been suggested, therefore, that a test be conducted to determine OCT at a constant depth of cut and feed rate, varying the cutting speed and measuring the tool wear, the main component of the cutting speed, and the roughness of the machined surface simultaneously.[4] The optimum cutting speed (OCS) can be defined as corresponding to OCT for a given combination of cutting parameters. Figure 6.2 shows an example of the results where OCT was determined to be 875°C. As seen, OCT can be defined as corresponding to the minimum stabilized cutting force. To support this point, Figure 6.3 shows an example of the determination of OCT by the minimum stabilized cutting force. It is also seen from this figure that, as stated above, OCT does not depend on the particular cutting regime(s) selected for the experiment.

Makarow[5] suggested that OCT can be found without an actual cutting experiment. If the microhardnesses of the tool and workpiece materials are considered as functions of temperature, OCT can be thought of as the

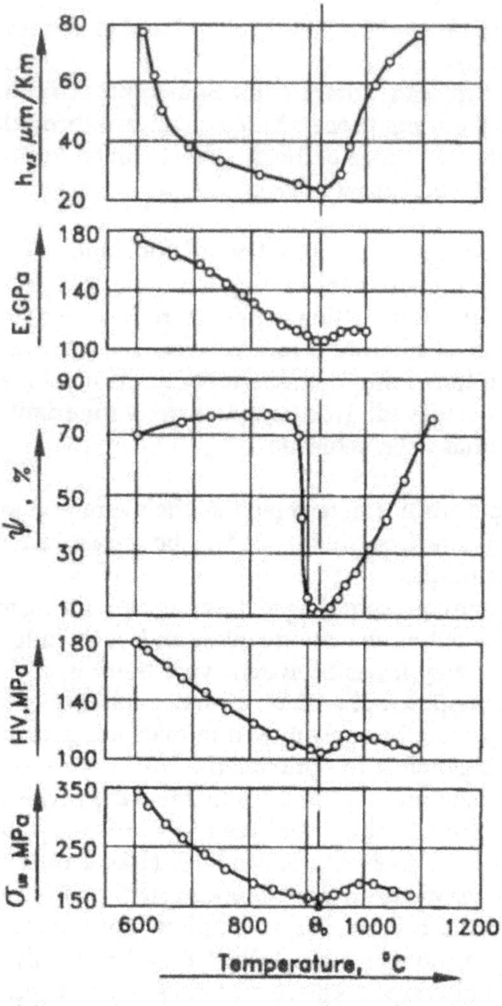

Figure 6.1 Effect of the cutting temperature on the mechanical properties (ultimate stress σ_{UTS}, elongation ψ, hardness *HV*) of the workpiece material and tool volumetric wear per length of the cutting path $h_{V,F}$. Workpiece material: pure iron; tool material: carbide P10 (79% WC, 15% TiC, 6% Co).

temperature corresponding to the maximum difference of these microhardnesses, as shown in Figure 6.4.

An interesting practical interpretation of OCT was suggested by Vinogradov.[7] In his study, the cutting speed was the main concern, so that the changes in the tool life, *T*; chip compression ratio, ζ; cutting force, R_c; and cutting temperature, Θ_c, with the cutting speed were the main focus. It was found in studies of cutting difficult-to-machine materials that particular points on the curve $T = f(v)$ correspond to specific points on curves $R_c = f(v)$, $\zeta = f(v)$, and $\Theta_c = f(v)$ (Figure 6.5). Furthermore, the specific interaction

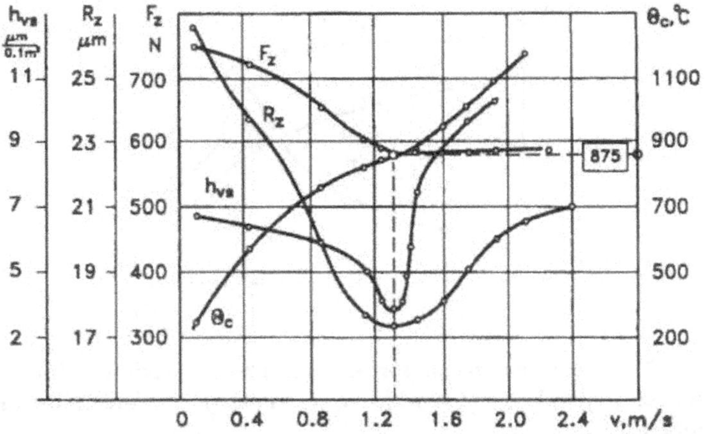

Figure 6.2 Experimental determination of OCS using longitudinal turning of 4340 steel. Tool material: carbide P10 (79% WC, 15% TiC, 6% Co). Cutting regime: f = 0.15 mm/rev, t_1 = 1 mm. Tool geometry: $\gamma_n = 10°$, $\alpha_n = 8°$, κ_r = 45 deg, $\kappa_{r1} = 25°$, r_n = 1 mm.

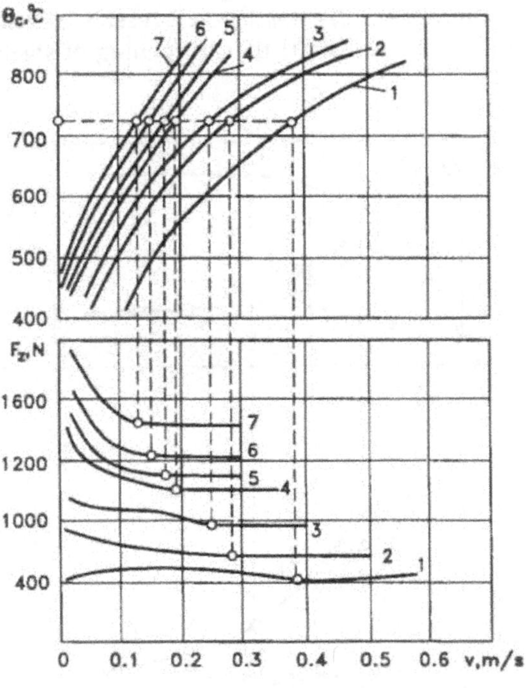

Figure 6.3 Experimental determination of OCS by the minimum stabilized cutting force. Workpiece material: nickel-based high alloy (0.08% C, 1% Cr, 56% Ni, 1% Co, 1% Al). Tool material: carbide M30 (92% WC, 8% Co). Tool geometry: $\gamma_n = 12°$, α_n = 12°, $\kappa_r = 45°$, κ_{r1} = 45 deg, r_n = 1 mm. Cutting regime: t_1 = 1 mm; *f*, mm/rev = 1, 0.074; 2, 0.11; 3, 0.15; 4, 0.25; 5, 0.30; 6, 0.34; 7, 0.39.

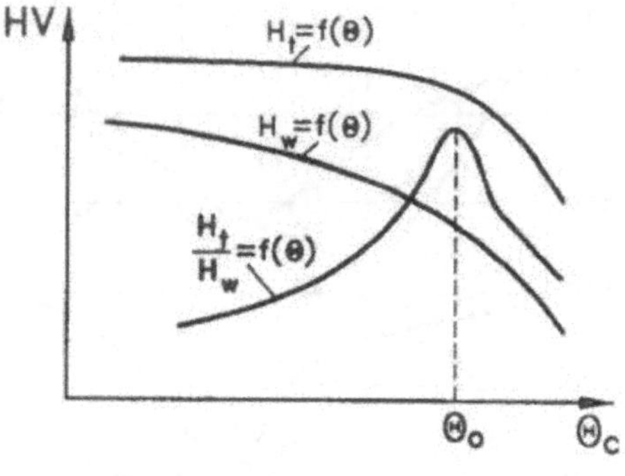

Figure 6.4 OCT is the temperature corresponding to the maximum difference of the microhardnesses of tool H_t and workpiece H_w materials considered as functions of temperature.

processes between the workpiece material and tool correspond to these points.

It has been also shown in many experimental studies of metal cutting that the curve $T = f(v)$ generally has less expressive extremes (Figure 6.5b, continuous line). It is also known[8,9] that in turning of steels and alloys, the

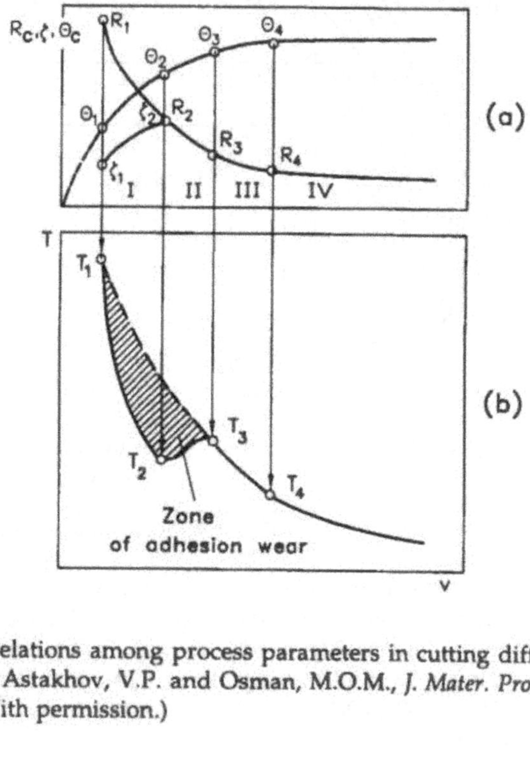

Figure 6.5 Correlations among process parameters in cutting difficult-to-machine materials. (From Astakhov, V.P. and Osman, M.O.M., *J. Mater. Process. Technol.*, 62, 175–179, 1996. With permission.)

relationships $R = f(v)$, $\zeta = f(v)$ have extremes, and the minimum tool life (point T_2 in Figure 6.5b) corresponds to the maximum cutting force and chip compression ratio. In cutting of some alloys with special properties, the curves $R = f(v)$, $\zeta = f(v)$ do not have expressive extremes, but the left part (from point T_2) of the curve $T = f(v)$ (Figure 6.5b, dashed line) has a greater slope than its right part.

The maximum of the tool life (point T_3) corresponds to the point R_3 on the curves $R = f(v)$, $\zeta = f(v)$. In the vicinity of point R_3 these curves have maximum curvature. Farther to the right of this point, they gradually transform into the curves having a moderate slope that reflects the small influence of the cutting speed (within the range considered) on the chip compression ratio and cutting force. The maximum of the tool life (point T_3) also corresponds to point Θ_3 on the curve $\Theta_c = f(v)$ (Figure 6.5a), which is the boundary point between two parts of the curve $\Theta_c = f(v)$.

In order to analyze the correlation among the cutting phenomena, the entire range of the cutting speed is divided into four distinctive regions (I to IV).[7] Region I is characterized by an intense increase in cutting temperature when cutting metals are prone to seizure. When the cutting speed is small, the thickness of the seizured layer is stable, which results in low cutting temperatures and small chip compression ratios. With increasing cutting speed, the cutting forces become great enough to make the seizured part unstable. As a result, adhesion-type tool wear takes place. However, when cutting metals are not prone to seizure, region I is also characterized by adhesion wear, and the wear rate depends upon the combination of the cutting regime, tool and workpiece material properties, and the presence of the cutting fluid.

Upon a further increase in the cutting speed (region II), the intensity of adhesion wear, the chip compression ratio, and the cutting force decrease significantly. As a result, the tool life increases. Whereas metals that are not prone to seizure, tool life continues to decrease but at a smaller rate than that in region I. Therefore, both regions I and II correspond to the predominance of adhesion wear.

An even further increase in the cutting speed (region III) significantly decreases the rate of adhesion wear while that due to diffusion becomes predominant. As a result, the tool life decreases. Further increasing the cutting speed (region IV) results in an additional increase in the cutting temperature, which leads to further increase of the diffusion wear. Because the cutting temperature in this region is high, seizure does not occur and no adhesion wear occurs.

The phenomena described above correspond to the characteristic points in the curves $\zeta = f(v)$, $R_c = f(v)$, and $T = f(v)$ (Figure 6.5). The presence of the cutting fluid does not change the characteristics of these relationships but shifts their maxima relative to those obtained without cutting fluid. Experimental verification of the above-considered phenomena is shown in Figures 6.6 and 6.7.

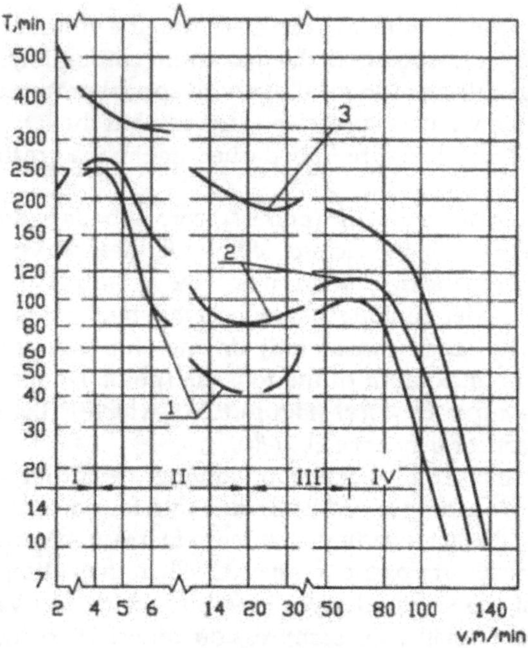

Figure 6.6 Tool life vs. cutting speed. Workpiece material: high alloy (0.2% C, 25% Cr, 20% Ni, 2%, Si). Tool material: carbide M30 (92% WC, 8% Co). Tool geometry: $\gamma_n = 12°$, $\alpha_n = 12°$, $\kappa_r = 45°$, $\kappa_{r1} = 45°$, $r_n = 1$ mm. Cutting regime: $t_1 = 2.5$ mm, $f = 0.06$ mm/rev. (From Astakhov, V.P. and Osman, M.O.M., *J. Mater. Process. Technol.*, 62, 175–179, 1996. With permission.)

In the tool life/cutting speed curves (Figure 6.6) obtained with and without cutting fluids, the cutting speed varied within the limits of 2 to 150 m/min. The maximum width of the flank wear area $VB_{Bmax} = 0.6$ mm was used was used as a tool-life criterion.

Referring to Figure 6.6, the four characteristic regions corresponding to certain cutting-speed ranges can be recognized.[7] Region I corresponds to turning with very small cutting speeds (2 to 4 m/min); within this region, the tool life increases with increasing cutting speed from 130 to 220 min in dry turning and from 215 to 255 min with the water-mix cutting fluid. Such a difference in the tool life can be explained by: (1) a change in adhesion under conditions of forming absorption films at the tool/workpiece contact area; (2) deep penetration of the cutting fluid to the contact areas (for a cutting-speed range of 2 to 3 m/min).

Region II (corresponding to a cutting-speed range of 4 to 19 m/min) is characterized by a sharp decrease in tool life with increasing cutting speed; its cause is the transition of the seizured layer from stable to unstable conditions. The tool wear initially takes place by crumbling out fine particles and

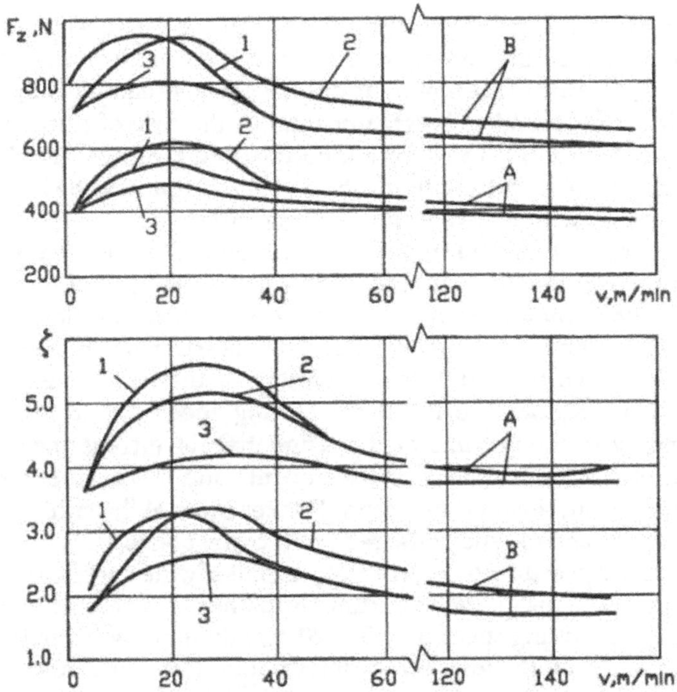

Figure 6.7 Cutting speed vs. tangential component of the cutting force F_z and chip compression ratio ζ. Workpiece material: stainless steel (0.12% C, 18% Cr, 10% Ni, 1% Ti). Tool material: carbide M30 (92% WC, 8% Co). Tool geometry: $\gamma_n = 12°$, $\alpha_n = 12°$, $\kappa_r = 45°$, $k_{r1} = 20°$, $r_n = 1$ mm. Cutting regime: $t_1 = 1.5$ mm; (A) f = 0.06 mm/rev, (B) f = 0.16 mm/rev. (1) dry cutting; (2) with water-mix cutting fluid; (3) with oil-based cutting fluid. (From Astakhov, V.P. and Osman, M.O.M., *J. Mater. Process. Technol.*, 62, 175–179, 1996. With permission.)

then, with increasing the cutting speed from 10 to 20 m/min, by rough particles of the tool carbide. The deformation of the layer being cut takes place non-uniformly, and a chip has parts of different thicknesses along its length (as discussed in Chapter 4).

With a further increase in cutting speed, up to 60 to 70 m/min (region III), the tool life, when turning without cutting fluid, increases from 40 to 100 min. On the other hand, when turning with cutting fluid, the tool life increases at a lower rate (from 90 to 118 min). The increase in the tool life within this speed range is explained by a gradual transition from the adhesion type of tool wear into the diffusion type. The deformation of work material takes place uniformly, and the thickness of chip becomes uniform along its length. Region IV corresponds to machining at a high cutting speed. The decrease in the tool life within this region is explained by high cutting temperatures.

When using the oil-based cutting fluid in the cutting-speed range of 2 to 80 m/min, the tool life/cutting speed curve does not have the expressive maxima (i.e., the rate of the tool wear does not change significantly). The tool life in this case is found to be greater than in the case of turning with the water-mix cutting fluid. Thus, for a cutting-speed range of 12 to 30 m/min when using oil-based cutting fluid, the tool life is found to be 2.7 to 5 times higher than that with the water-mix cutting fluid.

The comparison of data from Figures 6.6 and 6.7 shows that the characteristic regions of the relationships cutting speed/tangential force and cutting speed/chip compression ratio have a particular correlation with the characteristic regions of the relation cutting speed/tool life. Specifically, in cutting without cutting fluid, the beginning of the tool life increases (the transition from region II to III) at the cutting speed v = 18 to 20 m/min, corresponding to the maxima of the tangential force component and chip compression ratio. When using the water-mix and oil-based cutting fluids, the transition from region II to region III takes place at the cutting speed v = 18 to 20 m/min, and the end of the region III corresponds to the cutting speed v = 80 m/min. According to Figure 6.6, particularly the curve corresponding to machining with the oil-based cutting fluid, the characteristic region corresponding to the cutting speed v = 18 to 20 m/min is not very obvious, which can be explained by the lower chip deformation (smaller chip compression ratio) and smaller cutting force when compared with cutting at the same speed without the cutting fluid or with the water-mix cutting fluid.

The established basic correlations among the process variables in metal cutting provide the possibility for quick determination of the cutting speed corresponding to the maximum machined area of workpiece before the chosen wear limit of tool is achieved, i.e., corresponding to OCT. To do this, the experimental determination of relations $\zeta = f(v)$ (for ductile materials) or $F_z = f(v)$ (for brittle materials) should be completed in order to determine the maximum value of the chip compression ratio, ζ_{max}, or of the tangential force, R_{Tmax}. Cutting tests conducted on a broad spectrum of engineering materials[7] show that the optimum cutting speed (OCS), v_{opt} (defined earlier as the cutting speed corresponding to OCT under given cutting conditions) can be determined, using experimentally obtained relations $\zeta = f(v)$ or $F_z = f(v)$, as follows: (1) in cutting of ductile materials, the optimum cutting speed is the speed corresponding to the chip compression ratio $\zeta_{v0} = (0.7 \text{ to } 0.8)\zeta_{max}$, and (2) in cutting of brittle materials, the optimum cutting speed is the speed corresponding to cutting force $F_{zv0} = (0.7 \text{ to } 0.9)\ F_{zmax}$.

It follows from above considerations that $v_o > v_{\zeta max}$ or $v_o > v_{Rmax}$ where $v_{\zeta max}$ and v_{Rmax} are the cutting speeds corresponding to the maximum chip compression ratio and tangential force, respectively.

The OCS for the case considered in Figure 6.6 lies in the speed range corresponding to the beginning of a small influence of the cutting speed on the tangential cutting force and chip compression ratio that corresponds to the definition of OCT.

6.1.3.2 Similarity numbers

Silin[4] has introduced a set of similarity numbers for use in metal cutting research. Some of these are as follows:

- The A-criterion is one of the most fundamental criteria in the study of metal cutting. As it was first derived and studied by Silin, it may be referred as the Silin criterion. It is represented by

$$A = \frac{a_1 b_1 (c\rho)_w \theta_c}{F_z} \tag{6.5}$$

  and characterizes the heat quote of the chip relative to the entire amount of heat generated in the deformation zone. In this equation a_1, b_1 are the uncut chip thickness and its width, respectively, m (see Appendix); $(c\rho)_w$ is the volume-specific heat of workpiece material, J/(m^3 °C); θ_c is the cutting temperature, °C; F_z is the tangential cutting force, N.
- The B-criterion is represented by:

$$B = \tan\varphi \tag{6.6}$$

  which characterizes the deformation process in metal cutting. In this equation, φ is the shear angle.
- The Peclet number (Chapter 4), calculated from:

$$Pe = \frac{va_1}{w_w} \tag{6.7}$$

  characterizes the relative influence of the cutting regime va_1 with respect to the thermal properties of workpiece material w_w. In this equation, v is the velocity of a heat source (the cutting speed, m/s); w is the thermal diffusivity of workpiece material (m^2/s),

$$w_w = \frac{k_w}{(c\rho)_w} \tag{6.8}$$

  where k_w is thermal conductivity of the workpiece material, J/(m/s °C)
- The D-criterion:

$$D = \frac{a_1}{b_1} \tag{6.9}$$

  characterizes the uncut chip cross-section.

- The E-criterion:

$$E = \frac{\rho_1}{a_1} \quad (6.10)$$

characterizes the influence of the cutting edge radius ρ_1 (m) with respect to the uncut chip thickness, a_1.

- The F-criterion:

$$F = \frac{k_t}{k_w}\beta\varepsilon_{tn} \quad (6.11)$$

characterizes the influence of tool geometry with respect to the thermal conductivities of tool and workpiece materials. In this equation k_t and k_w are thermal conductivities of tool and workpiece materials (J/(m/s °C)), respectively; β is the tool wedge angle; ε_{tn} is the acute angle in the reference plane between the major (side) and minor cutting edges (see Appendix) which is referred to as the cutting tool nose angle.

- The M-criterion:

$$M = \frac{b_\Sigma}{b_1} \quad (6.12)$$

characterizes the relative load on the cutting edges where b_Σ is the total length of the cutting edges engaged in cutting, *m* (see Appendix).

Having introduced these criteria, Silin[4] attempted to solve the major problems in metal cutting using the theory of similarity. Although this was a great attempt, the results did not comply satisfactorily with experiment. As a result, even the main idea of the approach was blatantly disregarded by further researchers. At this point, the author would like to present his vision of the problem.

It is understood that the theory of similarity could not "improve" the cutting model selected for simulations. Consequently, the results obtained by Silin in his attempt to derive the relationships among cutting parameters suffered from the all the drawbacks of the single-shear plane model discussed in the previous chapters. Moreover, in order to improve coherence between the theoretical and experimental results, Silin adopted a number of additional assumptions that made his approach even more difficult to justify. As a result, his very sound approach was not widely recognized. This is unfortunate in metal cutting studies, because, for experimental studies, it seems to be a very useful and promising method. This approach allows for generalization of the results of experiments bridging the gap between the results obtained in modeling of the cutting process and practical cutting operations.

6.1.3.3 Use of similarity numbers in metal cutting

6.1.3.3.1 *Energy balance.* Because practically all of the mechanical energy associated with chip formation ends up as thermal energy, the heat balance equation is of prime concern in metal cutting studies. This equation can be written as

$$F_z v = Q_\Sigma = Q_c + Q_w + Q_t \tag{6.13}$$

where Q_Σ is the total thermal energy generated in the cutting process, Q_c is the thermal energy transported by the chip, Q_w is the thermal energy conducted into the workpiece, and Q_t is the thermal energy conducted into the tool. In contrast to the known consideration of Equation (6.13), Silin[4] analyzed Equation (6.13) in its complete form:

$$Q_\Sigma = \left(Q_{cd} + Q_{cf}\right) + \left(Q_{wd} + Q_{wf}\right) + \left(Q_{tr} + Q_{tf}\right) \tag{6.14}$$

Here Q_{cd} is the thermal energy in chip due its plastic deformation occurring in chip formation; Q_{cw} is the thermal energy in the chip due to friction in its sliding over the tool rake face; Q_{wd} is the thermal energy conducted into the workpiece from the deformation zone; Q_{wf} is the thermal energy conducted into the workpiece due to plastic deformation and friction on the tool flank; Q_{tr} is the thermal energy conducted into the tool through the tool/chip contact area on the tool rake face; and Q_{tf} is the thermal energy conducted into the tool through the tool flank face.

Applying the theory of similarity, Silin[4] obtained the following equations for the terms of Equation (6.14). Chip thermal energy can be expressed as:

$$Q_{cd} = \lambda_w b_1 \theta_d \varphi_1 \varphi_2 \tag{6.15}$$

where:

$$\varphi_1 = \frac{1}{1 + \dfrac{0.0225\, FD^{0.3}}{Pe(1-\sin\gamma)^{0.4}}}$$

$$\varphi_2 = 1 - \frac{2}{PeB} + \frac{1.125}{\sqrt{PeB}} \frac{\exp\left(-\dfrac{PeB}{4}\right)}{\operatorname{erf}\sqrt{\dfrac{PeB}{4}}} \tag{6.16}$$

$$Q_{cf} = \frac{\lambda_w b_1 \theta_d PeB \varphi_3 \left[\cos\gamma + \sin\gamma - B(\cos\gamma - \sin\gamma)\right]}{\operatorname{erf}\sqrt{\dfrac{PeB}{4}}(\cos\gamma + B\sin\gamma)} \tag{6.17}$$

where:

$$\varphi_3 = \frac{1}{1+\dfrac{0.325\,FD^{0.3}\left(1-\sin\gamma\right)^{0.25}}{\sqrt{Pe}}} \tag{6.18}$$

In these equations, γ is the tool rake angle, θ_d is the temperature in the deformation zone, and:

$$\operatorname{erf} x = \frac{2}{\sqrt{\pi}}\int_0^x e^{-x^2}dx \tag{6.19}$$

The integral (Equation (6.19)) cannot be integrated but has been tabulated because one needs it quite frequently in multiple applications of the normal distribution (for example, see Table A7 in Kronenberg[9]). To simplify the calculations, the following approximations can be used:

$$\begin{aligned}
&\operatorname{erf}\left(0.5\sqrt{PeB}\right) = 0.52(PeB)^{0.35} && \text{if } PeB \le 5\\
&\operatorname{erf}\left(0.5\sqrt{PeB}\right) = 0.85(PeB)^{0.05} && \text{if } 5 < PeB \le 20\\
&\operatorname{erf}\left(0.5\sqrt{PeB}\right) = 1 && \text{if } PeB > 20
\end{aligned} \tag{6.20}$$

The temperature in the deformation zone θ_d can be expressed in terms of the similarity numbers as:

$$\theta_d = \frac{\tau_f}{(c\rho)_w}\operatorname{erf}\sqrt{\frac{PeB}{4}} \tag{6.21}$$

where τ_f is the shear flow stress in the deformation zone (Pa).

Combining Equations (6.15), (6.17), and (6.21), one may obtain:

$$Q_c = \frac{\tau_f a_1 b_1 v}{B}\left[\varphi_1\varphi_2\operatorname{erf}\sqrt{\frac{PeB}{4}} + B\varphi_3\frac{\cos\gamma+\sin\gamma-B\left(\cos\gamma-\sin\gamma\right)}{\cos\gamma+B\sin\gamma}\right] \tag{6.22}$$

Analyzing Equation (6.22), one may notice that the thermal energy transported by the chip increases with the shear flow stress of workpiece material τ_f, uncut chip cross-section a_1b_1, and cutting speed v, while it decreases with B.

It was proven in Chapter 2 that chip formation is a cyclic process; therefore, B cannot be considered as a similarity number as it is not constant. As was shown in Chapter 2, the position of the plane of maximum combined stress, which is actually the shear plane in the cutting of ductile materials, changes within a chip formation cycle from 45° at the beginning (pure compression) to its lowest value at the end of this cycle. Therefore, the thermal energy transported by the chip has its minimum at the beginning of a chip formation cycle and then increases to its maximum at the end of this cycle. In the example shown in Figure 2.14, $B_{max} = 1$ at the beginning and $B_{min} = 0.577$ at the end of chip formation cycle.

As discussed in Chapter 2, the shear flow stress of the workpiece material τ_f also varies over a chip formation cycle. As opposed to B, τ_f is maximum at the beginning of a cycle while assuming its minimum value at the end of a cycle. Variation of τ_f over a chip formation cycle as a function of the similarity numbers is yet to be defined. The maximum flow shear stress τ_{fmax} can be calculated through the ultimate strength of workpiece material σ_{UTS} as:[8]

- $\tau_{fmax} = \sigma_{UTS}$ for annealed and austenitic steels and alloys.
- $\tau_{fmax} = 0.9\,\sigma_{UTS}$ for normalized steels and alloys.
- $\tau_{fmax} = 0.8\,\sigma_{UTS}$ for heat-treated steels.

The minimum shear flow stress τ_{fmin} in a chip formation cycle depends on the bending moment as seen from Equation (2.22).

The above consideration suggests that the exact value of Q_c is determined as:

$$Q_c = \int_{B_{min}}^{B_{max}} \int_{\tau_{min}}^{\tau_{max}} Q_c\left(B\left(\frac{1}{f_c}\right), \tau\left(\frac{1}{f_c}\right)\right) d\left(\frac{1}{f_c}\right) \tag{6.23}$$

where f_c is the frequency of chip formation (Figure 2.15).

As an approximation, the average values of B and τ_f can be used. The average value of B_a in any number of complete chip formation cycles is the same as the average B_a in the first half-cycle when its change is approximated by a sinusoidal function:

$$B_a = \frac{1 + B_{min}}{\pi} \tag{6.24}$$

In this equation, B_{min} can be determined using experimentally obtained values for the chip compression ratio ζ as:

$$B_{min} = \frac{\zeta^2 - 2\zeta \cos\gamma + 1}{\zeta \cos\gamma} \tag{6.25}$$

Analysis of a large body of experimental results in cutting of steels shows that the average shear flow stress to be substituted into equations for thermal energy can be selected to the first approximation as $\tau_{fa} = 0.87\ \tau_{fmax}$.

Thermal energy conducted into the workpiece can be expressed as:

$$Q_{wd} = \frac{0.665\lambda_w b_1 \theta_d Pe\varphi_4}{\sqrt{PeB_a}} \tag{6.26}$$

where:

$$\varphi_4 = \frac{1}{1+\dfrac{0.27\,FD^{0.3}E^{0.3}}{\sqrt{Pe}\,\sin^{0.18}\alpha}} \tag{6.27}$$

$$Q_{wf} = \frac{0.625\lambda_w b_1 \theta_d PeB_a^{1.5} ME\varphi_5}{\operatorname{erf}\sqrt{\dfrac{PeB_a}{4}}\,\sin^{0.55}\alpha} \tag{6.28}$$

where:

$$\varphi_5 = \frac{1}{1+\dfrac{0.35\,FD^{0.3}\sin^{0.1}\alpha}{\sqrt{Pe}E^{0.2}}} \tag{6.29}$$

Combining Equations (6.21), (6.26), and (6.28), one may obtain:

$$Q_w = \tau_{fa} a_1 b_1 vM\left(\frac{0.665\varphi_4 \operatorname{erf}\sqrt{\dfrac{PeB_a}{4}}}{\sqrt{PeB_a}} + \frac{0.625\,E\varphi_5 B_a^{1.5}}{\sin^{0.55}\alpha}\right) \tag{6.30}$$

Analysis of Equation (6.30) shows that the thermal energy conducted into the workpiece is proportional to the average shear flow stress, uncut chip cross-section a_1b_1, and cutting speed v. However, the influence of the cutting speed on Q_w is not as strong as on Q_c due to the Peclet number in the denominator. As with Q_c, Q_w varies within a chip formation cycle, although this variation is not as significant as with Q_c.

As with Q_c, Q_w can be calculated using the experimentally defined chip compression ratio.

The thermal energy conducted into the tool can be expressed as:

Table 6.1 Energy balance in machining (steel 1045)

v (m/s)	Pe	Q_c (J/s)	Q_c/Q (%)	Q_w (J/s)	Q_w/Q (%)	Q_t (J/s)	Q_t/Q (%)	Q (J/s)	$F_z v$
0.10	1.76	47.9	50.2	38.4	40.2	9.2	9.6	95.5	98.4
0.20	3.52	93.7	55.7	63.7	37.8	11.0	6.6	168.4	174.8
0.50	8.80	272.3	70.3	100.3	25.9	14.7	3.8	287.3	295.4
1.00	17.60	501.6	76.2	136.9	20.8	19.7	3.0	658.3	669.3
2.00	35.20	1177.1	82.8	217.5	15.3	27.0	1.9	1421.6	1494.4
4.00	70.40	2306.2	86.3	336.7	12.6	29.4	1.1	2572.3	2689.6

$$Q_{tr} = \frac{0.144\, \lambda_w b_1 \theta_d Pe^{0.47} F^{0.81} D^{0.24}}{B^{0.2}(1-\sin\gamma)^{0.935}\, \mathrm{erf}^{0.75}\sqrt{\frac{PeB_a}{4}}} \tag{6.31}$$

$$Q_{tf} = \frac{0.45\, \lambda_w b_1 \theta_d Pe^{0.29} F^{0.96} D^{0.285} M B_a^{0.82} E^{0.57}}{\sin^{0.315}\alpha\, \mathrm{erf}^{0.53}\sqrt{\frac{PeB_a}{4}}} \tag{6.32}$$

Combining Equations (6.21), (6.31), and (6.32), one may obtain:

$$Q_t = \frac{0.54\, \tau_{fa} a_1 b_1 v F^{0.86} D^{0.26} M^{0.47} E^{0.27}\, \mathrm{erf}^{0.35}\sqrt{\frac{Pe\,B_a}{4}}}{Pe_{0.615}(1-\sin\gamma)^{0.5}\sin^{0.15}\alpha B_a^{0.72}} \tag{6.33}$$

Equation (6.33) shows that the thermal energy conducted into the tool increases with τ_{fa}, a_1, b_1, and v, as well as with the similarity numbers F, D, E, and M. As for other terms of Equation (6.14), it can be calculated using the experimentally obtained chip compression ratio. Table 6.1 presents the results of calculations of the energy balance in machining steel 1045 using experimentally obtained chip compression ratios. For comparison, the total process energy calculated using experimentally obtained F_z is also shown. The following parameters were used in the calculations and experiments.

- Workpiece material: steel 1045, $k_w = 40.1$ J/(m/s °C); $(c\rho)_w = 5 \cdot 10^6$ J/(m³/s), $w_w = 8 \cdot 10^{-6}$ m²/s.
- Carbide P10 (79% WC, 15% TiC, 6% Co); tool geometry: $\gamma_n = 10°$, $\kappa_n = 8°$, $\kappa_r = 45°$, $\kappa_{r1} = 25°$, $r_n = 1$ mm.
- Cutting regime: f = 0.18 mm/rev = $0.18 \cdot 10^{-3}$ m/rev, $t_1 = 2$ mm = $2 \cdot 10^{-3}$ m.

Figure 6.8 shows the energy balance.

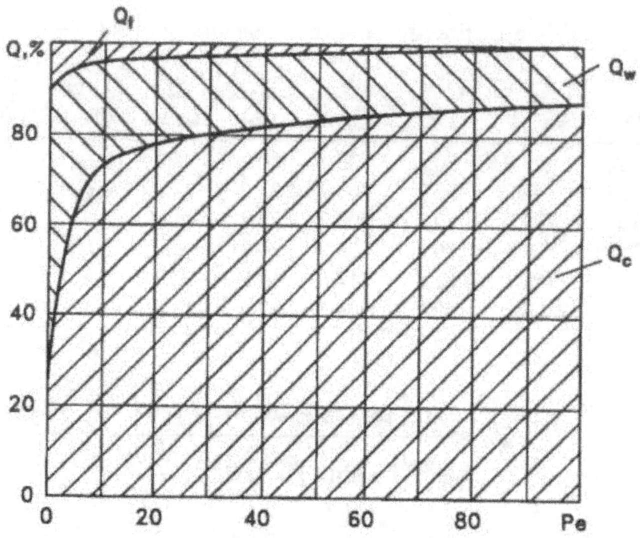

Figure 6.8 Energy balance in machining steel AISI 1045.

6.1.3.3.2 *Optimum cutting speed and machinability.* Generally, to establish the optimum cutting speed, the following equation should be considered:

$$A = n_0 Pe^{n_1} F^{n_2} D^{n_3} E^{n_4} M^{n_5} P^{n_6} U^{n_7} \tag{6.34}$$

where n_0–n_7 are experimentally defined constants; P and U are additional similarity numbers to characterize the geometry of a cutting operation:

$$P = \frac{L_w}{d_w} \quad \text{and} \quad U = \frac{B_t}{H_t} \tag{6.35}$$

where L_w and d_w are the length and diameter of the workpiece, respectively; B_t and H_t are the width and height of the tool shank.

In studies of machinability, identical tools are used in experiments; therefore, at this stage numbers P and U are excluded from further considerations. Furthermore, in experimental studies when the specific tool and workpiece are selected for testing, it is often sufficient at the first stage of the study to consider the following relationship:

$$Pe = n_0 A^{m_1} \quad or \quad \frac{v\, a_1}{w_w} = n_0 \left(\frac{a_1 b_1 (cp)_w \theta}{F_z} \right)^{m_1} \tag{6.36}$$

The equation for the optimum cutting speed v_o is obtained by substituting the optimum cutting temperature θ_o in Equation (6.36), which yields:

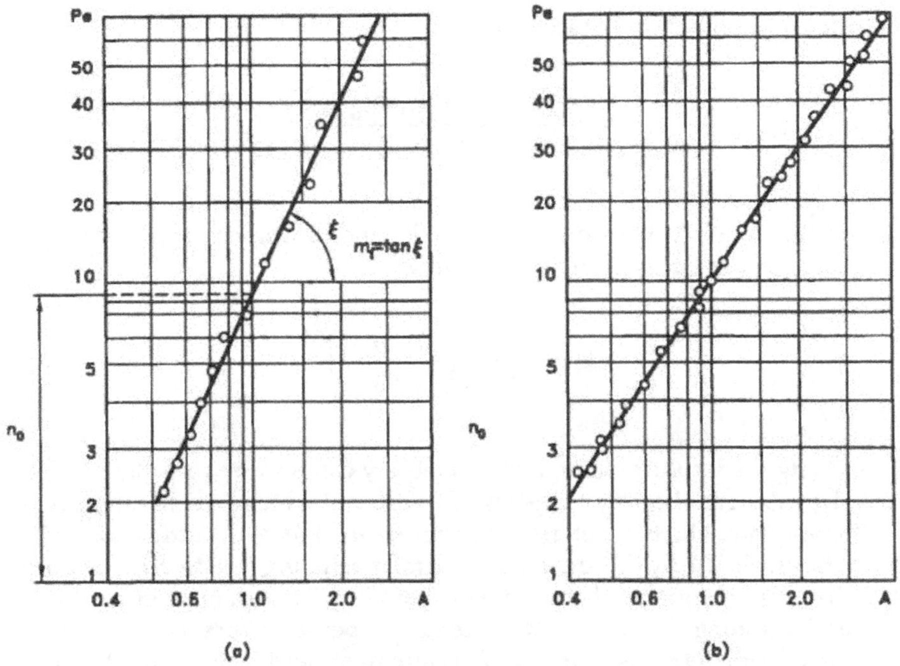

Figure 6.9 Experimental determination of the constants of Equation (6.37). **(a)** Workpiece material: stainless steel AISI 303. Tool material: carbide P01 (66% WC, 30% TiC, 4% Co). Tool geometry: $\gamma_n = 12°$, $\alpha_n = 10°$, $\kappa_r = 45°$, $\kappa_{r1} = 25°$, $r_n = 1$ mm). Similarity numbers: $F = 1.48$, $D = 0.0126$–0.1500, $E = 0.06$–0.76. **(b)** Workpiece material: steel AISI 1045. Tool material carbide P10 (79% WC, 15% TiC, 6% Co). Tool geometry: $\gamma_n = 12°$, $\alpha_n = 12°$, $\kappa_r = 45°$, $\kappa_{r1} = 15°$, $r_n = 1$ mm. Similarity numbers: $F = 2.31$, $D = 0.022$–0.122, $E = 0.018$–0.105.

$$v_o = \frac{n_0 w_w}{a_1}\left(\frac{a_1 b_1 (c\rho)_w \theta_o}{F_z}\right)^{m_1} \tag{6.37}$$

Constants n_0 and m_1 in Equation (6.37) are determined experimentally. In the tests, the force F_z and the cutting temperature θ are measured simultaneously. If the test results are plotted on a double logarithmic A vs. Pe diagram (the same module along both axes) as shown in Figure 6.9, then $n_0 = Pe$ when $A = 1$ and $m_1 = \tan\xi$.

For the data shown in Figure 6.9a, Equation (6.37) becomes:

$$v_o = \frac{9.1 w_w}{a_1}\left(\frac{a_1 b_1 (c\rho)_w \theta_o}{F_z}\right)^{2.3} \tag{6.38}$$

and for data shown in Figure 6.9b:

$$v_o = \frac{8.35\, w_w}{a_1}\left(\frac{a_1 b_1 (cp)_w \theta_o}{F_z}\right)^{2.0} \tag{6.39}$$

Using the same set of experimental results, the cutting temperature may be calculated from Equation (6.36) as follows:

$$\theta_c = \frac{F_z}{a_1 b_1 (cp)_w}\left(\frac{v a_1}{n_0 w_w}\right)^{m_1} \tag{6.40}$$

Using the results obtained, machinability can now be considered. Commonly, machinability is a property of a material which governs the ease or difficulty with which a material can be machined using a cutting tool.[11] The term is widely use by those concerned with machining, yet detailed inquiries would reveal a measurable vagueness about its precise definition, or even its general meaning. Unlike most material properties, there is no generally accepted parameter used for its measurement, and it is evident that, in practice, the meaning attributed to the term "machinability" tends to reflect the immediate interest of the user. If the tool life is of prime concern, then machinability is understood as a property affecting the tool wear; if surface finish problems are concerned, then it is thought of as "finishability"; if chip-breaking creates problems, then machinability is considered as "chip breakability", etc. In most fields of science and technology, great care is devoted to the definition of relevant parameters, but, in machining, machinability tends to remain a term which means "all things to all men."[11]

The foregoing analysis leads to a new approach to machinability determination using the following procedure. It is only possible to describe the machinability procedure in general terms, as conditions will vary with each situation. The method to follow is the same as that used for good machine tool operation except that great care and observation must be exercised and certain measurements must be taken during machining. In particular, cutting force and temperature measurements should be taken simultaneously. Before starting the test, it should be ascertained that the machine tool (lathe) and workpiece fulfill all the requirements of the ANSI/ASME B94.55M-1985 standard. Cutting force and cutting temperature measurements should be made using methodology discussed further in this chapter. Five to seven different cutting feeds should be selected for the study. The depth of cut should be kept the same for all tests. The number of tests corresponds to that of the selected feeds. In each test, the cutting speed is varied, and the cutting force and cutting temperature are measured. The readings should be plotted on the cutting force and cutting temperature (ordinates) vs. cutting speed

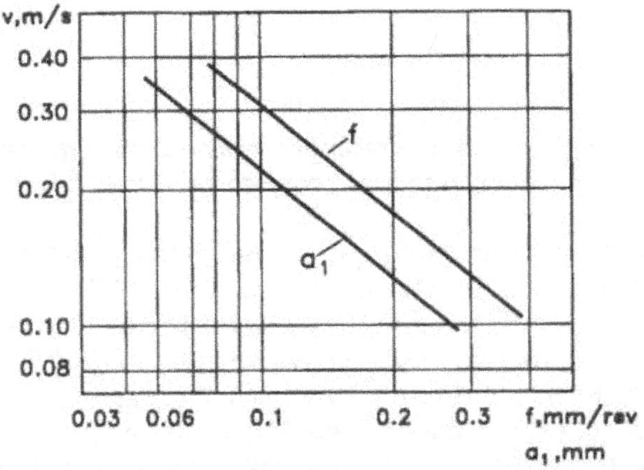

Figure 6.10 The optimum cutting speed vs. cutting feed and true uncut chip thickness.

(abscissa) diagram as shown in Figure 6.3. Such diagrams should show at least 15 experimental points for each curve so that the cutting speed at which the cutting force assumes the minimum stabilized value can be assessed with sufficient accuracy. Under no circumstances should the cutting force and/or cutting temperature be determined by extrapolating corresponding curves. The optimum cutting speed is defined for each feed using points on each curve $F_z(v)$ where F_z assumes the minimum stabilized value corresponding to the optimum cutting temperature θ_o. To set the test properly (calibration stage), it is recommended to determine the optimum cutting temperature θ_o using the point where F_z assumes the minimum stabilized value when $f = 0.2$ to 0.4 mm/rev. Plotting the coordinates (v_{o1}, a_1), (v_{o2}, a_2), etc. obtained from Figure 6.3 on a double logarithmic, the true uncut chip thickness vs. cutting speed (same module along both axes) will produce an a_1–v curve as shown in Figure 6.10. As such, the true uncut chip thickness is calculated using the feed and tool geometry as discussed in the Appendix. This a_1–v curve may be considered linear within a certain range of the uncut chip thickness. The equation for this linear proportion of the curve is

$$v_o = C_v a_1^{k_1} \tag{6.41}$$

where $k_1 = \tan\xi$ defines the slope of the curve; C_v is a constant defined using any selected uncut chip thickness a_{11} and corresponds to v_{o1} by:

$$C_v = a_{11}^{-k_1} \cdot v_{o1} \tag{6.42}$$

and the following expression is obtained:

$$C_v = a_{11} \cdot v_{o1}^{-\frac{1}{k_1}} \tag{6.43}$$

The discussed method does not require expensive and time-consuming tool-life tests to define machinability as required by the ANSI/ASME B94.55M-1985 standard.

Analysis of a large body of experimental data shows that the simplified relationship between similarity numbers (Equation (6.36)) provides satisfactory results over a wide range cutting speed variation. However, it is valid only for the cutting tools of similar geometry when cutting feed $f \geq 0.08$ mm/rev and the depth of cut $t_1 \geq 1$ mm. To obtain a more general relationship than Equation (6.36) of experimental relationships between process parameters in metal cutting it is necessary to use a more sophisticated relationship between similarity numbers than that of Equation (6.36). Equation (6.44) is the proper relationship:

$$A = \frac{c_A Pe^{n_8}}{F^{n_9} D^{n_{10}} E^{n_{11}} \sin^{n_{12}} \alpha} \tag{6.44}$$

where n_8–n_{12} are experimentally defined constants. The design of experiments is very suitable and reliable for the determination of these constants.

6.1.3.3.3 *Quality of the machined surface and tool wear.* The quality of the machined surface is a complex parameter which, in general, is very difficult to specify in a meaningful way. It may be considered included in the machining residual stress and surface finish. The residual stress is unavoidably gained by the machined surface. The problem appears not to be as severe when cutting conventional engineering materials while it becomes of prime concern when cutting high alloys. Therefore, more and more the residual stress problem is selected as the subject of experimental studies.

The mechanisms which create residual stresses in machined and ground surfaces are not well understood; however, tentative conclusions can be drawn from the experimental evidence available in the literature. It is suggested that the sources or causes of residual stresses be separated into two groups identified as thermal and mechanical.

The thermal mechanism is usually related to grinding and considered to be the same as that which occurs in quenching of a part which is being hardened by heat treating; different rates of cooling create residual stresses resulting in warpage of the part. Those areas which cool faster end up with residual tensile stresses; essentially the same thing happens in grinding.

The mechanical sources of residual stresses would appear to be much more complex than thermal sources. The technical literature reports both tensile and compressive stresses resulting from ordinary metal cutting operations, and there can be no doubt that both have been observed at

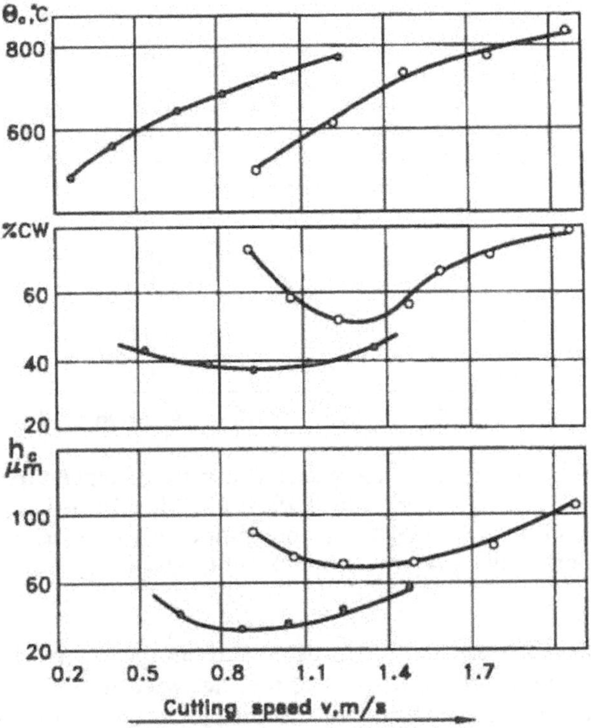

Figure 6.11 Influence of the cutting speed on cutting temperature θ_c, percent cold work (%CW), depth of cold worked layer h_c. Workpiece materials: ∘, steel AISI 304; •, high alloy (0.15% C, 10% Cr, 22% Ni, 8% Ti, 1% Mo). Tool material: carbide P01 (66% WC, 30% TiC, 4% Co).

relatively low temperatures which could not have been the source of tensile stresses of the magnitude which are reported.

Figure 6.11 shows the influence of the cutting speed on the cutting temperature θ_c; percent cold work, %CW; and depth of the cold-worked layer, h_c, in cutting of two Ni-based alloys. All the obtained curves have extremes (minimums); therefore, there is an optimal regime corresponding to a given value of cold working under given cutting conditions (cutting tool material and geometry, chosen cutting fluid and its flow rate, etc.). Our experience with Ni-based alloys shows that they are extremely sensitive to even small changes in the cutting regime and in the tool geometry. To support this point, Figure 6.12 illustrates the heavy dependence of the residual stress on the cutting speed in cutting of a Ni-based alloy. It is seen that under certain cutting speeds, the residual stress is at its maximum. It is also seen that there is no visible correlation between the residual stress and the cutting temperature alone. Obviously, the cutting temperature plays a significant role in the formation of residual stresses. However, without cutting

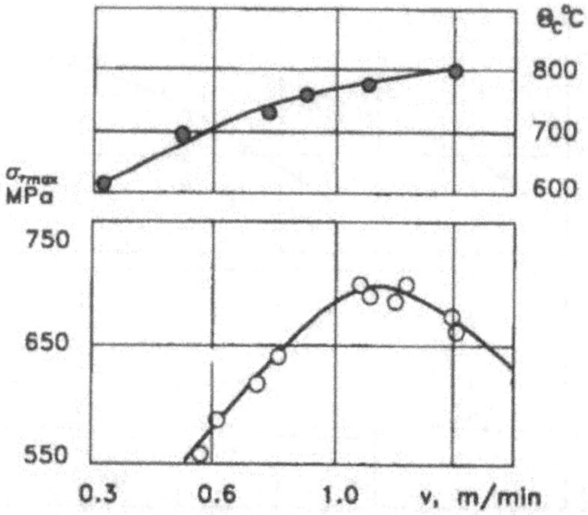

Figure 6.12 Influence of the cutting speed on the cutting temperature θ_c and residual stress. Workpiece material: high alloy (0.08% C, 68% Ni, 1% Cr). Tool material: carbide P01 (66% WC, 30% TiC, 4% Co). Tool geometry: $\gamma_n = -5°$, $\alpha_n = 12°$, $\kappa_r = 45°$, $\kappa_{r1} = 15°$, $r_n = 0.4$ mm. Cutting regime: depth of cut $t_1 = 0.7$ mm, $f = 0.10$ mm/rev.

forces and phase transformations, the cutting temperature seems to be useless for defining the optimum cutting regime when residual stresses are of prime concern. Figure 6.13 shows the influence of the tool geometry on residual stresses.

To study residual stresses experimentally, an additional set of similarity criteria is introduced:

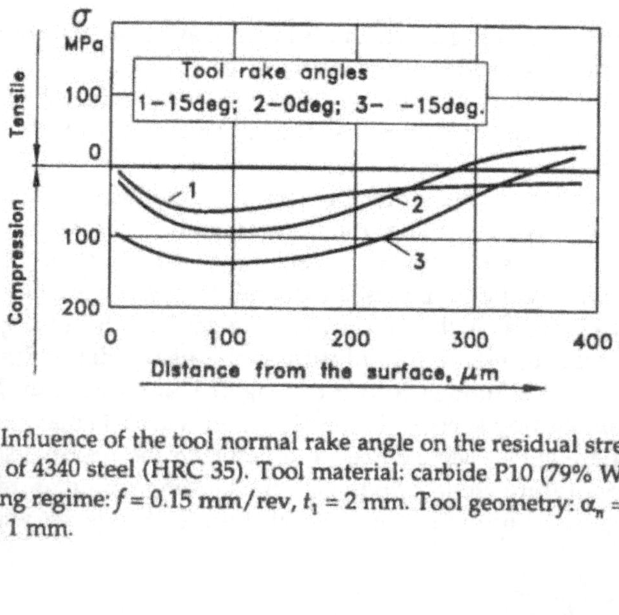

Figure 6.13 Influence of the tool normal rake angle on the residual stress. Longitudinal turning of 4340 steel (HRC 35). Tool material: carbide P10 (79% WC, 15% TiC, 6% Co). Cutting regime: $f = 0.15$ mm/rev, $t_1 = 2$ mm. Tool geometry: $\alpha_n = 8°$, $\kappa_r = 45°$, $\kappa_{r1} = 25°$, $r_n = 1$ mm.

$$G=\frac{\alpha_r E_w}{(c\rho)_w};\quad P=\frac{R_w}{a_1};\quad Z=\cos\alpha;\quad K=\frac{h_r}{a_1};\quad L=\frac{\sigma_{UTS_w}}{\sigma_{yw}} \tag{6.45}$$

Here, α_r is the coefficient of linear expansion of workpiece material (1/°C); E_w is the modulus of elasticity of workpiece material (Pa); R_w is the radius of the machined surface (m); $h_r \geq a_1$ is the variable coordinate directed toward the center of rotation of the workpiece along the radius. The coordinate origin is shifted from the workpiece surface by factor $h_r/a_1 = 1$; σ_{UTS_w} and σ_{yw} are the ultimate and yield strength of workpiece material, respectively (Pa).

Generally, the following relationship of the similarity numbers should be determined experimentally

$$\frac{\sigma_r}{\sigma_{UTS_w}}=\frac{c_r F^{n_{13}} D^{n_{14}} E^{n_{15}} G^{n_{16}} P^{n_{17}} Z^{n_{18}} L^{n_{19}}}{Pe^{n_{20}} K^{n_{21}}} \tag{6.46}$$

where σ_r is residual stress; c_r, n_{13}–n_{21} are constants to be determined experimentally.

When experiments are carried out using a given workpiece and cutting tool with defined geometry and the depth of cut is kept constant, the similarity numbers F, G, P, Z, and L are treated as constants. As such, Equation (6.46) becomes:

$$\frac{\sigma_r}{\sigma_{UTS_w}}=\frac{c_r D^{n_{14}} E^{n_{15}}}{Pe^{n_{20}} K^{n_{21}}} \tag{6.47}$$

Furthermore, if the tests are run at a constant cutting feed, f, using a sharp cutting tool ($\rho_1 = const$), then F and D are also constants. As such, Equation (6.47) takes the form:

$$\sigma_r=\frac{c_r \sigma_{UTS_w}}{Pe^{n_{20}} K^{n_{21}}} \tag{6.48}$$

The constants in Equation (6.48) are determined using experimental curves $\sigma_r/\sigma_{UTS_w} = f(Pe, K)$. For example, for the machining of high-alloy (0.12% C, 11% Cr, 2% Ni, 2% Mo) tool material—carbide M10 (94% WC, 6% Co), depth of cut $t_1 = 1$ mm, $f = 0.23$ mm/rev — Equation (6.48) becomes:

$$\sigma_r=\frac{600\,\sigma_{UTS_w}}{\left(\dfrac{v a_1}{a_w}\right)^{1.8}\left(\dfrac{h_r}{a_1}\right)^{11}} \tag{6.49}$$

The depth of the cold-worked layer h_c is a function of the following similarity numbers:

$$\frac{h_c}{a_1} = f\left(Pe, F, D, E, E, \sin\gamma, L\right) \tag{6.50}$$

Consequently, an experimental study should be carried out to establish the following relationship:

$$\frac{h_c}{a_1} = \frac{c_c\, F^{n_{22}}\, D^{n_{23}}\, E^{n_{24}}\, L^{n_{25}}\left(1-\sin\gamma\right)^{n_{26}}}{Pe^{n_{27}}} \tag{6.51}$$

where c_c, n_{22}–n_{27} are constants to be determined experimentally.

When experiments are carried out using a given workpiece and cutting tool with defined geometry, Equation (6.51) simplifies to:

$$\frac{h_c}{a_1} = \frac{c_c\, D^{n_{23}}\, E^{n_{24}}}{Pe^{n_{27}}} \tag{6.52}$$

and when F and D are constants throughout the tests, Equation (6.52) becomes:

$$\frac{h_c}{a_1} = \frac{c_c}{Pe^{n_{27}}} \tag{6.53}$$

Many methods have been developed to measure surface finish from very simple visual or touch comparative methods where a surface is compared with one of a series of standard surfaces using sophisticated equipment which will measure the root mean square (RMS) average of a surface (R_a) or the center line average (CLA) of a surface (R_z). Probably the most accepted measure of surface finish is CLA value.

Although it is well known that CLA depends on many process parameters, analysis shows that when a given combination of tool and workpiece is considered the dimensionless number $S = R_a/f$ characterizing surface finish becomes a function of the following dimensionless numbers:

$$S = f\left(Pe, D, E, N = \frac{f}{r_n}\right) \tag{6.54}$$

where r_n is the tool radius, m. Experience shows that when cutting without seizure — that is, when the part of the deformation zone on the tool rake face is negligibly small (appears, and is considered in the literature, as the absence of the built-up edge, though it is not the case, as discussed in Chapter 2), the constants c_h, n_{28}–n_{31} of the following equation should be determined experimentally:

$$\frac{R_z}{f} = \frac{c_h \left(\frac{a_1}{b_1}\right)^{n_{28}} \left(\frac{f}{r_n}\right)^{n_{29}}}{\left(\frac{v a_1}{w_w}\right)^{n_{30}} \left(\frac{\rho_1}{a_1}\right)^{n_{31}}} \tag{6.55}$$

The tool wear processes of cutting tools which are relevant to the cutting tool failure have been reviewed.[8,9,11–14] The wear mechanisms, including abrasive, adhesive, diffusion, chemical, etc., are described in the mentioned references in detail. Of the four listed principal types of wear, adhesive wear is the only one which never can be eliminated. Three basic types of adhesive wear can be distinguished — severe wear, moderate wear, and burnishing.[15]

Adhesive wear obeys a Holm-Arcard relationship[16] of the type:

$$wear\ volume = \frac{wear\ coefficient \times load \times distance\ of\ sliding}{hardness} \tag{6.56}$$

Equation (6.56) indicates that, when the load, workpiece material, and sliding distance are given, two ways of minimizing adhesive wear are possible — to use tools made of hard tool materials or to achieve a low wear coefficient. The hardness range which is available in practical tool materials is relatively limited. Therefore, a practical way to achieve minimum wear is to select the optimum cutting speed as discussed above. As experience has shown, the burnishing type of adhesive wear takes place. Burnishing, or material removal on a molecular scale, represents the least possible amount of adhesive wear. It is worthwhile to mention here that the transition from one wear regime to another, as sliding conditions are gradually changed, often occurs quite abruptly. The sudden change from severe to moderate wear, which is generally accompanied by a change from a system giving large wear particles to one yielding small oxide particles, has been studied extensively,[17,18] but such factors as the critical temperature[19] or oxide thickness[20] have been invoked.

Makarow suggested[6] that the relative surface wear W_s is the most suitable characteristic of the tool life when dimensional accuracy is of major importance. This wear is defined as:

$$W_s = \frac{WB_r}{L_p f} = \frac{W_1}{f} \tag{6.57}$$

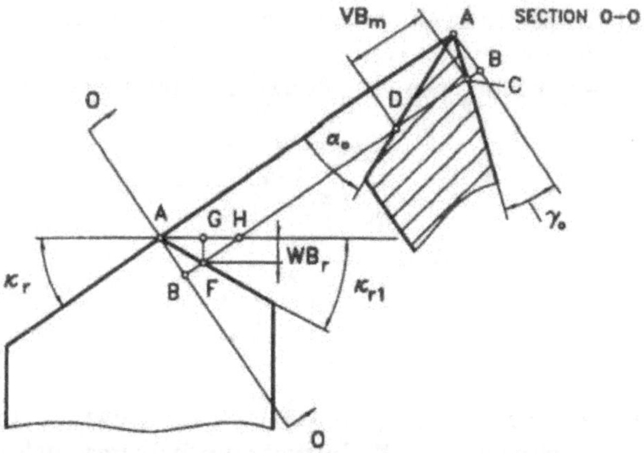

Figure 6.14 Model used to derive the relationship between the radial and flank wear.

Here, WB_r is the radial tool wear (m); $L_p = vT_t$ is the total length of the tool path passed over the workpiece during the time period corresponding to the tool life T_t (dimensional accuracy) and at cutting speed v; W_l is the linear tool wear.

Experience shows that the tool radial wear which defines dimensional accuracy is a more suitable criterion for the tool life than that defined by the ANSI/ASME B94.55M-1985 standard (*Tool Life Testing With Single-Point Turning Tools*) for which the criteria most commonly used for sintered carbide tools are the three types of flank wear. The defined criteria are difficult to measure objectively, as according to all the flank criteria the flank wear has to be measured using data not existing in a worn tool. Figure 6.14 shows the relation between the radial tool wear and the flank wear WB_m:

$$AB = BD\tan\alpha_o = (BC + CD)\tan\alpha_o = (AB\tan\gamma_o + VB_m)\tan\alpha_o \tag{6.58}$$

or

$$AB = \frac{VB_m \tan\alpha_o}{1 - \tan\alpha_o \tan\gamma_o} \tag{6.59}$$

On the other hand,

$$AB = AH\sin\kappa_r = (AG + GH)\sin\kappa_r = (WB_r\cot\kappa_{r1} + WB_r\cot\kappa_r)\sin\kappa_r \tag{6.60}$$

Finally,

$$WB_r = \frac{VB_m}{(\cot\alpha_o + \tan\gamma_o)(\cot\kappa_r + \cot\kappa_{r1})\sin\kappa_r} \tag{6.61}$$

In experimental studies of tool wear, additional criteria should be considered:

$$G_1 = \frac{(c\rho)_w \theta_c}{\tau_{f_n}\left(1+e_f\right)},\ P_1 = \frac{\sigma_{UTS_w}}{\sigma_{UTS_t}},\ U_1 = \frac{\theta_{cb}}{\theta_c} \tag{6.62}$$

where e_f is the standard percentage elongation of workpiece material (%); σ_{UTS_t} is the ultimate strength of tool material (Pa); θ_{cb} = 1490°C is the melting point of cobalt (the matrix for sintered carbides).

Silin[4] showed that when the test is conducted using the optimum cutting regime, the constants of the following expression should be defined experimentally:

$$W_s = C_w \left[\frac{(c\rho)_w \theta_c}{\tau_{f_n}\left(1+e_f\right)}\right]^{n_{32}} \left(\frac{\sigma_{UTS_w}}{\sigma_{UTS_t}}\right)^{n_{33}} \left(\frac{\theta_{cb}}{\theta_c}\right)^{n_{34}} \tag{6.63}$$

6.2 *Temperature measurements in metal cutting*

It is generally recognized that the temperatures in metal cutting are extremely important, as they affect the shear flow stress, chip formation process, chip breakability, and tool wear.[4–9,11–14] As a result, a number of methods have been developed for the measurements of temperatures in metal cutting. However, the results of temperature measurements have not been widely applied in correlations between the obtained temperatures and cutting parameters.

6.2.1 *Conventional thermocouples*

The simplest electrical method of temperature measurement uses the conventional thermocouple.[21] When two dissimilar metals are joined together to form an electrical circuit, as shown in Figure 6.15a, an electromagnetic field (emf) will be generated and can be registered by a potentiometer. This emf is a primary function of the junction temperature. This phenomenon is referred to as the Seebeck effect. If two metals are connected to an external circuit in such a way that a current is drawn, the emf may be altered slightly owing to the phenomenon called the Peltier effect. Furthermore, if a temperature gradient exists along either or both materials, the junction emf may undergo an additional slight alternation known as the Thomson effect. There are, then, three emfs present in a thermoelectric circuit: the Seebeck emf, caused by the junction of dissimilar metals; the Pelter emf, caused by a current flow in the circuit; and the Thomson emf, which results from a temperature gradient in the connected materials. The Seebeck emf is of prime concern, as it is dependent on the junction temperature. This dependence is known for any junction formed by most common metals and, therefore, a

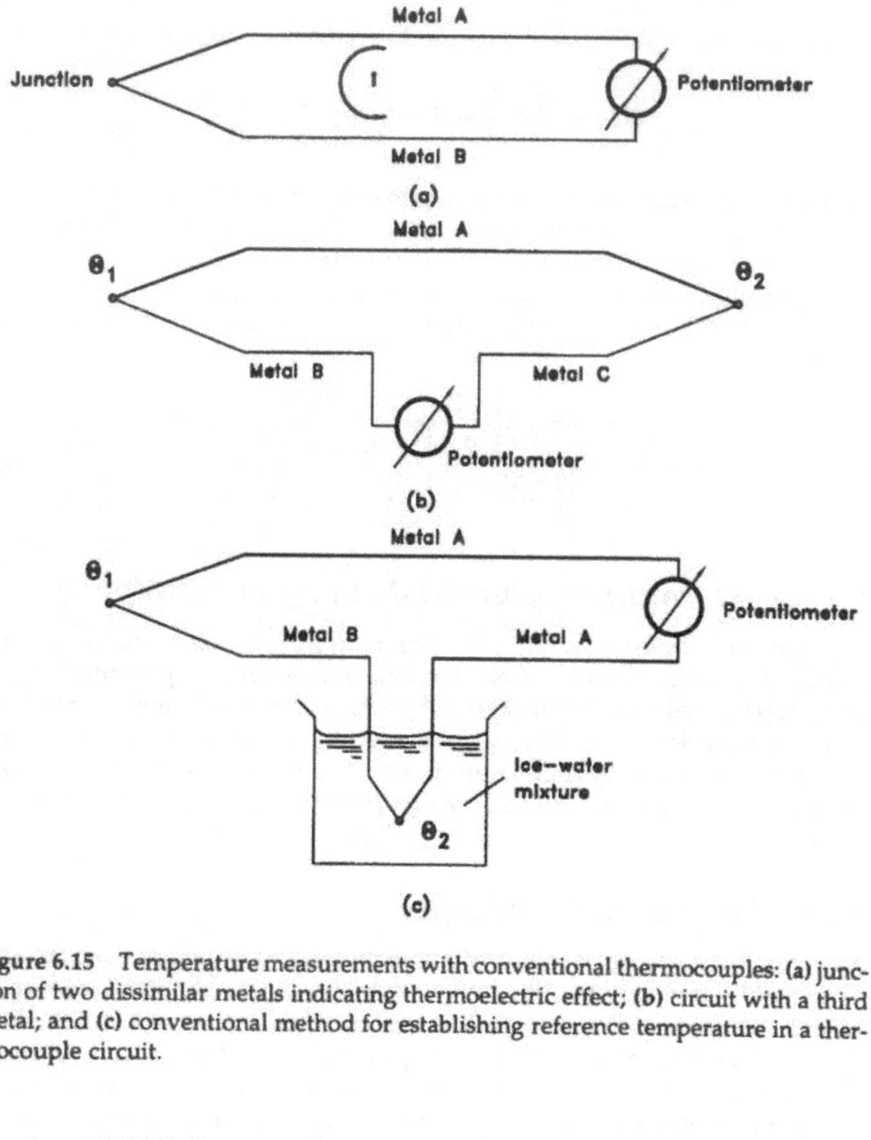

Figure 6.15 Temperature measurements with conventional thermocouples: **(a)** junction of two dissimilar metals indicating thermoelectric effect; **(b)** circuit with a third metal; and **(c)** conventional method for establishing reference temperature in a thermocouple circuit.

junction of dissimilar metals can be used to measure temperature if the generated emf is carefully measured. The main problem arises here when an experimenter attempts to measure the potential, as another thermal emf will be generated at the junction of the materials, joined to measure temperature, to a potentiometer of any type. This second emf will depend on the temperature of connection, and provisions must be made to take into account this additional potential.

Two important properties (sometimes referred to as laws[12]) of circuits with conventional thermocouples are useful in applications. First, if a third metal is connected in the circuit as shown in Figure 6.15b, the net emf of the

Table 6.2 Thermal emf in absolute millivolts for some commonly used metals and alloys in conjunction with platinum

Metal or alloy	emf (mV)
Cromel (90% Ni, 10% Cr)	+2.4
Iron	+1.8
Molybdenum	+1.2
Tungsten	+0.8
Copper	+0.76
Aluminum	–0.40
Nickel	–1.50
Alumel (1% Si, 2% Al0.17 Fe, 2% Mn, 94.83% Ni)	–1.70
Constantan (58% Cu, 40% Ni, 2% Mn)	–3.40
Copel (56.5% Cu, 43.5% Ni)	–3.60

Note: Reference junction at 0°C, terminal junction at 100°C.

circuit is not affected as long as the new connection is at the same temperature as the initial junction, that is when $\theta_1 = \theta_2$. Second, when this is not the case (i.e., when $\theta_1 \neq \theta_2$), then the circuit develops an emf of E_{t1} at the first junction and E_{t2} at the second. The law of intermediate temperatures states that the circuit will develop a resultant emf $E_{tr} = E_{t1} + E_{t2}$ when operating between different temperatures θ_1 and θ_2.

It may be observed that all thermocouple circuits must involve at least two junctions. If the temperature of the one junction is known, then the temperature of the other junction may be easily calculated using the thermoelectric properties of the metals. The known temperature is referred to as the reference temperature. A common arrangement for establishing the reference temperature is the ice bath as shown in Figure 6.15c. An equilibrium mixture of ice and saturated-air distilled water at standard atmospheric pressure produces a known temperature of 0°C. When the ice-water mixture is kept in a Dewar flask, the reference temperature may be maintained for extended periods of time.

Practically all metals and alloys can be used as thermocouple materials. It is common to express the thermoelectric emf in terms of potentials generated with a reference junction at 0°C and a terminal junction at 100°C when a material is coupled with platinum. Table 6.2 shows thermal emfs for some commonly used materials. To obtain a higher output emf, the materials for a thermocouple should be selected so that the first would have the maximum positive emf and the second would have the maximum negative emf, as selected from Table 6.2.

In metal cutting studies, standard cromel-alumel and cromel-copel thermocouples are used. Figure 6.16 presents emf-temperature relations for these thermocouples. If tool or workpiece material is selected as one of these

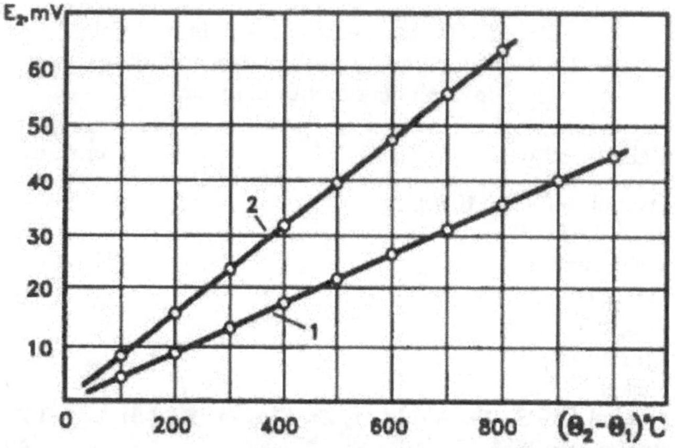

Figure 6.16 emf-temperature relations for a cromel-alumel (1) and cromel-copel (2) thermocouples.

thermocouple materials, then the second thermocouple material is recommended to be copel or constantan. If the workpiece material is aluminum, an aluminum alloy, or nickel or nickel-based alloys, then cromel is selected to be the second metal. The emf characteristics of standard thermocouples are well known, and the thermocouple (made for the study using rolls of commercial thermocouple wire) should be calibrated directly against the known thermocouple standards to determine its actual emf-temperature relations.

In metal cutting studies, standard cromel-alumel and cromel-copel thermocouples are referred to as artificial thermocouples; those having the terminal junction formed by constantan or copel thin wire and tool (workpiece) material are referred to as semi-artificial; those having the terminal junction formed by tool and workpiece material are referred to as natural thermocouples.

Usachev probably was the first who used an artificial thermocouple to measure cutting temperature.[22] The principle of measurement is shown in Figure 6.17. The thermocouple is placed in a small hole made in the cutting tool. The hole diameter should be as small as possible to reduce disturbances which may appreciably change the temperature field. Experience shows that temperatures can be measured with sufficient accuracy when the terminal thermocouple junction is pressed against the cutting insert with a force of no less than 50 N. As this is not always possible, it is recommended that the terminal end is welded to the inserts using condenser welding when HSS inserts are used. By placing thermocouple holes in different positions of the insert, the temperature field can be determined. A useful method to obtain such a field using a single thermocouple is as follows. A thermocouple is initially placed in the most remote (from the cutting edge) point of interest and then, by regrinding the flank and the rake tool faces in the defined

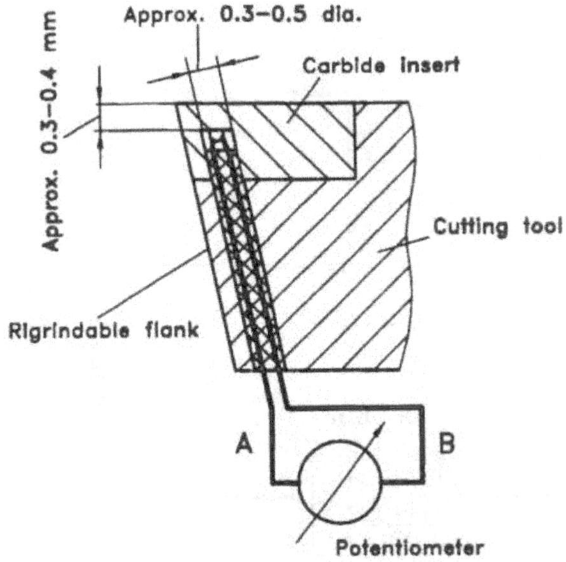

Figure 6.17 Embedded thermocouple technique for mapping temperatures in cutting tool used for turning operation.

sequence, the relative location of the thermocouple is moved to any desirable point relative to the cutting edge.

The output of thermocouples is in the millivolt range and may be measured by a digital millivoltmeter. The voltmeter is basically a current-sensitive device; hence, the meter reading will be dependant on both the emf generated by the thermocouple and the total circuit resistance, including the resistance of connecting wires. Therefore, the complete system, including the thermocouple, connecting wires, and millivoltmeter should be calibrated directly to furnish a reasonably accurate temperature determination.

A large number of commercially available electronic voltmeters are suitable for thermocouple measurements. Among them, those providing a digital output which can be used for direct computer processing of the data are most suitable if the cutting temperature is measured simultaneously with the cutting force, as discussed above. In these devices, the problem with the reference junction can be alleviated with a special circuit having a thermistor or other hardware-compensation devices

6.2.2 *Tool-work (natural) thermocouple*

The most widely used method for measuring temperatures in metal cutting is with a tool-work or natural thermocouple. With this method, the average integral temperature at the tool/chip interface is measured and has the sense of the cutting temperature, as discussed in Section 6.1. Therefore, if the

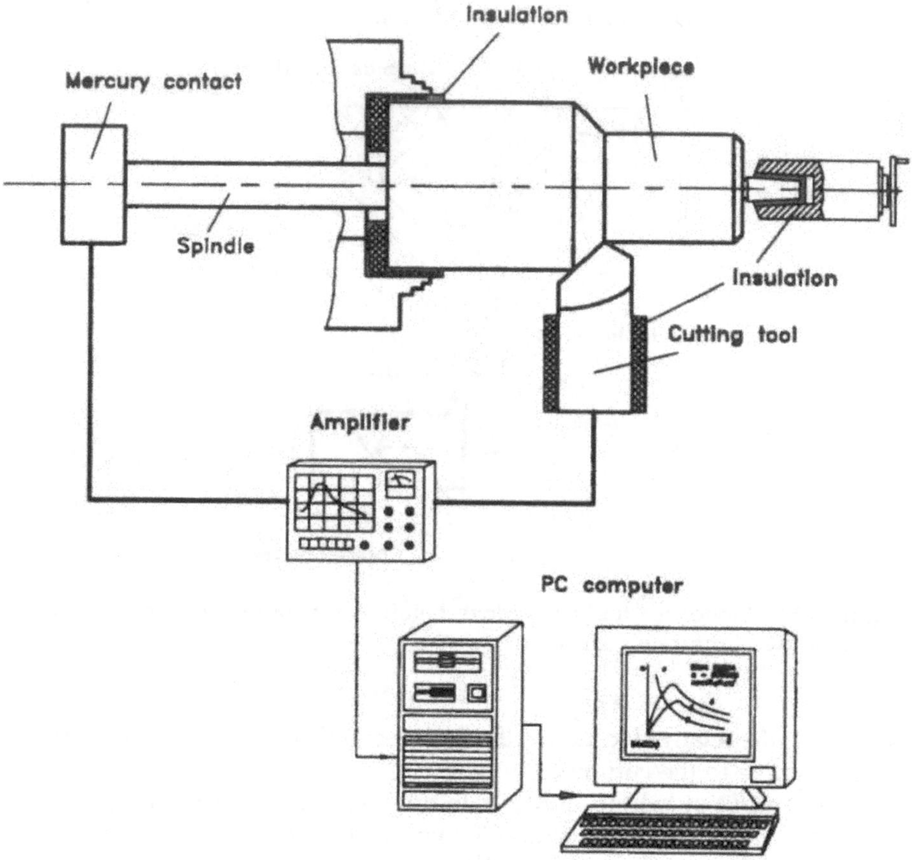

Figure 6.18 Workpiece-tool (natural) thermocouple circuit.

results of Section 6.2 are to be applied to experimental studies, this method of temperature measurement should be selected.

Figure 6.18 shows the principle of this method. Because the tool and workpiece materials are different, they are used as materials *A* and *B* to form a natural thermocouple. The hot junction of this thermocouple is the tool/chip interface. The components of the thermocouple are isolated from the machine to eliminate noise in the output signal.

The major problem with natural thermocouples is their proper calibration. The simplest and most widely used method for calibrating these thermocouples involves comparing the emf produced by the tool with that produced by a standard thermocouple when both are placed in a bath of a molten metal.[12,13] Because the calibration is critical to obtaining accurate results, a few severe problems associated with such calibration have to be pointed out:

- Different circuits are used in calibration and in actual cutting. Commonly, the samples that represent the tool and chip are not even in contact (see Figure 12.2 in Shaw,[12] Figure 7.3a in Stephenson[13]). Even though they may be in contact, the actual contact area is not the same as in reality and the contact pressure is well below that in actual cutting. Besides, the contact between the tool flank and the workpiece taking place in actual cutting is completely ignored, and a mercury contact connecting the rotating spindle with the amplifier is not present in the calibration circuit.
- A lead bath is used in calibrations to obtain the uniform maximum calibration temperature; however, the temperature of molten lead is far below those expected in cutting. Because the calibration curve for a natural thermocouple may not be linear (see Figure 12.3 in Shaw[12]), it is next to impossible to extrapolate the obtained results (i.e., the calibration curves obtained may contain significant errors).
- Because the specimen representing the tool in calibrations is restricted in length, it is very difficult to keep the cold junction of a natural thermocouple at constant temperature. As pointed out in Stephenson,[13] it is particularly true when small, indexable tool inserts are used. The same problems occur in obtaining a sufficiently long chip for workpiece materials of relatively low ductility.

As discussed in Section 6.1, if the theory of similarity is to be used in a study, the cutting temperature has to be measured precisely; therefore, it is extremely important to use an accurate calibration method. One of the possible methods for calibrating the natural thermocouples is based on the above-discussed first principle of thermoelectric circuit.[23] According to this principle, a third metal, connected in the circuit, does not change the net emf of the circuit if the new connection is at the same temperature as the initial junction (Figure 6.15b).

An arrangement for calibration is shown in Figure 6.19a. The workpiece (1) is cut by two geometrically similar cutters (2 and 3) made of different tool materials. Because both tools work in the same cutting regime and the tool material does not significantly affect the cutting temperature, it may be assumed that both tools have the same contact temperature so that the requirement of the second law is justified when the workpiece is considered as the third metal in the circuit. The arrangement also includes a two-position switch (4) and an analog-digital millivoltmeter (5) with built-in amplifier. The output of the millivoltmeter (5) is connected to a PC-type computer (6).

The calibration procedure includes two successive stages. In the first stage, a relatively low melting point material (for instance, aluminum) is used as workpiece material, and its melting point is stored in the computer program used in the calibration. The switch (4) is in the a position. When cutting starts, the cutting speed is gradually increased up to the point when the workpiece material begins to melt at the tool/chip contact of both cutters.

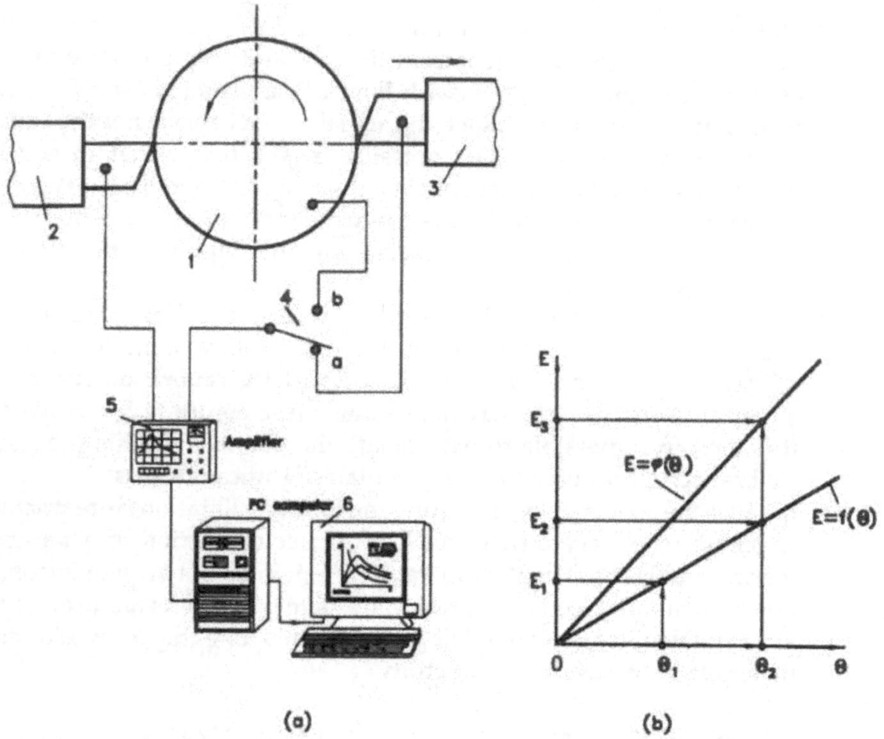

Figure 6.19 Arrangement for calibration of natural thermocouple.

The emf E_{m1} corresponding to this point is registered by the millivoltmeter and then is stored in the computer program. A few tests with different workpiece materials can be carried out to obtain the calibration curve $E_{mf} = f(\theta)$ as shown in Figure 6.19b.

At the second stage of calibration, the workpiece made of the metal used for temperature measurement is employed. In a given cutting regime, the emf of the natural two-cutter thermocouple is measured to be E_2, which the computer calculated using the corresponding temperature θ_2. After this measurement is done, the computer gives a signal to a servomechanism for fast withdrawal of the second cutter (3) in the direction shown by an arrow in Figure 6.19a, and simultaneously the switch (4) is moved to the b position. The cutter is still working in the same cutting regime. A new emf, E_3, is measured. Although the cutting temperature θ_2 does not change, $E_2 \neq E_3$, as the latter was measured using a new natural thermocouple "tool-workpiece". Using a few measuring points (cutting regimes) and corresponding differences in the emfs, the computer calculates the corrected calibration curve, $E_{mf} = \varphi(\theta)$, which is used in the further experiments.

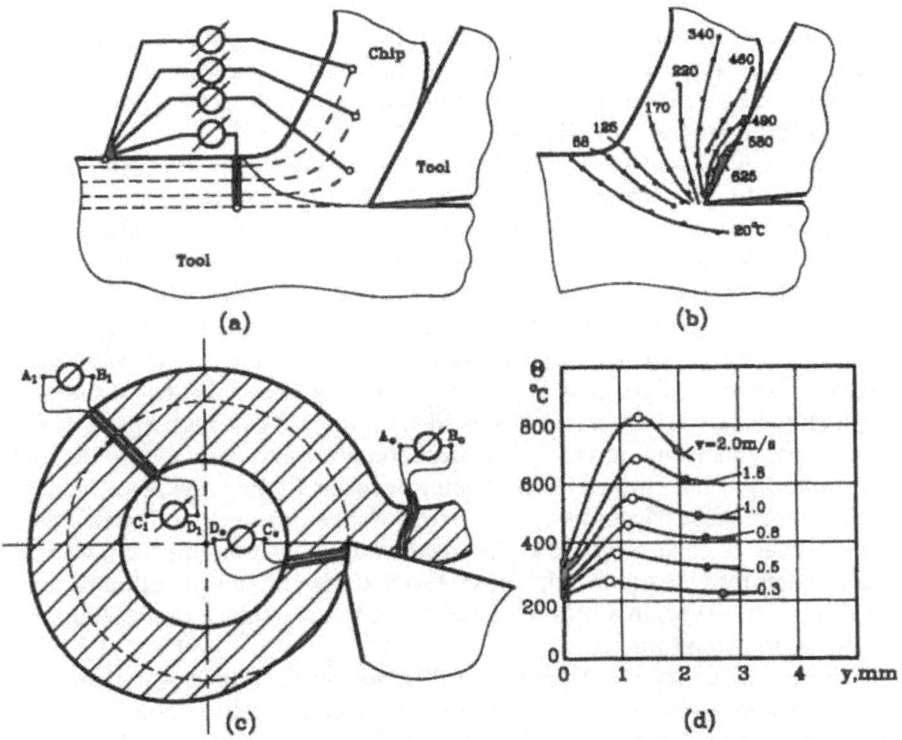

Figure 6.20 Temperature measurements with semi-artificial and running thermocouples. **(a)** Semi-artificial thermocouple technique for mapping temperatures in the deformation zone and chip. **(b)** Reconstructed temperature field for the orthogonal cutting test carried out on a shaper using following cutting parameters: the depth of cut a_1, 2 mm; tool rake angle, 12°, tool material, P10 carbide; workpiece material, steel AISI 1045. **(c)** Running thermocouple technique for mapping temperatures along the rake and flank contact areas. **(d)** Temperature distribution on the tool rake face as determined by running thermocouple measurements. The test conditions are the same as those used in the orthogonal cutting test (b).

6.2.3 Semi-artificial and running thermocouples

To measure temperature in the deformation zone and in the chip, semi-artificial thermocouples embedded in the layer to be removed can be used as show in Figure 6.20a. Using the results of such measurements, the temperature field in the deformation zone and in the chip can then be reconstructed. Insulated constantan wires of 0.12-mm diameter are embedded in holes of different depths in the layer to be removed. When the layer approaches the deformation zone, it deforms plastically so that the wires are gripped securely in the holes, forming in this way semi-artificial thermocouples. The cold junctions of the thermocouples were secured to the workpiece far enough from the deformation zone to keep the cold thermocouple junctions at room temperature. Arranged in this way, thermocouples produce the emf while the layer to be removed passes the deformation zone, becoming the chip, and then slides over the tool rake face.

Figure 6.20b shows the reconstructed temperature field obtained using this method in an orthogonal cutting test on a shaper. As seen, the results are in compliance with the conclusions reached in Chapter 4 — that the temperatures in orthogonal cutting do not affect the mechanical properties of workpiece material if the cutting speed is higher than that of heat expansion. The chip gains its temperature on sliding over the tool rake face, and the maximum temperature has been registered at the distances of 0.5 to 1.5 mm from the cutting edge, depending on the cutting speed. When cutting with high speeds and feeds, it was observed with the aid of a high-speed camera that, during its formation in the deformation zone and even within the tool/chip interface, the chip has a gray color peculiar to the workpiece material. Only when the chip loses its contact with the rake face and continues to move freely does it becomes red, as the heat gained by the chip contact layer propagates into the entire chip volume so that the chip temperature becomes uniform. However, this heat does not affect the resistance of the workpiece material in orthogonal cutting.

Figure 6.20c shows the running thermocouple technique for mapping temperatures along the tool/chip interface and along the flank/workpiece contact surface. A tube-workpiece is used for experiments. A hole of 0.8- to 1.2-mm diameter is drilled in the workpiece, and a thin-wall protective tube is inserted into the hole. Then, two insulated wires made of thermocouple materials (for instance, one from cromel and the other from copel) are inserted into the tube, and their ends A_1B_1 and C_1D_1 are connected to two amplifiers connected with a data acquisition board. In cutting, the tube (made of material similar to that of the workpiece) is cut into two portions as shown in Figure 6.20c. As such, two running artificial thermocouples are formed having outputs A_2B_2 and C_2D_2. The first thermocouple runs along the tool/chip interface, registering a temperature distribution in this region. The second runs along the tool/workpiece flank contact area which yields the temperature distribution along this area.

In the study leading to Figure 6.20d, longitudinal turning tests were conducted on a lathe at different cutting speeds. As was expected, the maximum temperature (○ in the figure) at the tool/chip interface depends on the cutting speed. This is readily explained by the increase of sliding speed. Because the chip sliding speed is calculated as the cutting speed over the chip compression ratio and the latter also decreases with the cutting speed, a non-proportional increase in the maximum temperature is observed in Figure 6.20d. The same tendency can be observed with the tool temperatures corresponding to the end point of the tool/chip interface (• in Figure 6.20d). Also, a most interesting observation can be made in Figure 6.20d about the temperatures of points corresponding to the distance $y = 0$ from the cutting edge.

As seen, these are relatively low. Although the cutting regime, tool geometry, and tool and workpiece materials are the same as those used in the orthogonal cutting tests (Figures 6.20a,b), the higher temperatures in the deformation zone are explainable by residual heat from the previous revolution as discussed in Chapter 3.

The experimental results, obtained with the discussed semi-artificial and running thermocouples, prove that:

- True orthogonal cutting cannot be used as a model for real cutting operations when the thermal phenomenon is to be studied.
- Because the residual heat on the current pass may affect the thermomechanical characteristics of workpiece material, it should be accounted for properly in the tests.

6.3 Cutting force measurements

As pointed out by Shaw,[2] presentation of the analysis of metal cutting operations on a qualitative basis requires certain observations before, during, and after a cut. The number of observations that can be made during the cutting process is rather limited. Among the others, one of the most important measurements of this type is the determination of cutting force components.

6.3.1 General

An understanding of the forces in machining processes is necessary, as they are related to such things as tool wear, vibrations, accuracy, power consumption. Because force magnitudes can be determined by a variety of techniques, it is important to know the methodology of measurement to ensure the correct results. This is extremely important when certain aspects of a new theory have been examined due to the fact that results are used to judge the theory.

In measuring cutting force, it has been recognized for a long time that, although each dynamometer must be specially designed to meet specific requirements, certain design criteria are general.[2,3] Unfortunately, recent books on metal cutting do not pay particular attention to this significant aspect, though they present the results of studies including cutting forces as proofs of the suggested theories.[4–6]

In general, the papers on cutting force measurement[24–45] do not provide sufficient information on the calibrations of the measuring setups and the measuring procedure. Significant issues such as the calibration of the setups for the force measurement, possible errors in the recording of the force signals caused by the cut-off frequency of a recorder, the cut-off frequency itself, sampling rate, number of points per FFT time slice, number of slices averaged, type of aliasing filters, etc. are not addressed sufficiently. It is understood that insufficient information regarding measuring and reporting of cutting force makes it very difficult to compare the results from various research. It may be partially explained by the restrictions on paper size imposed by many journals.

It should be mentioned here that there are a few works which provide the results of well-performed experiments on force measurements. For example Buryta et al.[39] presented the procedure of static and dynamic calibrations of the setup used, although it would be easier to understand the results if the type of piezoelectric impact hammer used in the experiments was reported, as well as the sampling rate, aliasing filters, cut-off frequency, number of slices, etc. Noori-Khajavi and Komandury[40] presented a well-conducted study with a theoretical background on spectral estimation. To the best of the author's knowledge, this is the first work in the field which discusses the aliasing filters, signal-to-noise ratio of sensor signals in the frequency domain, and selection of sampling rate. Unfortunately, the static and dynamic calibrations are not mentioned, and the inter-influences of the force channels were not reported. Berger et al.[41] presented a short theoretical background on bispectra and bicoherence. In their work, the conditions of the dynamic force measurements are well presented and the inter-influence of the force channel is well analyzed. However, there is no explanation for why the reported parameters of the spectrum analyzer were selected as reported, and there is no mention of the static and dynamic calibration procedures.

A special group of studies during the last 20 years claims that the metal cutting process is of a stochastic nature.[42–50] This conclusion is drawn from the force signature represented by the cutting force signal. The cutting process is defined as being a stochastically stationary process so that its prediction cannot be made on the basis of its theoretical analysis. This result may be explained as follows.

A new era in the cutting force dynamometry started when piezoelectric transducers became available (for example, the two-component load washer Kistler 9271A); however, the amazing sensitivity of such transducers was unusual for metal cutting researchers. Instead of one simple signal (recording) from a dynamometer which could be easily decoded using a static calibration curve, they started receiving an assembly of signals reflecting even small dynamic unevennesses of the machine, fixtures, cutting tool, auxiliary components, and environmental noise.

To process signals from these dynamometers, researchers had to catch up with the development of new experimental methodologies, never used before in metal cutting experiments, such as, for example, Fast Fourier Transform (FFT). Moreover, they had to master the theory behind the new instruments (FFT analyzers), as machine tools are in general far from being simple linear systems. Another thing that has never been really recognized is that machine tools as dynamic systems exist only when machining. The reason for this would be that only after the cutting forces have been applied to take up many clearances in the assembly, thus introducing friction, etc., has the dynamic system been established. This calls for costly real-time measurements with extensive use of triggering facilities. Unfortunately, rarely have both the static and the dynamic calibrations been performed, but when they were, the calibrations were carried out on a stationary machine tool without cutting or even idling. As a result, calibration is carried out on one system while using a completely different one, in terms of dynamics, when testing.

To represent signals in the frequency domain, new functions have been introduced, among them the most common being the power spectral density function.[51] This function describes the general frequency composition of the acquired signal in terms of spectral density of its mean square value. However, the results of power spectral density function measurements, known as autospectra, are not always properly interpreted in the cutting force measurements. The trouble came from misunderstanding the word "power" that has been kept in the signal analysis theory only for historical reasons. In the field of electrical engineering, from which it came, the unit really reflected the energy of the signal. In the theory of stochastic processes, this word means that the integrated function is squared so that the power spectral density function shows the rate of change of the mean square value with frequency. In metal cutting dynamics, particularly in cutting force measurement, this word was sometimes thought of as reflecting the mechanical power. As a result, the power spectral density function was given dimensions such as g^2/Hz, or even W/Hz, instead of N^2/Hz and $(Nm)^2$/Hz when measuring forces and torques. Moreover, the number of Hz to scale the axes was never related to the instrument setup. Such a misuse of the power spectral density function resulted in a conclusion that the cutting process is fully stochastic in nature, and the cutting force changes show no distinct frequency rather than "white noise". Obviously, this conclusion contradicts the entire metal cutting practice which desperately requires proper dynamic force measurements to improve the quality and efficiency of machining.

6.3.2 *Procedure used in the current study*

Because the cutting force is known to be very sensitive to even the smallest changes in the cutting process, special focus was directed to the selection of the conditions of the tests and experimental methodology. Measuring the cutting force involves three successive stages: pre-process, measuring, and analysis of the results obtained in the test.

The workpiece materials which were used in the test were selected to represent the major group of workpiece materials used in industry. The composition and element limits for each workpiece material were chosen according to the requirements of standard ANSI/ASME B94.55M-1985 and were requested from the steel dealer. Special parameters such as the chemical composition, element counts, microstructure, and grain size were inspected using quantitative metallurgy. The latter is very important because of a real possibility of mistakes in certificates issued by steel dealers. In our research practice, we have experienced such mistakes.

For this particular study, we bought bars of steel AISI 4130. Following our standard procedure, we requested from a steel dealer a certificate showing the results of analysis on the chemical composition, structure, grain size, and hardness. According to the certificate, the steel should have had a ferrite-pearlite structure. The steel was used in an experiment where the influence of high cutting speed (thus high strain rate) on the chip structure was studied. An analysis of chip micrographs revealed that the chip structure contained primarily martensite (as shown in Figure 5.13b) so that the conclusion was drawn that the high strain rate and temperature resulted in a change of the structure of workpiece material from ferrite-pearlite to martensite as the result of a significant influence of the strain rate and temperature occurring during cutting. Because this conclusion is at odds with our experience and earlier experimental results, it was decided to examine the initial structure of the material. It was quite a surprise that, in contrary to the certificate provided, the initial structure contained mainly martensite (as shown in Figure 5.13a) as a result of improper heat treatment. This is a graphic example of how a wrong conclusion may be drawn if sufficient care is not devoted to the early stages of the preparation, carrying out, and analysis of an experiment. Although workpieces having different microstructures may have approximately the same hardness, the cutting force required to fracture the workpiece materials may differ significantly, as discussed in Chapter 4.

As discussed in Chapter 4, end-tube turning cannot be used to simulate orthogonal cutting. In the present study, the special specimens were prepared to truly simulate the orthogonal cutting. After being machined to the configuration shown in Figure 6.21, the specimens were tempered at 180 to 200°C to remove the residual stresses. The hardness and microhardness of each specimen were determined over the entire working part. Cutting tests were conducted only on the bars where the hardness was within the limits ±10%.

A retrofitted Schaerer HPD 631 lathe was used. The drive unit motor was replaced with a 15-kW variable speed AC motor, and the feed motor was replaced with a 5-kW variable speed AC motor. The motors are individually controlled by AC invertors. The AC invertors are designed to provide the required volts/hertz ratio, allowing the AC motors to run at their optimum efficiency and providing rated torque capability through the motor's rated base speed. The control section of the AC invertors consists of a control board with a 16-bit microprocessor and keypad interface with an 8-bit microprocessor.

It is instructive to mention here that the accuracy of workpiece location in the machine and the machine itself may affect both the steady-state and dynamic cutting forces in such a way that it becomes difficult to distinguish between the cutting signature due to the dynamic response of the cutting process and that due to noise originating from inaccuracy of workpiece

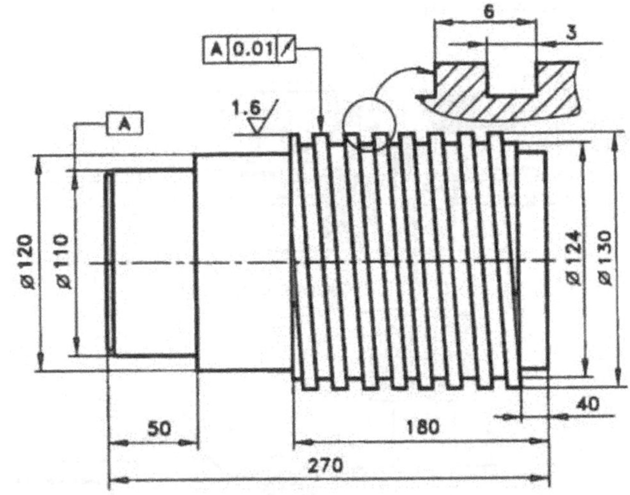

Figure 6.21 Configuration of the workpieces used in the experiments.

rotation. In particular, it becomes a problem when a relatively light cutting regime is used. Nevertheless, the accuracy of the workpiece rotation has never been considered as a factor in cutting force studies. The significant discrepancy in the reported results for cutting force measurements may be partially explained by this inaccuracy.[52]

A general-purpose tool holder CTJNR2520L16 was used. The geometry parameters of the tool were chosen to ensure real orthogonal conditions and were controlled according to American National Standard B94.50-1975. Tolerances for all angles were ±0.5°. The roughness, R_a, of the face and flank of drills did not exceed 0.25 μm and was measured according to American National Standard ANSI B46.1-1978. Each cutting edge was examined at a magnification of 15× for visual defects such as chip and/or cracks.

A two-component dynamometer made similarly to Kistler Type 9271A was used. Based on the standard mounting as specified by the supplier (Kistler), the load washer (Kistler Type 9065) was installed in the dynamometer and pre-loaded to 120 kN. At this pre-load, the range for force measurements was from –20 to +20 kN; threshold was 0.02 N; sensitivity was –1.8 pC/N; linearity was ≤±1.0% FSO; overload was 144 kN; cross-talk was ≤0.02 N/N; resonant frequency was ≈40 kHz; temperature error was +30 N/°C.

A schematic diagram of the setup used in the experiments is shown in Figure 6.22. The load washer was connected to the charge amplifiers (Kistler, Model 5004). The charge amplifier (Type 5004) was a mains-operated DC amplifier of very high input impedance with capacitive negative feedback, intended to convert the electric charge from a piezoelectric transducer into a proportional voltage on the low impedance amplifier output. The calibration factor setting (adjustment of transducer sensitivity at the amplifier) makes

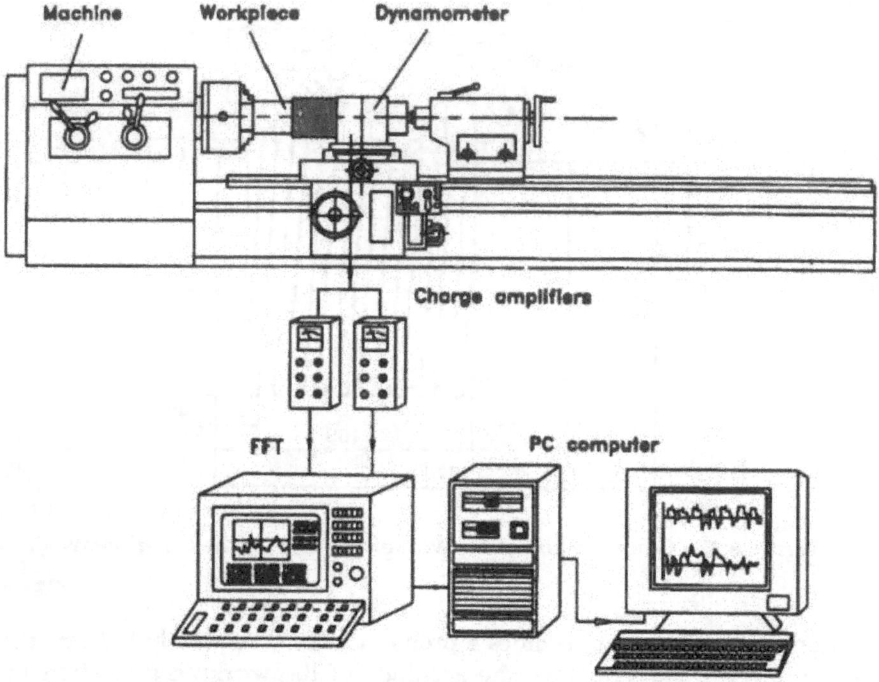

Figure 6.22 Schematic of the experimental set-up.

possible standardized amplifier sensitivities of, for example, 1, 2, 5, etc. mV per mechanical unit (N). The carefully designed calculating disc enables the reciprocal value of sensitivity to be shown directly as a measuring range. With the charge amplifiers, the specifically piezoelectric part of the measuring system ends. Charge calibrators (Type 541A) were connected instead of transducers, allowing the entire measuring chain to be calibrated with an appropriate charge signal.

Cables used in the connections were specially made for Kistler equipment. In addition to an extremely high insulation resistance, they offer freedom from disturbing charge signals when the cables are moved in the smallest possible capacity; 1619 cables protected by metallic tubing were used.

The outputs of the charge amplifiers were connected to the FFT analyzer (B&K, Model 2032). The dual channel signal analyzer (Type 2032) is flexible, easy to use, and a fully self-contained, two-channel FFT analysis system with 801 lines of resolution. The analyzer has a real-time speed of >5 kHz (>10 kHz in a single channel). It is flexible, because calibration, display scales, post-processing, etc. are user definable, and functions such as signal-to-noise ratio, cross-spectra, autospectra, etc. are computed directly without the need for intermediate processing. It is easy to use because operation is largely self-exploratory with all relevant control settings clearly shown on the display screen, and because complete measurement and display setups can be stored for later recall and use. It is self-contained because it has a fully instrumented front-end, built-in digital zoom, a built-in zooming signal generator, and IEC/IEEE interface. Its 801-line resolution is of special importance, as more modes of vibration can be identified and characterized in signal analysis than with a conventional 250 or 400-line analyzer.

The setup was calibrated statically and dynamically. In addition, the validity of measurements were examined. The objective of the static calibration was to establish a relationship between the measured value of the dynamometer and the actual value of the forces to be measured. In spite of the fact that the piezoelectric transducers are generally used to sense dynamic quantities, the static calibration is possible because of the high insulation resistance of the load washer and because of the high input impedance provided by the charge amplifier. Thus, for any value of the load less than the maximum value, in both the tangential and radial directions, the time duration of the signal is sufficiently large to permit a reading on a digital voltmeter. Static calibration of the dynamometer was carried out by applying various loads of known magnitudes, measuring the output of the dynamometer, and establishing a graphical relation between the measured quantities and the applied forces. In calibration of the dynamometer, the known loads were applied using a vertical hydraulic loading machine. To minimize the error introduced in static calibration, the charge amplifiers were set to "long" mode so that the time constant of the system becomes large, and the rate of charge decay during calibration was reduced. The load was provided by a loading machine and was incrementally increased to a value of 10 kN. At each step, a reading was recorded. Similar readings were recorded during the unloading cycle. The latter was performed to estimate the hysteresis of the system. It was observed that the difference in the readings obtained during loading and unloading were sufficiently small to neglect the hysteresis. A typical calibration chart is shown in Figure 6.23.

Dynamic calibration of the dynamometer-workpiece-machine tool system has been carried out to:

- Avoid measuring the vibration of the tool instead of the force fluctuations.
- Determine the frequency band, over which the dynamometer can be used for reliable measurements.
- Determine the resonant frequencies of the system so as to make sure that the response of the cutting process is not in the neighborhood of the dynamometer resonant frequencies.
- Determine the range of frequencies of the cutting forces which could be measured accurately without distortion.

Because in the present study only the static components of the cutting force were measured, the dynamic calibration of the setup has been performed

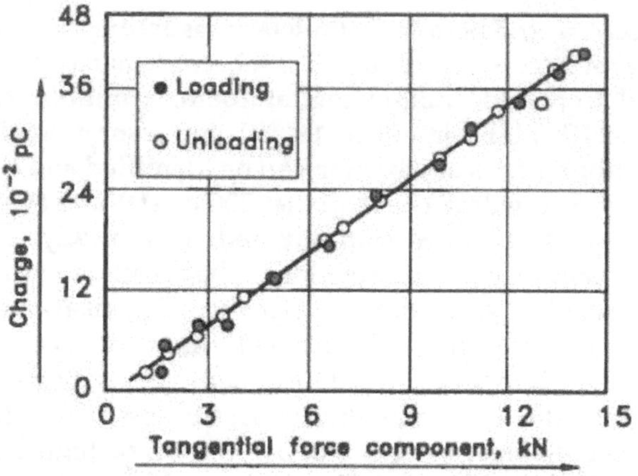

Figure 6.23 Typical calibration chart (channel of the tangential force component, static calibration).

using a simplified procedure. At this point, it is worthwhile to discuss an issue that is sometimes troublesome to many researchers — namely, the fact that dynamic calibration is a mandatory step even if only the static cutting force is to be measured. The reason for this is the nature of the signal from the piezoelectric transducers used in the dynamometer. Because the static cutting force is obtained by averaging the signal from the transducer, it is very important to know exactly what has been averaged. Although the natural frequency of a Kistler load cell is about 40 kHz, it cannot be considered as the natural frequency of the dynamometer. Even if the natural frequency of the dynamometer is known, it is not of prime importance, as the dynamometer is installed into the machine, and the natural frequency of the machining system where the cutting tests are to be conducted is of prime concern, as it may dramatically affect the measured results. Therefore, the whole measuring setup should be dynamically calibrated, when all accessories used in the test are installed. Even with this arrangement, the results of the dynamic calibration can be far from reality, as the machining system exists only during actual machining. Therefore, it may be concluded that the methods for the precise dynamic calibration of the machining system have yet to be developed. Fortunately enough, when only the static component of the cutting force is of prime concern, a simplified procedure of calibration is sufficient.[52]

A hammer B&K 8203 with a built-in cell has been used to supply an impulse excitation to the machine dynamometer-workpiece system, while the FFT has been used to resolve the signals coming out from the hammer and the load washer. First, an autospectrum of the input channel was acquired to make sure the hammer is capable of providing enough input energy over the frequency range of interest. Because the response is a transient function, the autospectrum was scaled to show the energy spectral density of the signal. The results show that when equipped with a metal tip, the hammer provided uniform input over a broad frequency range from 0 Hz to 6 kHz (Figure 6.24). A drop of less than 20 dB was tolerated following the common recommendation for resonance measurements.[53] Second, the frequency response function (FRF), $H_1(f)$, of the machine dynamometer-workpiece system was determined in order to find out the frequency band in which the FRF function remained constant. This frequency band determines the range of frequencies of the resultant force system, which can be measured without the introduction of any signal distortion due to the system nonlinearities. Figure 6.25a shows that the FRF (direct correlation) is very close to unity in the frequency range of 0 to 1100 Hz, and Figure 6.25b shows that the FRF (cross-correlation) has a constant value of 0.02 over the same range.

The validity of the measurements was checked by calculating the coherence function $\gamma^2(f)$ for the measured spectra, as shown in Figures 6.26a,b. The coherence gives a measure of the degree of linear dependence between any two signals over the frequency axis. It is calculated from the two autospectra and cross-spectrum. As seen from these figures, this function is close to unity conforming to the linear independency of the measurements.

The measurement mode is the most important element in determining how the measurements are carried out. There are four measurement modes available with the FFT analyzer: (1) spectrum averaging mode; (2) spectrum averaging, zero pad mode; (3) 1/n octave spectrum averaging mode; and (4) signal-enhancement mode. The first mode (spectrum averaging) was used in the experiments because it is suitable for signal and system analysis, whereas other modes are implemented for other types of measurements.

The analyzer can capture time records consisting of 128, 256, 512, 1024, or 2048 samples. When the time records are Fourier transformed, the resultant complex spectra may have 64, 128, 256, 512, and 1024 independent lines each. However, due to anti-aliasing filters, only 51, 100, 201, 401, or 801 lines will have the correct amplitude. It is an advantage to have as many samples in the time record as possible, as this gives the best frequency resolution. Therefore, the use of 801 spectral lines has been selected to obtain the best frequency resolution, as there is no large change in signal amplitude or frequency within the signal record.

The FFT analyzer cannot execute the Fourier transformation continuously. It has an internal computer dedicated to performing these calculations, but it must look at a time block of data. This is the purpose of the sampling zone in the FFT analyzer. The sampling zone holds the amplified and the filtered signal in short-term memory while voltage readings are taken. The voltage readings are taken to convert the time waveform into a table of numbers. When complete, the time block of waveform data resides in the computer memory as a table of numbers that contains both amplitude and phase information. This is necessary because the Fast Fourier Transform is a digital mathematical process that operates with numbers. The typical

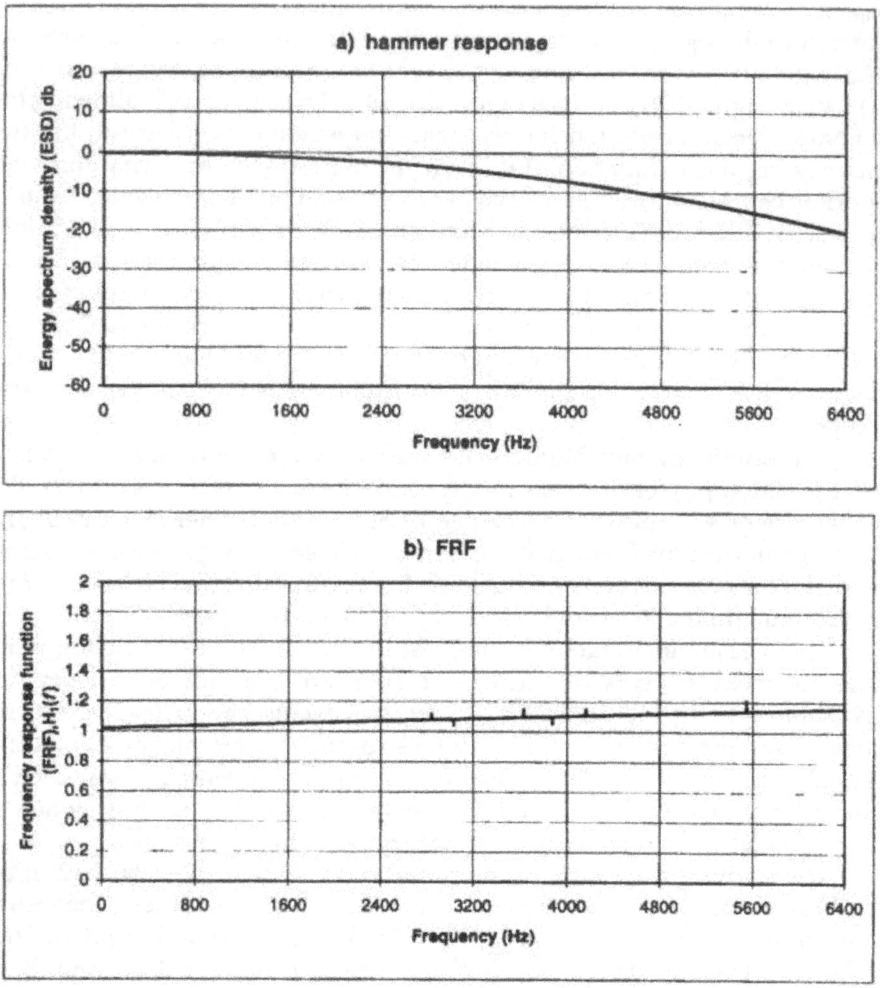

Figure 6.24 Calibration of the hammer for the frequency response function (FRF) measurements: **(a)** hammer response, and **(b)** FRF values.

spectrum analyzer takes about 50,000 readings per second, on the input waveform. The analyzers measure this quickly to overcome the problem of tracking high-frequency signals. According to the Nyquist criterion,[51] an analyzer that has a frequency range of 0 to 20,000 Hz must measure the voltage of the input signal faster than twice this frequency (or 40,000 Hz) to be able to see changes at 20,000 Hz. In the experiments, the sampling frequency was set to 1600 Hz to see the change at 800 Hz, which was the range of interest.

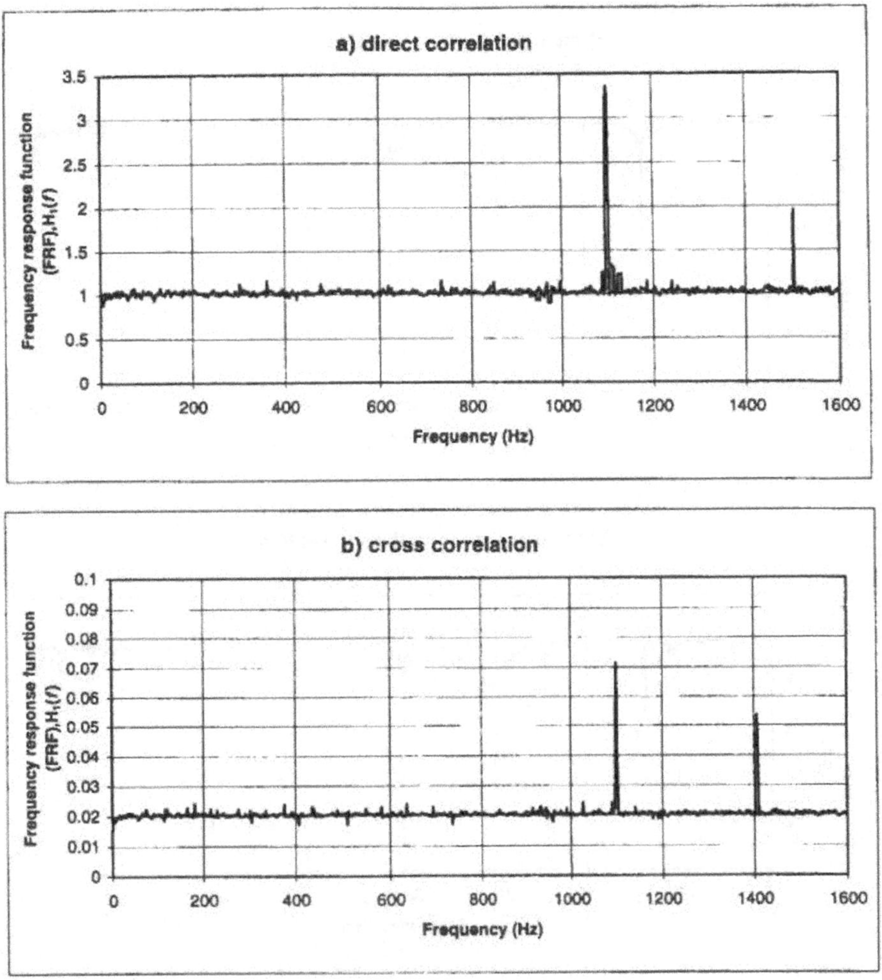

Figure 6.25 Frequency response function (FRF) when the input excitation was along the direction of the power component of the cutting force: **(a)** direct correlation, and **(b)** cross-correlation.

There are three modes of averaging which can be selected to obtain suitable autospectra: (1) exponential, (2) linear, and (3) peak averaging. Each mode has its application depending on the type of the signals. Exponential averaging places more emphasis on the latest spectrum. Linear averaging places equal emphasis on all of the averaged spectra where the peak averaging records the largest amplitude of each spectral line. In the measuring setup, exponential averaging has been used to continuously monitor signals which may be slowly varying.

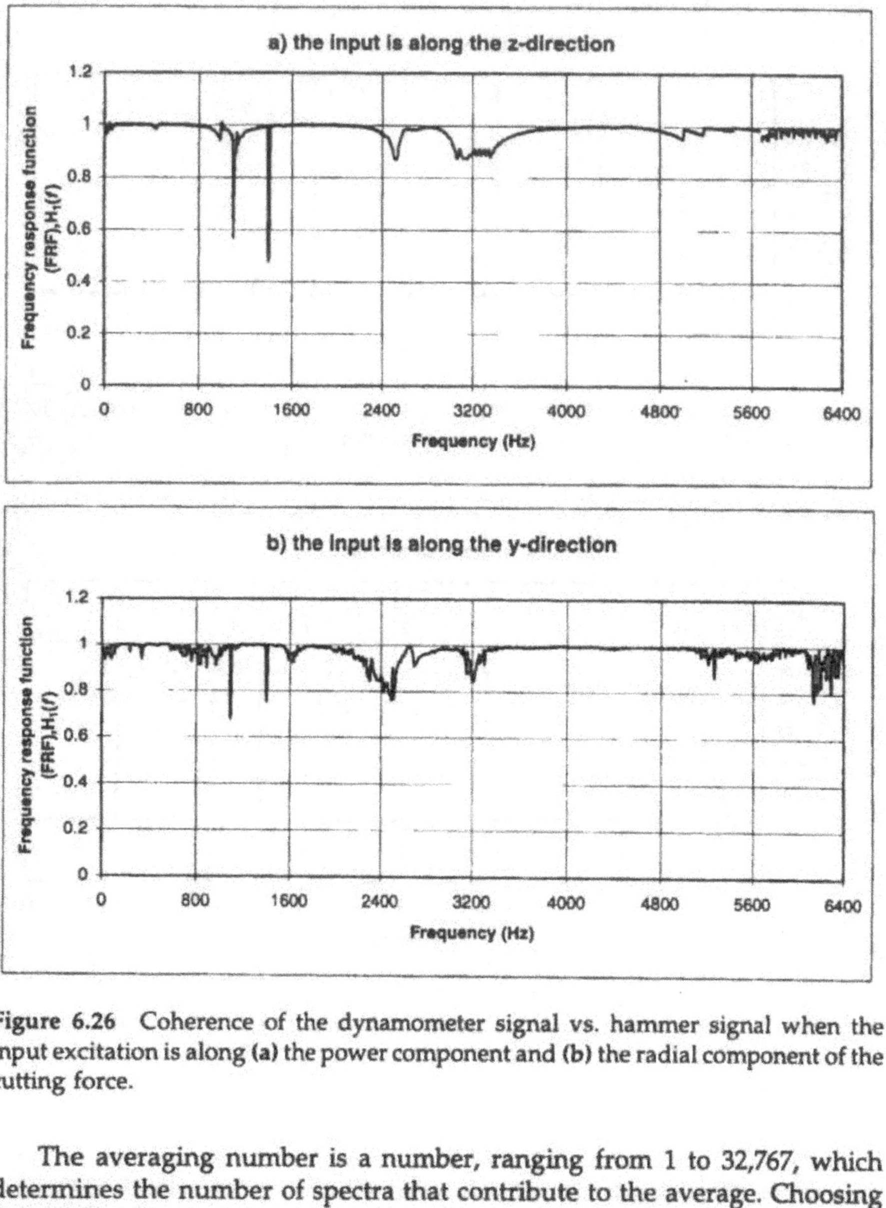

Figure 6.26 Coherence of the dynamometer signal vs. hammer signal when the input excitation is along **(a)** the power component and **(b)** the radial component of the cutting force.

The averaging number is a number, ranging from 1 to 32,767, which determines the number of spectra that contribute to the average. Choosing this number depends on: (1) the desired accuracy — the greater the accuracy, the higher the required averaging number, (2) the amount of the noise in the measurements, and (3) the type of signal being analyzed. For stationary signals, only a few averages are necessary to obtain a good estimate; however, averaging of 15 has been used in the measurements to increase the accuracy and to eliminate the noise as much as possible. This averaging number results in an error of less than 0.5 dB, and the process does not take too long.[51]

Aliasing is a mirror of the high-frequency signals into the lower frequency range where the high-frequency signals can form false peaks in the frequency domain. This is a consequence of the digitizing process. They appear as vertical lines in the frequency domain of the FFT spectrum analyzer. They are ghost images of high-frequency data input to the analyzer. Aliasing has been eliminated using two methods. First, the input data were measured at more than the twice the highest frequency of interest. Second, used in conjunction with the first method, built-in low-pass filters appropriate for each frequency range with a cutoff frequency at about 80% of the Nyquist frequency and displaying the results unaffected by the filter, were applied to the analyzer. Typically for 2048 points, 801 frequency lines were displayed.

Windowing the data is one of the most important parameters. In the FFT analyzer, the digital sampling takes place during discrete blocks of time. The FFT analyzer assumes that the signal which exists in that block of time also exists for all the time before and after the sampling. Processing by time blocks is necessary for real-time data acquisition. Therefore, it is not possible for the analyzer to wait for an extended time period to check if the signal is periodic. The Fourier transformation, however, should make an assumption of periodicity to start the acquisition process. In doing so, it can produce a leakage error if the actual signal does not match up at the beginning and end of the current time block. This leakage is undesirable, as it may hide low-amplitude signals, causing distortions in the resulting spectrum. To avoid the leakage, input data are weighted with a mathematical function that favors the data in the center of the time block and reduces the data at both ends to zero. There are many useful window functions which can be used such as rectangular, Hanning, Kaiser-Bessel, and flat top.

A better choice of the window function for stationary signals, as for the case of the signal of the cutting force, is one which is equal to zero at each end of the time blocks and whose amplitude varies smoothly. The Hanning window is an excellent window function which meets these requirements and was used to process the experiment data.[51]

References

1. Schenck, H., Jr., *Theories of Engineering Experimentation*, 2nd ed., McGraw-Hill, New York, 1968.
2. Sedov, L.I., *Similarity and Dimensional Methods in Mechanics*, MIR, Moscow, 1982.
3. Astakhov, V.P. and Osman, M.O.M., An analytical evaluation of the cutting forces in self-piloting drilling using the model of shear zone with parallel boundaries. Part 2. Applications, *Int. J. Machine Tools Manuf.*, 36(12), 1335, 1996.
4. Silin, S.S., *Similarity Methods in Metal Cutting* (in Russian), Mashinostroenie, Moscow, 1979.
5. Makarow, A.D., *Wear and Tool Life* (in Russian), Mashinostroenie, Moscow, 1966.

6. Makarow, A.D., *Optimization of Cutting Processes* (in Russian), Mashinostroenie, Moscow, 1976.
7. Vinogradov, A.A., *Physical Foundations of Drilling of Hard-To-Machine Materials* (in Russian), Naukova Dumka, Kiev, 1985.
8. Zorev, N.N., *Metal Cutting Mechanics*, Pergamon Press, Oxford, 1966.
9. Kronenberg, M., *Machining Science and Application: Theory and Practice for Operation and Development of Machining Processes*, Pergamon Press, London, 1966.
10. Kreyszig, E., *Advanced Engineering Mathematics*, John Wiley & Sons, New York, 1993, Appendix 5.
11. Mills, B. and Redford, A.H., *Machinability of Engineering Materials*, Applied Science Publishers, London, 1983.
12. Shaw, M.C., *Metal Cutting Principles*, Clarendon Press, Oxford, 1984.
13. Stephenson, D.A. and Agapionu, J.S., *Metal Cutting Theory and Practice*, Marcel Dekker, New York, 1997.
14. Trent, E.M., *Metal Cutting*, Butterworth-Honeymoon, Oxford, 1991.
15. Rabinowicz, E., The least wear, *Wear*, 100, 533, 1984.
16. Archard, J.F., Contact and rubbing of flat surface, *J. Appl. Phys.*, 24, 981, 1953.
17. Archard, J.F. and Hirst, W., The wear of metals under unlubricated conditions, *Proc. R. Soc. London, Ser. A*, 3, 397, 1956.
18. Peterson, M.B., Florek, J.J., and Lee, R.E., Sliding characteristics of metals at high temperatures, *ASLE Trans.*, 3, 101, 1960.
19. Quinn, T.F.J. and Sullivan, J.L., A review of oxidation wear, *Proc. Int. Conf. Wear Metals*, American Society of Mechanical Engineers, New York, 1977, p. 110.
20. Foley, R.T., Peterson, M.B., and Zapf, C., Frictional characteristics of cobalt, nickel and iron as influenced by their surface oxide films, *ASLE Trans.*, 6, 29, 1963.
21. Holman, J.P., *Experimental Methods for Engineers*, McGraw-Hill, New York, 1994, chap. 8.
22. Usachev, Y.G., Phenomena occurring during the cutting of metals (in Russian), *Izv. Petrogradskogo Politekhnicheskogo Inst.*, XXIII(1), 245, 1915.
23. Resnikov, A.N. and Resnikov, L.A., *Thermal Processes in Manufacturing Systems* (in Russian), Mashinostroenie, Moscow, 1990, chap. 4.
24. Lin, A.C. and Lin, S.Y., A coupled finite element model of thermo-elastic-plastic large deformation for orthogonal cutting, *ASME J. Eng. Industry*, 114, 218, 1992.
25. Wang, D.H., Ramulu, M., and Arola, D., Orthogonal cutting mechanisms of graphite/epoxy composite. Part II. Multi-directional laminate, *Int. J. Mach. Tools Manuf.*, 35(12), 1639, 1995.
26. Olgac, N. and Guttermuth, J.R., A simplified identification method for autoregressive models of cutting force dynamics, *ASME J. Eng. Industry*, 110, 288, 1988.
27. Song, X., Strain-hardening and thermal-softening effects on shear angle prediction: new model development and validation, *ASME J. Eng. Industry*, 117, 28, 1995.
28. Stephenson, D.A., Material characterization for metal-cutting force modeling, *ASME J. Eng. Mater. Technol.*, 111, 210, 1989.

29. Stevenson, R. and Stephenson, D.A., The mechanical behaviour of zinc during machining, *ASME J. Eng. Industry*, 117, 172, 1995.
30. Thangaraj, A. and Wright, P.K., Computer-assisted prediction of drill-failure using in- process measurements of trust force, *ASME J. Eng. Industry*, 110, 192, 1988.
31. Zhang, G.M. and Kapoor, S.G., Dynamic modeling and analysis of the boring machine system, *ASME J. Eng. Industry*, 109, 219, 1987.
32. Marui, E., Kato, S., Hashimoto, M., and Yamada, T., The mechanism of chatter vibration in a spindle-workpiece system. Part 2. Characteristics of dynamic cutting force and vibration energy, *ASME J. Eng. Industry*, 110, 242, 1988.
33. Lee, S.J. and Kapoor, S.G., Cutting process dynamics simulation for machine tool structure design, *ASME J. Eng. Industry*, 108, 68, 1986.
34. Elbestawi, M.A., Papazafiriou, T.A., and Du, R.X., In-process monitoring of tool wear in milling using cutting force signature, *Int. J. Mach. Tools Manuf.*, 31, 55, 1991.
35. Budak, E., Altintas, Y., and Armarego, E.J.A., Prediction of milling force coefficients from orthogonal cutting data, *ASME J. Eng. Industry*, 118, 216, 1996.
36. Lee, L.C., Lee, K.S., and Gan, C.S., On the correlation between dynamic cutting force and tool wear, *Int. J. Mach. Tools Manuf.*, 29(3), 295, 1989.
37. Adolfsson, C. and Stahl, J.E., Cutting force model for multi-toothed cutting processes and force measuring equipment for face milling, *Int. J. Mach. Tools Manuf.*, 35(12), 1715, 1995.
38. El-Wardany, T.I., Gao, D., and Elbestawi, M.A., Tool condition monitoring in drilling using vibration structure analysis, *Int. J. Mach. Tools Manuf.*, 36(6), 687, 1996.
39. Buryta, D., Sowerby, R., and Yellowley, I., Stress distribution on the rake face during orthogonal cutting, *Int. J. Mach. Tools Manuf.*, 34(5), 721, 1994.
40. Noori-Khajavi, A. and Komanduri, R., Frequency and time domain analyses of sensor signals in drilling. I. Correlation with drill wear, *Int. J. Mach. Tools Manuf.*, 35(6), 775, 1995.
41. Berger, B.S., Minis, I., Deng, K., Chen, Y.S., Chavali, A., and Rokni, M., Phase coupling in orthogonal cutting, *J. Sound Vibration*, 191(5), 976, 1996
42. Chandrashekhar, S., Osman, M.O.M., and Sankar, T.S., An experimental investigation for the stochastic modeling of the resultant force system in BTA deep-hole machining, *Int. J. Prod. Res.*, 23(4), 657, 1985.
43. Chandrashekhar, S., Sankar, T.S., and Osman, M.O.M., A stochastic characterization of the tool-workpiece system in BTA deep hole machining. Part 1. Mathematical modeling and analysis, *Adv. Manuf. Process. J.*, 2, 37, 1987.
44. Chandrashekhar, S., Sankar, T.S., and Osman, M.O.M., A stochastic characterization of the tool-workpiece system in BTA deep hole machining. Part II. Response analysis and evaluation of the tool tip motion, *Adv. Manuf. Process. J.*, 2, 71, 1987.
45. Osman, M.O.M. and Sankar, T.S., Short-time acceptance test for machine tools based on the random nature of the cutting forces, *ASME J. Eng. Industry*, 94, 1020, 1972.

46. Sankar, T.S. and Osman, M.O.M., Profile characterization of manufactured surfaces using random function excursion technique. Part 1. Theory, *ASME J. Eng. Industry*, 97, 190, 1975.
47. Frazao, J., Chandrashekhar, S., Osman, M.O.M., and Sankar, T.S., On the design and development of a new BTA tool to increase productivity and workpiece accuracy in deep hole machining, *Int. J. Adv. Manuf. Technol.*, 1(4), 3, 1986.
48. Chandrashekhar, S., Frazao, J., Sankar, T.S., and Osman, M.O.M., On the stochastic description of the tool tip motion and its influence on the surface texture in BTA deep-hole machining considering the dynamic interaction between the cutting tool and workpiece, *Adv. Manuf. Process. J.*, 1, 393, 1986.
49. Chandrashekhar, S., Ahmed, Z., Osman, M.O.M., and Sankar, T.S., On the prediction of roundness error in BTA deep-hole machining based on the stochastic description of the cutting tool motion, *Int. J. Prod. Res.*, 24(4), 879, 1985.
50. Osman, M.O.M., Xistris, G.D., and Chahil, G.S., Measurement and stochastic modeling of torque and trust in twist drilling, *Int. J. Prod. Res.*, 17(5), 571, 1979.
51. Bendat, J.S. and Piersol, A.G., *Random Data: Analysis and Measurement Procedures*, Wiley-Interscience, New York, 1971.
52. Hayajneh, M.T., A Generalized Approach for Mechanics of Chip Formation in Steady-State and Dynamic Orthogonal Metal Cutting Using a New Model of Shear Zone with Parallel Boundaries and Its Verification to Cutting-Forces Prediction in Self-Piloting Machining, Ph.D. thesis, Concordia University, Montreal, 1998.
53. Randall, R.B., *Frequency Analysis*, K. Larsen & Son, Glostrup, Denmark, 1987.

Appendix

Cutting tool geometry

To apply any theory of metal cutting to predict the cutting process, it is necessary to know the geometry of the cutting tool used. International Standard ISO 3002-1977 defines cutting tool geometry in the tool-in-hand and tool-in-use systems in which the angles of the cutting tool are defined in a series of reference planes. In the author's opinion, however, it is not sufficient, as the standard does not account for the inaccuracies of tool position on the machine. Therefore, another system, which may be referred to as the tool-in-machine system (setting system),[1] has to be introduced in addition to the standard systems. This appendix aims to introduce the complete system of tool geometry.

A.1 Tool-in-hand system

A.1.1 Planes

The working part of the cutting tool basically consists of two surfaces intersecting to form the cutting edge. The surface along which the chip flows is known as the rake face or more simply as the face, and that surface which is ground back to clear the new or machined surface is known as the flank surface or simply as the flank. In the simplest yet common case the rake and flank surfaces are planes.

Figure A1 shows the definition of the main reference plane P_r as perpendicular to the assumed direction of primary motion and the tool-in-hand coordinate system. In this figure, v_f is the assumed direction of the cutting feed. Because angles of the cutting tool are defined in a series of reference planes, the standard defines a system of these planes in the tool-in-hand system, as shown in Figure A2. The system consists of five basic planes defined relative to the reference plane P_r. Perpendicular to the reference plane P_r and containing the assumed direction of feed motion is the assumed

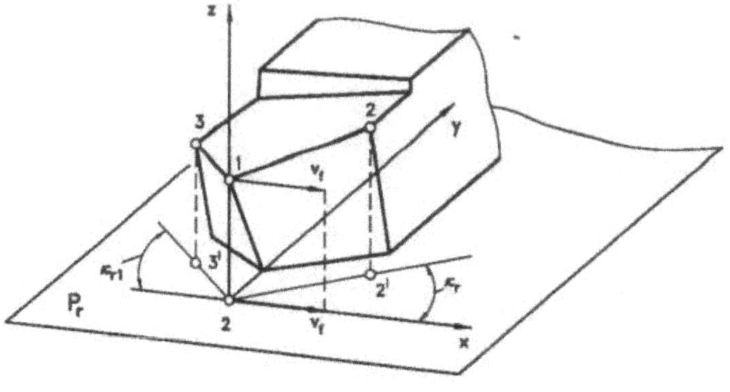

Figure A1 Definition of the main reference plane P_r

working plane P_f. The tool cutting edge plane P_s is perpendicular to P_r and contains the side (main) cutting edge (12 in Figure A1). The tool back plane P_p is perpendicular to P_r and P_f. Perpendicular to the projection of the cutting edge into the reference plane is the orthogonal plane P_o. The cutting edge normal plane P_n is perpendicular to the cutting edge.

Similarly, an additional system of planes can be attributed to the minor cutting edge and contains the following planes: P'_s, P'_o, P'_n as shown in Figure A3.

A.1.2 Angles

The geometry of a cutting element is defined by certain basic tool angles and thus precise definitions of these angles are essential.[2] A system of tool angles is shown in Figure A4. Rake, wedge, and clearance (flank) angles are specified by γ, β, and α, respectively, and these are identified by the subscript of the plane of intersection. The definitions of basic tool angles in the tool-in-hand system are as follows:

- κ_r is the tool cutting edge angle; it is the acute angle that P_s makes with P_f and is measured in the reference plane P_r and the acute angle between the projection of the main cutting edge onto the reference plane and the x-direction (Figure A1). κ_r is always positive and is measured in a counter-clockwise direction from the position of P_f.
- κ_{r1} is the tool minor (end) cutting edge angle; it is the acute angle that P'_s makes with P_f and is measured in the reference plane P_r and the acute angle between the projection of the minor (end) cutting edge into the reference plane and the x-direction (Figure A1). κ_{r1} is always positive (including zero) and is measured in a clockwise direction from the position of P_f.
- ψ_r is the tool approach angle; it is the acute angle that P_s makes with P_p and is measured in the reference plane P_r as shown in Figure A4.

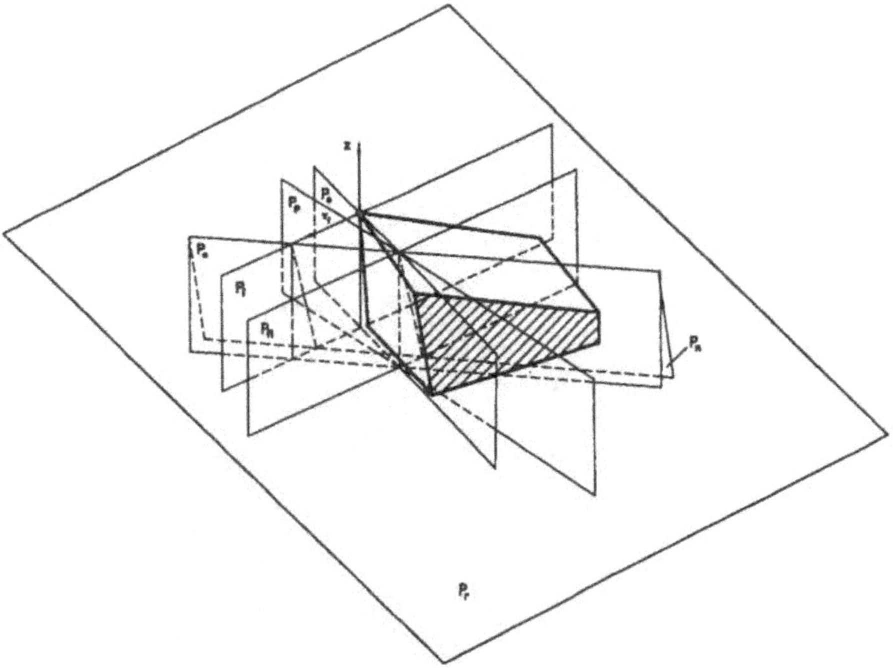

Figure A2 Standard system of reference planes in the tool-in-hand system (major cutting edge).

The rake angles are defined in the corresponding planes of measurement. The rake angle is the angle between the reference plane (the trace of which in the considered plane of measurement appears as the normal to the direction of primary motion) and the intersection line formed by the considered plane of measurement and the tool rake plane. The rake angle is defined as always being acute and positive when looking across the rake face from the selected point and along the line of intersection of the face and plane of measurement. The viewed line of intersection lies on the opposite side of the tool reference plane from the direction of primary motion in the measurement plane for γ_f, γ_p, γ_o, or a major component of it appears in the normal plane for γ_n. The sign of the rake angles is well defined (Figure A4).

The clearance (flank) angles are defined in a way similar to the rake angles, though here if the viewed line of intersection lies on the opposite side of the cutting edge plane P_s from the direction of feed motion, assumed or actual as the case may be, then the clearance angle is positive. Angles α_f, α_p, α_o, α_n are clearly defined in the corresponding planes. The clearance angle is the angle between the tool cutting edge plane P_s and the intersection line formed by the tool flank plane and the considered plane of measurement as shown in Figure A4.

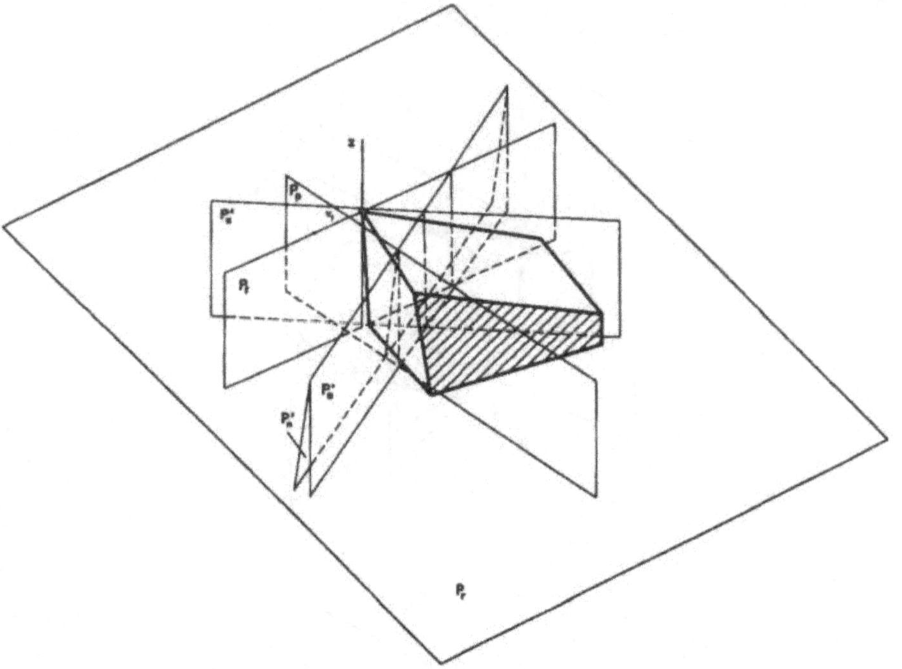

Figure A3 Standard system of reference planes in the tool-in-hand system (minor cutting edge).

The wedge angles β_f, β_p, β_o, β_n are defined in the planes of measurements. The wedge angle is the angle between the two intersection lines formed as the corresponding plane of measurement intersects with the rake and flank planes. For all cases, the sum of the rake, wedge, and clearance angles is 90°:

$$\gamma_p + \beta_p + \alpha_p = \gamma_n + \beta_n + \alpha_n = \gamma_o + \beta_o + \alpha_o = \gamma_f + \beta_f + \alpha_f = 90° \quad \textbf{(A1)}$$

For the minor (side) cutting edge, the flank angle α_{o1} is specified as the angle between the the tool minor (side) cutting edge plane P'_s and the intersection line formed by the tool minor flank plane and the plane of measurement P'_o as shown in Figure A4.

The orientation and inclination of the cutting edge are specified in the tool cutting edge plane P_s. In this plane, the cutting edge inclination angle λ_s is the angle between the cutting edge and the reference plane. This angle is defined as always being acute and positive if the cutting edge, when viewed in a direction away from the selected point at the tool corner being considered, lies on the opposite side of the reference plane from the direction of primary motion. This angle can be defined at any point of the cutting edge. The sign of the inclination angle is well defined in Figure A4.

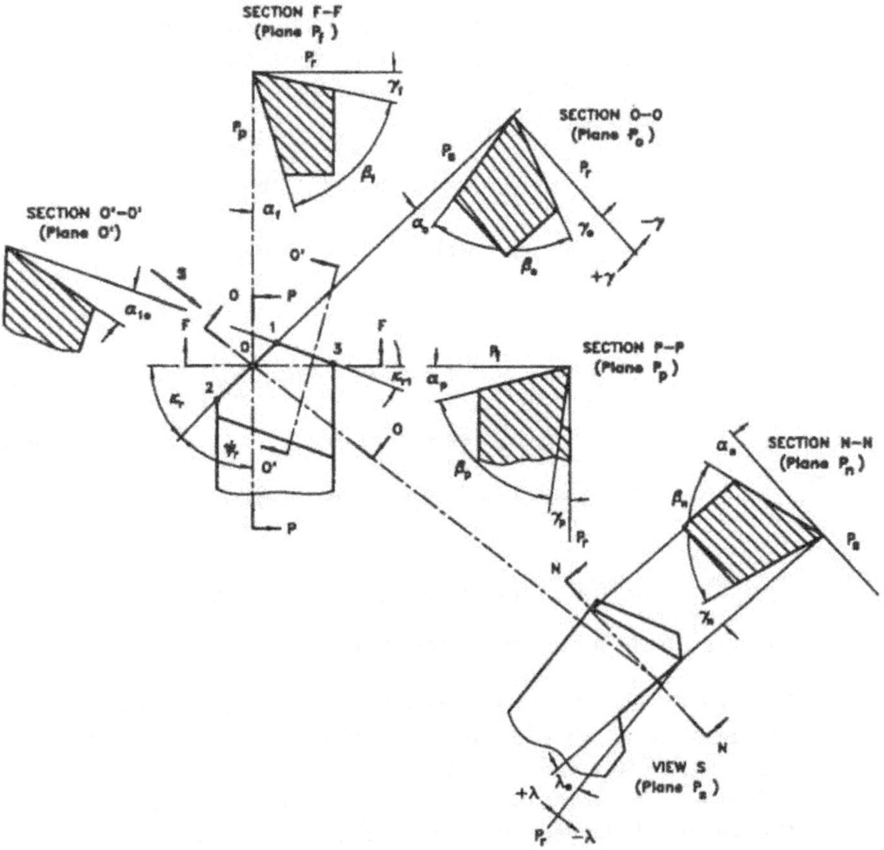

Figure A4 Tool angles in the tool-in-hand system.

Simple relationships exist among the considered angles in the tool-in-hand system. These relationships have been derived assuming that the tool side rake angle γ_f, the tool back rake angle γ_p, and the tool cutting edge angle κ_r are the basic angles for the tool face, and the tool side clearance angle α_f, tool back clearance angle α_p, tool cutting edge angle κ_r are the basic angles for the tool flank:[2,3]

$$\tan\lambda_s = \sin\kappa_r \tan\gamma_p - \cos\kappa_r \tan\gamma_f \quad \text{(A2)}$$

$$\tan\gamma_n = \cos\lambda_s \tan\gamma_o \quad \text{(A3)}$$

$$\tan\gamma_o = \cos\kappa_r \tan\gamma_p + \sin\kappa_r \tan\gamma_f \quad \text{(A4)}$$

$$\cot\alpha_n = \cos\lambda_s \cot\alpha_o \quad \text{(A5)}$$

It must be stated, however, that some of these relationships apply only when the cutting edge angle κ_r is less than 90°. Nowadays, it is becoming common practice to use cutting tools having κ_r greater than 90°. For these tools, the following relationships are valid:

$$\cot\alpha_o = \cos\kappa_r \cot\alpha_p + \sin\kappa_r \cot\alpha_f \quad \textbf{(A6)}$$

$$\tan\lambda_s = -\sin\kappa_r \tan\gamma_p - \cos\kappa_r \tan\gamma_f \quad \textbf{(A7)}$$

$$\tan\gamma_n = \cos\lambda_s \tan\gamma_o \quad \textbf{(A8)}$$

$$\tan\gamma_o = -\cos\kappa_r \tan\gamma_p + \sin\kappa_r \tan\gamma_f \quad \textbf{(A9)}$$

$$\cot\alpha_n = \cos\lambda_s \cot\alpha_o \quad \textbf{(A10)}$$

$$\cot\alpha_o = -\cos\kappa_r \cot\alpha_p + \sin\kappa_r \cot\alpha_f \quad \textbf{(A11)}$$

A.2 Tool-in-machine system (setting system)

It is assumed in the tool-in-hand system that:

- The tool tip (point 1 in Figure A2) and the axis of rotation of the workpiece are located on the same horizontal plane (reference plane).
- The geometrical axis of the cutter down through the tool tip (point 1 in Figure A1) is perpendicular to the axis of rotation of the workpiece.
- The vector of the feed v_f is parallel to the axis of rotation of the workpiece and thus perpendicular to the geometrical axis of the cutter.

In reality, however, the position of the cutting tool in the tool holder of a machine is far from perfect due to the fact that, in practice, major attention is paid to the working parts of the cutting tools and, consequently, much less attention is paid to the proper design and tolerance of the tool shanks, although the shank is defined as the relative location of the tool working part and workpiece. As a result, the tool geometry intended by the tool-in-hand system, should be re-considered to account for changes in actual angles. A set of the cutting tool angles changed to account for the inaccuracies in tool positioning in the machine constitutes the tool-in-machine system.[3]

Figure A5a shows the case where the geometrical axis of the cutter is perpendicular to the axis of rotation of the workpiece. Here, the tool cutting edge angle, tool minor (end) cutting edge angle, and tool approach angle are as in the tool-in-hand system. Figures A5b,c show two cases where the tool, installed in the machine, is rotated with respect to the point O on the cutting edge. If the rotation is in the clockwise direction (Figure A5b), then the tool cutting edge angle, tool minor (end) cutting edge angle, and tool approach angle are as in the tool-in-machine system and are calculated as:

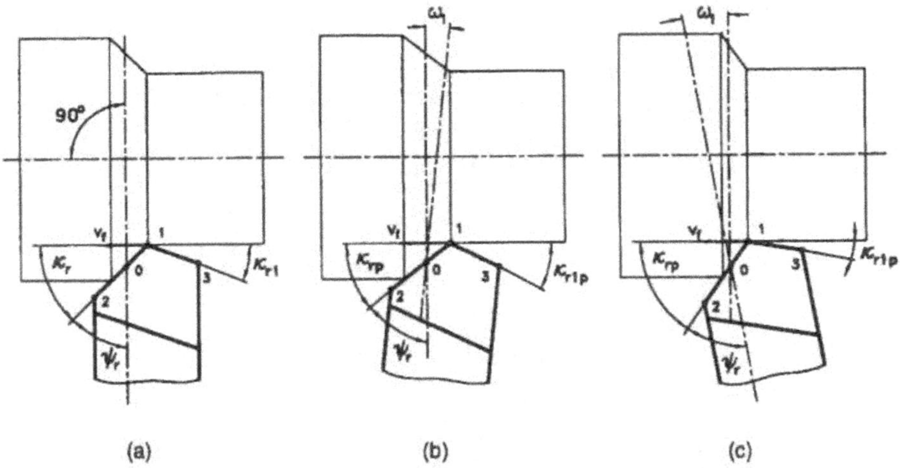

Figure A5 The tool cutting edge and minor (end) cutting edge angles depend on the setting angle.

$$\kappa_{rp} = \kappa_r - \omega_1, \quad \kappa'_{rp} = \kappa'_r + \omega_1, \quad \psi_{rp} = \psi_r + \omega_1 \tag{A12}$$

If the rotation is in the counterclockwise direction, then

$$\kappa_{rp} = \kappa_r + \omega_1, \quad \kappa_{r1p} = \kappa_{r1} - \omega_1, \quad \psi_{rp} = \psi_r - \omega_1 \tag{A13}$$

The location of the tool tip is assumed to be in the reference plane containing the axis of rotation of the workpiece, or even more accurately, on the line of intersection of the tool base plane and the reference plane contacting the axis of rotation of the workpiece. However, in practice, it is not always the case. The tool tip, after being installed in a machine, is often found to be shifted along the tool back plane relative to the mentioned tool reference plane (i.e., in the vertical direction). This shift causes changes in the major cutting angles that are now to be considered.

Consider the general case when $\lambda \neq 0$.[3] Figure A6a shows the intended tool position where the tool tip (1a) is located in the reference plane containing the axis of rotation of the workpiece. Figure A6b shows the case where the tool is located such that tip 1b is shifted relative to the reference plane distance H. It is seen from this figure that

$$\tan \tau_1 = \frac{\tan \lambda_s}{\sin \kappa_r} \tag{A14}$$

Because the vector of the cutting speed is perpendicular to the radius connecting the tool tip 1b with the axis of rotation of the workpiece, the angle

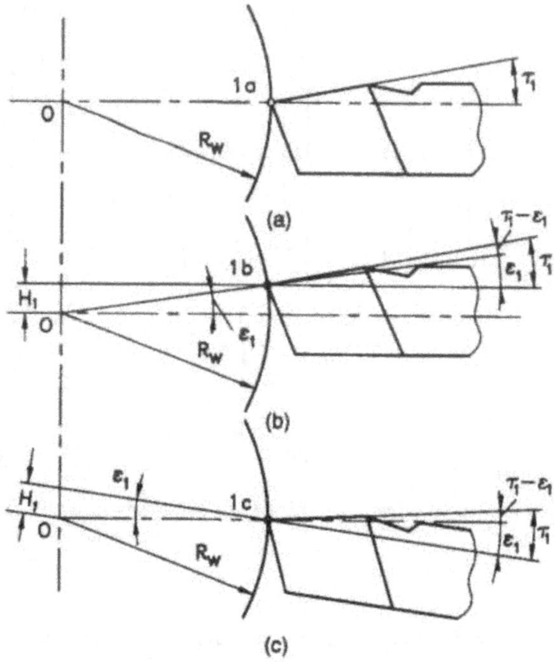

Figure A6 Influence of the vertical position of the cutting tip relative to the horizontal plane containing the axis of rotation of the workpiece.

ε_1 in this figure is the angle of rotation of the cutting speed vector relative to a perpendicular of the reference plane. The angle ε_1 depends on the radius R_w of the workpiece and the distance H_t. It can be calculated as:

$$\sin\varepsilon_1 = \frac{H_t}{R_w} \tag{A15}$$

It can be shown[3] that the case illustrated by Figure A6b is equivalent to that shown in Figure A1.6c where the tool tip 1c is located as in Figure A6a, but the cutting edge is rotated relative to the tool tip by the angle ε_1. The tool cutting edge angle can be calculated in this case as:

$$\tan\kappa_{rp} = \frac{\tan\kappa_r \cos(\tau_1 - \varepsilon_1)}{\cos\tau_1} = \frac{\tan\kappa_r \cos\left(\arctan\dfrac{\tan\lambda_s}{\sin\kappa_r} - \varepsilon_1\right)}{\arctan\dfrac{\tan\lambda_s}{\sin\kappa_r}} \tag{A16}$$

Equation (A16) is general, as it includes all possible cases of relative position of the tool and workpiece. If $\lambda_s < 0$ and/or H_t is located below the reference plane then λ_s and/or H_t should be substituted in this equation with negative signs. An analysis of equation (A16) shows that:

- If λ_s and H_t are of opposite signs then $\kappa_{rp} < \kappa_r$ is always the case.
- If λ_s and H_t have the same signs, then if $|\tau_1 - \varepsilon_1| > |\tau_1|$ then $\kappa_{rp} > \kappa_r$.

Other angles in the tool-in-machine system when the tool is located relative the workpiece, as shown in Figure A6b, are calculated as:

$$\tan\lambda_{sp} = \sin\kappa_{rp}\tan(\tau_1 - \varepsilon_1) = \sin\kappa_{rp}\tan\left(\arctan\frac{\tan\lambda_s}{\sin\kappa_r} - \varepsilon_1\right) \quad \textbf{(A17)}$$

$$\tan\gamma_{rp} = \tan\kappa_{rp}\tan\lambda_{sp} + \frac{\tan(\gamma_r + \varepsilon_1)}{\cos\kappa_r} \quad \textbf{(A18)}$$

$$\tan\alpha_{rp} = \tan(\alpha_r - \varepsilon_1)\cos\kappa_{rp} \quad \textbf{(A19)}$$

A.3 Tool-in-use system

When the tool is being used, the actual direction of primary motion and the feed motion may differ from the assumed directions used in the tool-in-hand system. In the tool-in-use system (Figure A7), the workpiece reference plane P_{re} is perpendicular to the resultant cutting direction, while the working plane P_{fe} contains both the direction of primary motion and the direction of feed motion. Consequently, the working plane P_{fe} is perpendicular to the working reference plane P_{re}. The working back plane P_{pe} is perpendicular to both P_{re} and P_{fe} and thus completes an orthogonal set known as the working system.[2]

It is believed that simple relationships exist between angles in the tool-in-hand and tool-in-use systems.[1,2] As such, a number of relationships have been derived using different methods.[2] However, in the known studies, tool-in-use angles are considered to be rather constant. Compared to the tool-in-hand angles, only the effect of the resultant cutting speed angle η (Figure A7) is considered. As follows from Figure A7, this angle is calculated as:

$$\tan\eta = \frac{v_f}{v} \quad \textbf{(A20)}$$

For most practical machining operations it may be a reasonable approximation except, maybe, for tools having high inclination angles. However, because the difference between the calculated and real angle may reach 2°, this approximation is unacceptable when metal cutting mechanics are considered.

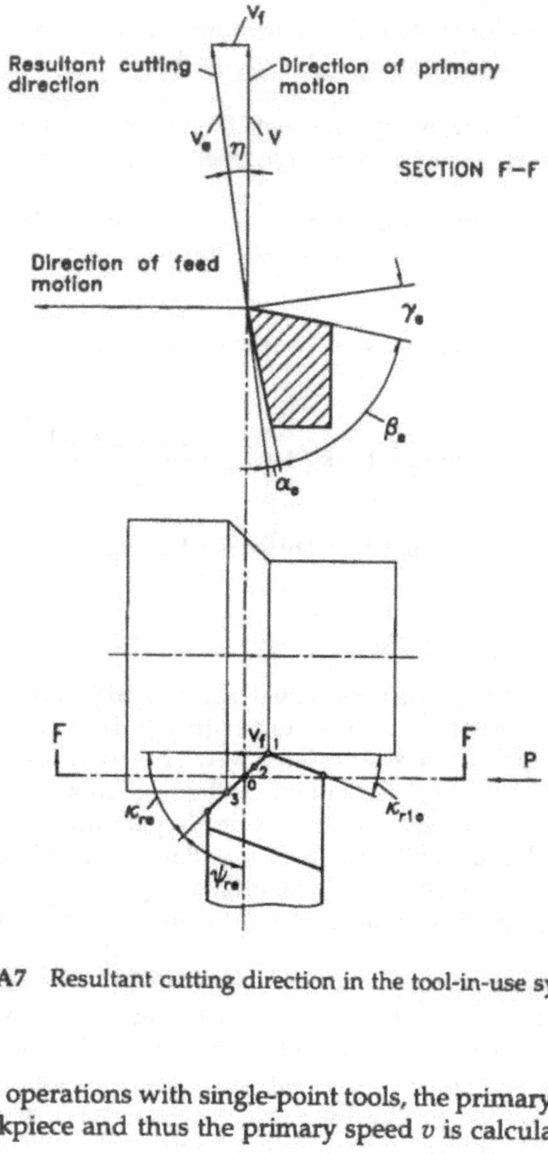

Figure A7 Resultant cutting direction in the tool-in-use system.

For turning operations with single-point tools, the primary motion is the rotation of workpiece and thus the primary speed v is calculated as:

$$v = \pi D_w n_w \tag{A21}$$

where D_w is the diameter of workpiece m , and n_w is the rotational frequency of the workpiece (rpm).

The feed motion is a motion provided to the tool, and in the simplest case, the speed of feed motion is calculated as:

$$v_f = f n_w \tag{A22}$$

where f is the cutting feed (m/rev).

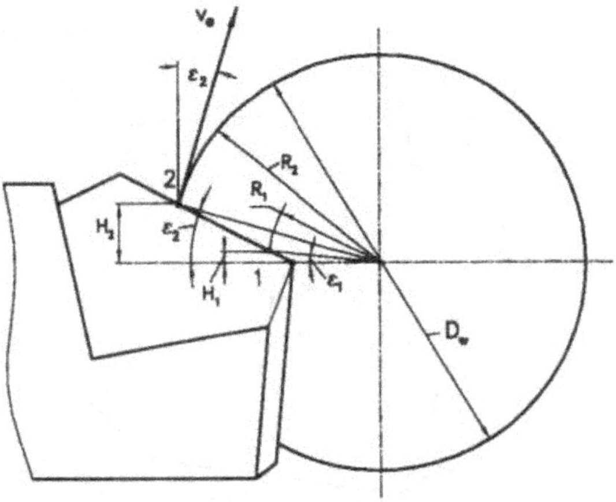

Figure A8 Resultant cutting direction varies along the cutting edge.

Substituting Equations (A21) and (A22) into Equation (A20), one may obtain the following final expression for the resultant cutting-speed angle η as follows:

$$\tan \eta = \frac{f}{\pi D_w} \tag{A23}$$

Because different points of the cutting edge lay at different distances from the workpiece rotational axis (that is, D_w for each point of cutting edge differs), the resultant cutting-speed angle η varies along the cutting edge.

Figure A8, which is view P on Figure A7 (enlarged and revolved 90° clockwise for clarity), illustrates that angle ε_1 also varies along the cutting edge. As seen from this figure and Equation (A15), for point 2 this angle is larger than that for point 1 located closer to the rotational axis of the workpiece.

The definitions of basic tool angles in the tool-in-use system are as follows. The tool cutting edge angle κ_{re} is measured in the plane containing the major cutting edge and v_f, while the minor tool cutting edge angle κ_{r1e} is measured in the plane containing the minor cutting edge and v_f in the sense shown in Figure A9.

The cutting edge inclination angle λ_{se} is the angle between the cutting edge and the plane perpendicular to the resultant cutting direction. Because the latter differs for different points of the cutting edge (as seen from Figure A8), each point of the cutting edge is characterized by its own λ_{sei} calculated as:

$$\begin{aligned} \sin \lambda_{sei} &= \cos \kappa_r \cos \lambda_s \sin \eta_i \\ &+ \sin \kappa_r \cos \lambda_s \sin \varepsilon_i \cos \eta_i \\ &+ \sin \lambda_s \cos \eta_i \cos \varepsilon_{1i} \end{aligned} \tag{A24}$$

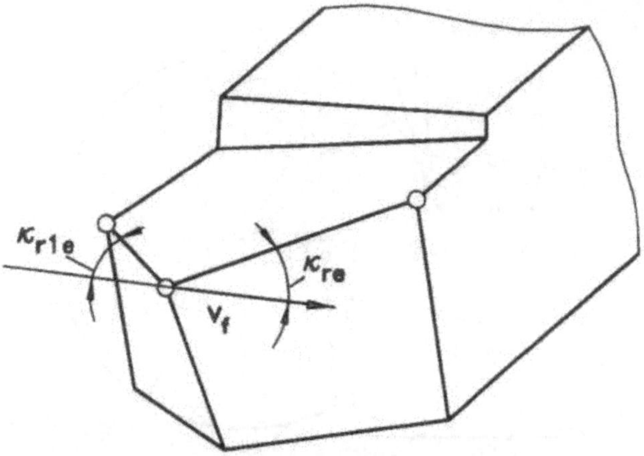

Figure A9 Model used in determining equivalent cutting edge and chip flow direction in non-free cutting.

In particular, when $\lambda_s = 0$ and, hence, $\varepsilon_1 = 0$,

$$\sin\lambda_{sei} = \cos\kappa_r \cos\lambda_s \sin\eta \tag{A25}$$

The rake angle is measured between the plane perpendicular to the resultant cutting direction and the chip flow direction. This angle differs for different points of the cutting edge, thus each point of the cutting edge is characterized by its own rake angle calculated as:

$$\sin\gamma_{ei} = \sqrt{1-\sin^2\gamma\cos^2\lambda_{sei}}\left(\begin{array}{l}\sin\left(\kappa_r + \arctan\dfrac{\tan\lambda_s}{\cos\gamma}\right)\\ \sin\eta + \cos\left(\kappa_r + \arctan\dfrac{\tan\lambda_s}{\cos\gamma}\right)\\ \cos\eta_i \sin\varepsilon_{1i}\\ \end{array}\right) \tag{A26}$$

$$+ \sin\gamma\cos\lambda\cos\eta_i\cos\varepsilon_{1i}$$

In particular, when $\lambda_s = 0$ and, hence, $\varepsilon_1 = 0$,

$$\sin\gamma_{ei} = \cos\gamma\sin\kappa_r \sin\eta + \sin\gamma\cos\eta \tag{A27}$$

The clearance (flank) angle is defined for each point of the cutting edge as:

$$\cos\alpha_e = \frac{2\pi \sin\kappa_r + \cos_{\kappa_r} \sin\varepsilon_{1i} - \tan\alpha \cos\varepsilon_{1i}}{\sqrt{\left(1+4\pi^2\right)\left(\sin^2\kappa_r + \left(\cos\kappa_r \sin\varepsilon_{1i} - \tan\alpha \cos\varepsilon_{1i}\right)^2\right)}} \quad \textbf{(A28)}$$

In particular, when $\lambda_s = 0$ and, hence, $\varepsilon_1 = 0$, Equation (A28) simplifies to:

$$\cos\alpha_e = \frac{2\pi \sin\kappa_r - \tan\alpha}{\sqrt{\left(1+4\pi^2\right)\left(\sin^2\kappa_r + \tan^2\alpha\right)}} \quad \textbf{(A29)}$$

A.4 Determination of the uncut chip cross-section for non-free and non-orthogonal cutting conditions

To apply the theoretical and experimental results obtained using orthogonal cutting as an approximation of the real cutting process, attention should be focused on the relationships between the uncut chip thickness in orthogonal cutting; orthogonal, non-free cutting; and non-orthogonal, non-free cutting. Another important aspect to be considered is the total length of the cutting edges involved in cutting, b_Σ (Chapter 6).

When orthogonal cutting is considered, the uncut chip cross-section area is determined as:

$$A_c = a_1 b_1 = f t_1 \quad \textbf{(A30)}$$

where a_1 and b_1 are uncut chip thickness and width, respectively; f is feed; t_1 is the depth of cut.

In orthogonal cutting, a_1 and b_1 are determined using simple relationships:

$$a_1 = f \sin\kappa_r, \; b_1 = \frac{t_1}{\sin\kappa_r} \quad \textbf{(A31)}$$

and both sides of the chip cross-section are perpendicular each other.

There have been a number of attempts to allow for the influence of the cutting action of the minor (end) cutting edge in determining the direction of chip flow. They are well summarized in Oxley.[4] Klushin[5] suggested determining the true uncut chip thickness a_{1T} in the plane perpendicular to the direction of chip flow and the true uncut chip width b_{1T} in a perpendicular direction and equal to the length of segment *AB* which joins the ends of the major and minor cutting edges engaged in cutting, as shown in Figure A10.

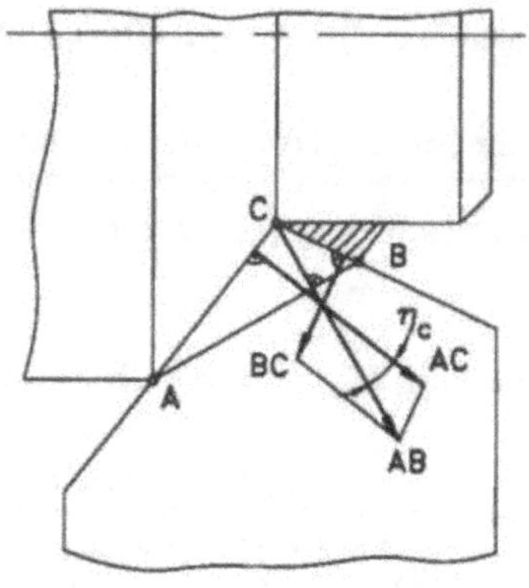

Figure A10 Tool cutting edge and minor cutting edge angles in the tool-in-use system.

In this figure, the directions *AC* and *BC* are orthogonal chip flow directions of the major and minor cutting edges, respectively, and direction *AB* is the resultant chip flow direction. The angle between *AC* and *AB* is referred to as the chip flow angle η_c. The segment *AB* is often referred to as the equivalent cutting edge.[4–6]

Figure A11a illustrates simple turning and the resultant apparent uncut chip cross-section. Figure A11b shows the determination of a_{1T} and b_{1T} for orthogonal, non-free cutting. As seen, the chip has a trapezoidal cross-section so that the width of the free side of the chip is always greater than that of the contact side. In particular, when the depth of cut is equal to the cutting feed, the chip has a triangular cross-section. Figure A11c illustrates the determination of a_{1T} and b_{1T} for non-orthogonal, non-free cutting.

Figure A12 presents four basic configurations in non-free cutting.[6] In the first configuration, shown in Figure A12a, the cutting tool is grounded with a nose radius r_n and set so that the depth of cut is greater than the nose radius. If the following relationships are justified:

$$t_1 \geq r_n(1-\cos\kappa_r), \quad f \leq 2r_n \sin\kappa_{r1} \tag{A32}$$

then the formulas for calculation of a_{1T}, b_{1T}, and b_Σ are as follows:

$$a_{1T} = \frac{f}{c'}\sin\arctan\frac{c'}{\left[1-a'(1-\cos\kappa_r)\right]\cot\kappa_r + a'(\sin\kappa_r + b')} \tag{A33}$$

$$b_{1T} = \frac{c't}{\sin\arctan\dfrac{c'}{\left[1-a'(1-\cos\kappa_r)\right]\cot\kappa_r + a'(\sin\kappa_r + b')}} \tag{A34}$$

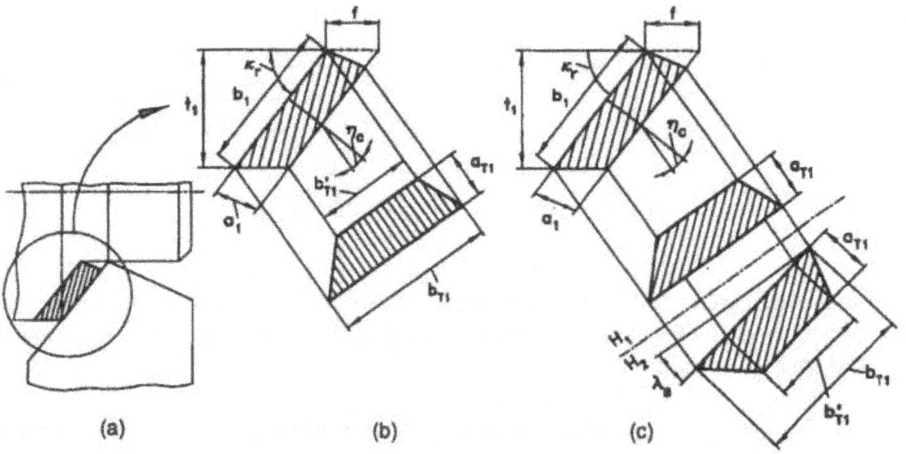

Figure A11 Determining the true uncut chip thickness a_{1T} and true uncut chip width b_{1T}: **(a)** model, **(b)** non-free orthogonal cutting, and **(c)** non-free, non-orthogonal cutting.

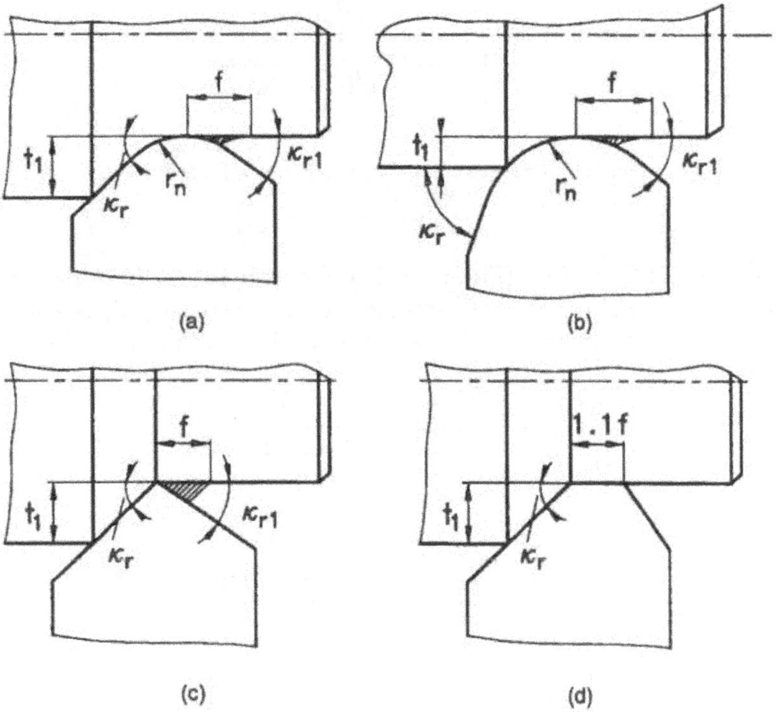

Figure A12 Four basic configurations in non-free cutting.

$$b_\Sigma = \frac{1}{\sin\kappa_r}\left[1 - a'\left(1 - \cos\kappa_r - \frac{\kappa_r}{\sin\kappa_r} - \frac{\arccos\sqrt{1-b'^2}}{\sin\kappa_r}\right)\right] \quad \text{(A35)}$$

where:

$$a' = \frac{r_n}{t_1} \quad b' = \frac{f}{2r_n} \quad c' = 1 - a'\left(-\sqrt{1-b'^2}\right) \quad \text{(A36)}$$

The second configuration (Figure A12b) is similar to the first except that only the radius part of the cutting edge is engaged in cutting. If the following relationships are justified:

$$t_1 < r_n(1 - \cos\kappa_r), \quad f \le 2r_n \sin\kappa_{r1} \quad \text{(A37)}$$

then the formulas for calculation of a_{1T}, b_{1T}, and b_Σ are as follows:

$$a_{1T} = \frac{f}{c'} \sin\arctan\frac{c'}{\sqrt{2a' - 1 + a'b'}} \quad \text{(A38)}$$

$$b_{1T} = \frac{c't}{\sin\arctan\dfrac{c'}{a'b' + \sqrt{2a' - 1}}} \quad \text{(A39)}$$

$$b_\Sigma = r_n\left[\arccos\left(1 - \frac{1}{a'}\right) + \arccos\left(\sqrt{1-b'^2}\right)\right] \quad \text{(A40)}$$

In the third configuration (Figure A12c), the nose radius is rather small so that only the straight parts of the major and minor cutting edges are considered. Then the formulas for calculation of a_{1T}, b_{1T}, and b_Σ are as follows:

$$a_{1T} = \frac{f}{d'}\sin\arctan\frac{d'}{1 + \cot\kappa_r - d'} \quad \text{(A41)}$$

$$b_{1T} = \frac{d't}{\sin\arctan\dfrac{d'}{1 + \cot\kappa_r - d'}} \quad \text{(A42)}$$

$$b_\Sigma = \frac{t_1}{\sin\kappa_r}\left[1 + \frac{\sin\kappa_r}{\sin\kappa_{r1}}(1 - d')\right] \quad \text{(A43)}$$

where:

$$d' = 1 - \frac{f}{t_1} \frac{1}{\cot \kappa_r + \cot \kappa_{r1}} \quad \text{(A44)}$$

In the fourth configuration (Figure A12d), the nose radius is rather small so that only the straight parts of the major and minor cutting edges are considered. To improve the machined surface finish, the minor cutting edge is positioned parallel to the axis of rotation of the workpiece so that $\kappa_{r1} = 0$. As such, the length of the minor cutting edge is usually selected to be equal to 1.1 f. The formulas for calculation of a_{1T}, b_{1T}, and b_{Σ} are as follows:

$$a_{1T} = f \sin \arctan \frac{1}{\cot \kappa_r + 2.2a'b'} \quad \text{(A45)}$$

$$b_{1T} = \frac{t_1}{\sin \arctan \dfrac{1}{\cot \kappa_r + 2.2a'b'}} \quad \text{(A46)}$$

$$b_{\Sigma} = t_1 \left(\frac{1}{\sin \kappa_r} + 2.2a'b' \right) \quad \text{(A47)}$$

Analysis of the above relationships shows that the true uncut chip thickness in non-free cutting increases with the cutting feed f and cutting edge angles κ_r and κ_{r1}, while it decreases with the ratio r_n/t_1. The influence of the ratio r_n/t_1 on the true uncut chip thickness is more significant when this ratio is small and the feed is high.

References

1. Boothroyd, G. and Knight, W.A. *Fundamentals of Machining and Machine Tools*, 2nd ed., Marcel Dekker, New York, 1989.
2. Watson, A.R., Geometry of drill elements, *Int. J. Mach. Tool. Des. Res.*, 25(3), 209, 1985.
3. Granovski, G.I. and Granovski, V.G., *Metal Cutting* (in Russian), Vishaja Shcola, Moscow, 1985.
4. Oxley, P.L.B., *Mechanics of Machining: An Analytical Approach To Assessing Machinability*, John Wiley & Sons, New York, 1989.
5. Klushin, M.I., *Metal Cutting* (in Russian), Mashgiz, Moscow, 1958.
6. Silin, S.S., *Similarity Methods in Metal Cutting* (in Russian), Mashinostroenie, Moscow, 1979.

Index

A

abrasive wear, 243
adhesive wear, 223, 225
  minimizing, 243
  three types of, 243
affine similarity, 216
angle of rotation, 278
angles, 272–276
assemblage
  mathematical definition of, 20–21
  vs. system, 15–26, 59
autospectra, 38, 40, 41, 42, 257, 260, 262, 263
averaging, modes of, 265
averaging number, 266

B

bending force, 29, 30
bending moment, 26, 27, 29, 32, 33, 231
  as plastic moment, 41, 42
  chip-breaking cycle, and, 55
  significance of, 39
bending stress, 29, 32, 47, 184
  analysis of, 39–44
  calculation of, 40–41, 43, 44
  in deformation zone, 34, 45
  verification of, 33
  vs. shear stress, 39, 44
black box diagram, 22
Bolza optimization problem, 58
boundary stress, 65–66
brittle vs. ductile fracture, 156, 163–167
built-up edge, 13, 243
  on rake face, 202–204
burnishing, 243

C

calibration chart, typical, 261, 262
carbon steel, cutting regime for, 138–139
cast iron, cutting of, 53
center line average (CLA), 242
chip
  basic types of, 51
  -breaking, 51–55
    cycle, four-stage, 55
  changing from gray to red, 254
  classifications of, 51
  contact layer, 32
  continuous fragmentary, 32
  continuous humpbacked, 49, 51, 55
  continuous uniform strength, 54
  control, lack of, 51
  deformation, 193, 226. *See also* deformation zone
  formed by shearing, 13
  fracture, 27
  grains, deformation of, 37
  irregular broken, 53
  plastic zone within, 75–76
  regular broken, 53

temperature, 253
thermal energy, 229, 232, 233
thickness, calculating, 237
three basic types of, 158
velocity, 92
chip compression ratio, 43, 44, 56, 91, 95, 96, 97, 98, 137, 138, 165, 220, 223, 225, 226, 233, 254
ductility, and, 47
measurement of, 45–46
chip formation, 26, 34, 76, 158
application of high load, and, 192
cause of, 45
cycle, 34–36, 82, 84, 86, 87, 88, 94, 184, 195, 196, 198, 200, 203, 205, 231
analysis of, by FEM, 191
frequency of, 38
frequency, 231
grain refining in, 166
in ductile materials, 29–30
mechanical energy of, 229
model of, 46, 47–51, 51–55, 198, 216, 218
process, 78, 82, 83, 161, 167, 185, 205, 245. *See also* seizure
system approach to, 26–55
properties of workpiece material and, 27
role of bending moment, 39
theory, 12–13
use of term, 51
zone, 4, 5, 33, 196
temperature in, 130
chip ratio, 91, 95
clearance face, nature of forces on, 5–6
coefficient of friction, 197
coherence function, 263
cold working, 239, 242
compression vs. cutting, 136, 139
compressive force, 29, 30, 135
continuity condition, 91
coordinate system, 215
cutting conditions, non-free, 283–287
cutting conditions, non-orthogonal, 283–287
cutting edge angle, 272, 276, 277, 284. *See also* angles
cutting force measurements, 255–267
three stages of, 257
cutting forces, 25
cutting speed angle, 279
cutting system, 26–69. *See also* system
arrangement of components, 29
cutting tool geometry, 271–287
cutting tool nose angle, 228
cutting vs. compression, 136

D

deep-hole drilling, 217
defects, 147, 173
deformation, velocity of, 80
deformation zone, 4, 30, 32–33, 45, 128, 129, 132, 140, 184, 196. *See also* strain
adjacent to tool rake face, 203
bending stress in, 39
bounded by plane surfaces, 74
development of, 192
displacement in, 93
experiment vs. theory, 159
forces acting within, 196–197
heat generated in, 137–138, 139
lower boundary of, 83
metallographic observation of, 158
modeling of, 192–197
parallel boundaries of, 88–89, 116
parallel-sided theory, 73–126
presence of bending stress, 33
primary, 204
primary and secondary, 51, 195–196
secondary, 32
shear flow stress in, 132
slip-line structure of, 33

strain in, 115
stress fields in, 87
structure of, 33–34
temperature field, and, 253
temperature in, 255
velocity fields in, 83. *See also* velocity discontinuity
velocity ratio in, 92
density function, 165
difractometer, 163–165, 166
dislocation theory, 127
displacement, finite, 108
displacement matrix, 67
ductile fracture, 171–174, 184, 195
criteria of, 169–171
vs. brittle fracture, 156, 163–169
dynamometer, 51, 205, 255, 259
dynamic calibration of, 261
natural frequency of, 262
signal, vs. hammer signal, 266
static calibration of, 261

E

elastic deformation, 171, 184
FEM, and, 184
elasticity, theories of, 127–128
electromagnetic field (emf), 245
end-tube turning, 258
energy balance in machining, 233
energy balance in machining steel, 234
engineering plasticity, 73–79
equicoheresive temperature (ECT), 173
equivalent cutting edge, 284
etchants, recommended, 161
experimental studies, methodology of, 213–270
similarity methods, 213–245
experimental verification, 134–141
discrepancies in results, 141–143
external vs. internal parameters, 2
extrusion, 170

F

failure theories, 145
Fast Fourier Transform (FFT), 256, 262
analyzer, 260, 263, 267
FEM. *See* finite element method
finite element analysis (FEA), 7
finite element method (FEM), 62, 87, 183–212
applications of, 189–191
calculated vs. actual data, 204–210
computational details, 184–189
convergence module, 186
data input and checking module, 185
initialization module, 185
load increment module, 185
loading module, 185
output module, 186
residual force module, 186
solution module, 186
stiffness module, 185–186
First Metal-Cutting Law. *See* Makarow's law
flank angle, 273. *See also* angles
flank, defined, 271
flank wear, 244
forging, 170
fractography, 166
fracture
brittle. *See* brittle fracture
ductile. *See* ductile fracture
mechanism of, 156–167
of polycrystallines, 156, 174
ductile vs. brittle, 156
plastic deformation, and, 174–178
regions of, 157
types of, 163–169
fracture strain, 134, 135, 143–145, 147, 150, 154–156, 174, 176
density, and, 147
determination of, 167–178

factors affecting, 155
hydrostatic stress, and, 153
zone of a parabolic increase in plasticity, 154
zone of insignificant dependence, 153
zone of significant dependence, 153
temperature, and, 154
frequency response function (FRF), 263, 264
FRF. *See* frequency response function

G

Griffith theory, 143, 171
grinding, 160

H

healing, of microdefects, 178
heat balance equation, 229
Hooke's law, 187
hypotheses, 214

I

incremental compression, 134–136, 140, 141
indentation, study of, 74
infinitesimal strain of a line element, 105–106
input vs. output parameters, 25, 56, 59
internal friction, 202
irrational tensor, 104

K

Kirpichev-Gukhman theorem, 218

L

Lagrange optimization problem, 58
limiting strains, 170

M

machinability, 236
machined surface, quality of, 238–270
machining system, components of, 3
machining zone, 202
macrohardness, 159
Makarow's law, 218–226
mathematics, and system engineering, 11–12
maximum normal stress theory, 145
Mayer optimization problem, 58
mechanical residual stress, 238–239
Merchant's force model, 128–129, 197
metal cutting
basic problems of, 6–8
brittle vs. ductile material, 226
coefficient of friction, and, 197
cutting speeds in testing, 142
definition of process, 44–45
energy necessary for, 134
experimental studies, 213–270
experimental verification of, 134–141
FEM, use of, 183–184. *See also* finite element method (FEM)
fracture, mechanism of, 156–167
heating rate, 139
history of, 1, 3–6
lack of model for, 184
necessary predictions, 132–134
optimum conditions for, 3
predicting performance, 7
reasons for studying, 2–3
role of engineering plasticity, 73
role of velocity diagram, 81
speed, 5
stochastic nature of, 256
strain in, 113–117
strain rate in, 117–122
system, 190–191

temperature, 2. *See also* temperature
optimum, 218–226
theory, need for, 2
total work per unit, 137
total work per unit volume, 136
velocity diagram, and, 88–94
vs. compression, 139
vs. cutting, 14
vs. incremental compression, 137
vs. shearing, 45
vs. shearing press operations, 14
metal removal tasks, 2
metallographic analysis, steps of, 160
metallurgy, modern, 158
microcrack nucleation, 176
microhardness, 93–94, 139–140, 159, 208, 219, 222, 258
scanning, 203, 204
tests, 206
millivoltmeter, 249, 251, 252
modeling process, 12
models, lack of predictive, 2
mountings, hot and cold, 160

N

nodal displacements, 68
nodal forces, 66
normal stress distribution, 199–212

O

OCS. *See* optimum cutting speed
OCT. *See* optimum cutting temperature
optical reflection microscope, 158
optimization problems, 58
optimum cutting speed (OCS), 219, 226, 234–238, 243
determination of, 221
optimum cutting temperature (OCT), 219, 220, 222, 226, 234–235, 237
orthogonal cutting, 12–13, 14, 75, 89, 96, 141, 142, 204, 205, 206, 209, 258, 283–287
FEM model of, 190
grid deformation in, 97
model of, 12–13
role of heat in, 142
test, 253, 254, 255

P

parallel-sided theory, 73–126
parameters of a process, 215
Peclet number, 142, 227, 232
Peltier effect, 245
penetration force, 27, 29, 30, 31, 32, 50, 51, 86, 88, 156
photo-elasticity technique, 204
plane of maximum stress, 35, 36
plane stress, 63, 65
plane-strain deformation, 73, 74
plane-strain theory, 74
planes, 271–272
plastic deformation, 4, 6, 13, 26, 33, 76–77, 112, 132, 135, 145, 154–156, 165, 170, 174, 195
analysis of, 95–117
as plane deformation, 96
bending and compression, and, 36
by slip, 119–120
discontinuities in stress, 84
ductile fracture, and, 171
due to compression, 194
energy fluxes of, 145–146
FEM, and, 185
fracture, and, 174–178
healing effect of, 27
incompressibility during, 85
instability of, 87
measure of, 95–98
nature of, 115–116
non-uniform, and FEM, 187
of fractured fragments, 161
of polycrystalline materials, 127, 143–156

rate of, 53
stages of, 171–172
theory of, 127
twinning, role of, 6
zone, 54
plastic moment, 41
plastic shear, 4
plastic strain, 6
density of material, and, 168
parameter, 152
plastic zone, 4, 75–76, 197
plasticity
history of science of, 73
measurement of, 146
metal cutting studies, and, 73–126
stress relaxation, and, 145–147. *See also* stress relaxation
theories of, 6, 90, 127–128, 129
polycrystallines, 59, 143–156
fracture of, 156. *See also* fracture: of polycrystallines
power spectral density function, 257
principal planes, 111
principle of minimum energy, 75
pure strain tensor, 104

Q

quick-stop device, 86, 159, 203, 205, 206
micrographs, 162, 163, 164
to observe built-up edge, 202

R

radial wear, 244
rake angle, 40, 89, 92, 97, 98, 191, 198, 199, 205, 273, 282. *See also* angles
chip compression ratio for, 96
effect on residual stress, 240
in cutting of cast irons, 28
positive vs. negative, 53
rake face, defined, 271
real virtual work equation, 94–95
relative displacement tensor, 100
relaxation time, as measure of plasticity, 146
residual stress, 238, 239, 240
resultant force, 27, 28, 29
on shear plane, 13
on tool face, 13
rolling, 170
root mean square (RMS), 242
rotation tensor, 105
rotational element, 102–105

S

scaling factor, 216
scanning electron microscopy, 166, 167
Seebeck effect, 245
segmentation, 16, 21, 22, 23, 59
seizure, 47–51, 161, 202, 224–225, 242
contact temperature, and, 202
increase in temperature, and, 223
setting. *See* system: tool-in-machine
shear angle, 13, 44, 75, 129
shear flow stress, 95, 129, 134, 143, 156, 231, 245
and standard tensile tests, 130–131
average, 232
increase due to strain hardening, 139
microhardness, and, 140
prediction of, 129–132
strain hardening, and, 132
thermal softening, and, 132
vs. standard material tests, 131
shear plane, 13, 75, 156, 195, 231
as a line, 75
model, 183–184, 195, 218
position of, 13
resultant force on, 13
shear strain, 130, 136, 137, 140. *See* strain: shear
average rate, 131

shear strength
  of chip fragments, 37
shear stress, 45, 47, 111, 120, 136, 139, 152, 157, 172, 183, 184, 192. *See also* shear flow stress
  as function of dislocation density, 131
  at fracture, 140–141
  at last stage of deformation, 205
  at tool/chip interace, 197–198
  average, computation of, 39–40
  calculated vs. experimental, 211
  caused by compression, 39
  distribution, 198–199, 204
  in cutting zone, 193
  in deformation zone, 195
  maximum, 5, 53, 77, 130, 191
    during tension vs. cutting, 130
  plane, 76
  plato-type distribution of, 198, 200
  reduction in, 137
  vs. bending stress, 39, 44
shear zone, 13, 89
  shear strain rate in, 118
  slip line as upper boundary of, 90
shearing, 165, 175, 195
  chip formation and, 26
  press operations, 14
sheet-forming processes, 170
similarity, basics of, 215–218
similarity criteria, 218
similarity numbers, 227–245
  A-criterion (Silin criterion), 227
  B-criterion, 227
  D-criterion, 227
  E-criterion, 228
  F-criterion, 228
  M-criterion, 228
  Peclet number. *See* Peclet number
similarity ratio, 216
similarity scale, 216
single shear plane model, 4, 76–77, 78, 79, 80, 81, 83, 88
slide plane, 32
sliding planes, 54
  number of, 47
slip lines
  as deformation zone boundaries, 90
  density of, 173
  difference in shape of, 172
  flow along, 74
  plastic zone, and, 76
  vs. sliding lines, 84
slip-line field, 74, 77, 78
slip-line fields, 33
slip-line solution, 75–76
slip-line theory, 120
strain
  calculating displacement from, 106
  compatibility equations, 106–108
  components, 106
  components, increments in, 112–113
  dilatational, 121
  direct, 101, 121
  energy, 68
  engineering shear, 101, 110, 121
  finite, 109–111
  infinitesimal, 98–101
  invariants, 111–112
  metal cutting, and, 113–117
  natural, 115
  plastic, 6
  principal, 111–112
  rate, 117–122
  rate components, 83
  rate tensor, 120–126
  rates, 6
  shear, 96–97, 98, 99, 103, 114, 115, 117
  tensor analysis, 97–98
  values of, 97
  volumetric, 121
strain hardening, 74, 128, 139, 173
  diagrams, 188
  exponent, 176

strain rate, 6, 25, 129, 139, 146, 163, 168, 170, 173, 178, 189, 258
  effect of sensitivity of, 148
  sensitivity of a metal, 155
strains, expression for, 46, 131
stress, 66
  bending. *See* bending stress
  boundary. *See* boundary stress
  distribution at tool/chip interface, 197
  distributions, 25
  limiting, 157
  maximum, 26–27
  maximum combined, 31, 35, 42, 45, 94, 161, 165, 203, 231
  normal, 91
  plane of maximum combined, 30
  plane. *See* plane stress
  shear. *See* shear stress
stress criterion, 145
stress gradient, of plane, 4
stress parameter, 152
stress relaxation, 172, 173, 174
  plasticity, and, 145–147
stress-strain curve, 130, 133, 137, 138
stress-strain relation, 73, 185, 187, 188, 189
stress-strain relationship, 74
system
  black box diagram of, 22
  components of, 215
  control of, 55–69
  cutting, 26–69
  definition of, 15
  dynamic behavior of, 17, 25
  energy, 145
  engineering, 11–71
  equilibrium conditions, 68
  functions as input to, 24
  future states of, 18–19
  input trajectory of, 16
  mathematical definition of, 15–20
  minimizing energy of, 67
  optimization of, 57, 58
  output trajectory of, 16
  past states of, 18, 19
    uniqueness of, 20
  planes, 272
  state of, 17
  theory, parts of, 24
  time, as variable, 25
  time trajectory of, 16
  tool-in-hand, 271–276, 279
  tool-in-machine, 271, 276–279
  tool-in-use, 271, 279–283, 284
  vs. assemblage, 15–26, 59

T

tangential force, 226
tangential velocity, 82–88
temperature
  blue shortness, 130
  effect on mechanical properties, 220
  grain boundary, 173
  influence of, 137–141
    ductility, 137
  measurement of, 245–255
  optimum cutting, 218–226
  reference, 247
tensile strength, 134
tensile stress, 157, 169
tensor, 103, 106, 111
tensors, 113
thermal residual stress, 238
thermal softening, 139
thermal strain hardening, 5, 130
thermocouples
  conventional, 245–249
    properties of, 246–247
  cromel-alumel/-copel, 248
  natural, 249–252
    calibration of, 250–252
  running, 253–255
  semi-artificial, 253–255
thermodynamics, second law of, 141
Thomson effect, 245

titanium alloys, 47, 174
tool approach angle, 272, 276. *See also* angles
tool geometry, 53
tool life, 223
  cutting speed, and, 224
  cutting speed curves, and, 224
  ductility of material, and, 47
  machinability, and, 236
  oil- vs. water-based cutting fluid, 226
  reduction in, 199
  seizure, and, 202–203
  testing, 219
tool materials modeling, 187–191
tool wear, 2, 5–6. *See also* tool life
  flank, 5–6
tool/chip interface, 98
tool/chip interface, modeling of, 197–202
tube end turning, 142
tungsten carbide, as tool material, 53
twinning, 119–120
  role in plastic deformation, 6
twins, 172

U

unit rectangular deforming region, 74

V

velocity diagrams, 79–81, 88, 200
  metal cutting, and, 88–94
  role in metal cutting, 81
velocity discontinuity, 84, 85, 86, 87–88
virtual work equation, 83
void fraction, 170
von Mises yield criterion, 74, 145, 151, 152, 188

W

wedge angle, 274. *See also* angles
workpiece
  accuracy of location, 258–259
  crack in front of cutting edge, 26, 27
  deformation zone. *See* deformation zone
  ductility of, 45–47. *See also* ductile fracture
  elastoplastic zone, 27, 29, 31
  fracture at cantilever support, 28
  interaction with tool, 14
  material, 5, 127–181
    as thermocouple, 247–248
    brittle vs. ductile, 226
    compression, and, 192
    etchants recommended for, 161
    initial density of, 177
    linear expansion of, 241
    mechanical properties of, 189
    modeling, 187–191
    strength of, and temperature, 138
    toughness of, 133, 144
    yield strength of, 195
  maximum stress in, 26
  plastic deformation of, 29. *See also* plastic deformation
  velocity, 92

X

X-ray, 165

Y

yield criteria, 74, 151. *See also* von Mises yield criterion
Young's modulus, 217

Milton Keynes UK
Ingram Content Group UK Ltd.
UKHW021622071024
449327UK00020BA/1148

9 780367 400149